Tangled Diagnoses

Tangled Diagnoses

Prenatal Testing, Women, and Risk

ILANA LÖWY

The University of Chicago Press
Chicago and London

The University of Chicago Press, Chicago 60637
The University of Chicago Press, Ltd., London
© 2018 by The University of Chicago
Published 2018
Printed in the United States of America

27 26 25 24 23 22 21 20 19 18   1 2 3 4 5

ISBN-13: 978-0-226-53412-1 (cloth)
ISBN-13: 978-0-226-53409-1 (paper)
ISBN-13: 978-0-226-53426-8 (e-book)
DOI: https:doi.org/10.7208/chicago/9780226534268.001.0001

Library of Congress Cataloging-in-Publication Data
Names: Löwy, Ilana, 1948– author.
Title: Tangled diagnoses: prenatal testing, women, and risk / Ilana Lowy.
Description: Chicago: The University of Chicago Press, 2018. |
    Includes bibliographical references and index.
Identifiers: LCCN 2017044387 | ISBN 9780226534121 (cloth: alk. paper) |
    ISBN 9780226534091 (pbk: alk. paper) | ISBN 9780226534268 (e-book)
Subjects: LCSH: Prenatal diagnosis. | Fetus—Diseases—Diagnosis.
Classification: LCC RG628 .L68 2018 | DDC 618.3/2075—dc23
LC record available at https://lccn.loc.gov/2017044387

♾ This paper meets the requirements of ANSI/NISO Z39.48-1992
(Permanence of Paper).

CONTENTS

"To See What Is about to Be Born"

Supervised Mothers, Scrutinized Fetuses

"Who is the wise man: the one who sees what is about to be born."[1] The ancient Hebrew expression "to see what is about to be born" was recorded in the early third century, but might have been coined earlier. It is a metaphor, meaning "to be able to foresee the future." By contrast, seeing what is about to be born in the concrete, not the metaphoric, sense—making the living fetus accessible to medical scrutiny—is a very recent development.[2] In its present form, it came into being in the late 1960s—and in one generation became inseparably associated with the routine monitoring of pregnancies of women in industrialized countries, and of middle- and upper-class women in "intermediary" or emerging countries. Each year, tens of millions of expectant mothers worldwide learn from their health care provider that their unborn child is doing fine—or not.

Birth has always been a symbol of hope and renewal, but until recently it was also seen as an extremely dangerous event for the birthing mother and the newborn. For nearly all of human history, maternal mortality has been very high. In industrialized countries, it receded to its present very low level only in the 1940s, yet remains appallingly high in many developing countries. Newborn mortality has been very high, too.[3] Sometimes, newborn babies died because of birth-related events or unknown/undetectable causes. In other cases, the newborn—and sometimes the stillborn child or spontaneously aborted (miscarried) fetus—was visibly malformed. Familiarity with misshapen, deformed, handicapped, and in extreme cases "monstrous" newborns is probably ageless, as are practices connected with these phenomena, such as the deformed babies' abandonment/exposure to the elements, and beliefs about the causes of their deformity. By contrast, the concept of birth defects—and its professional equivalent, congenital

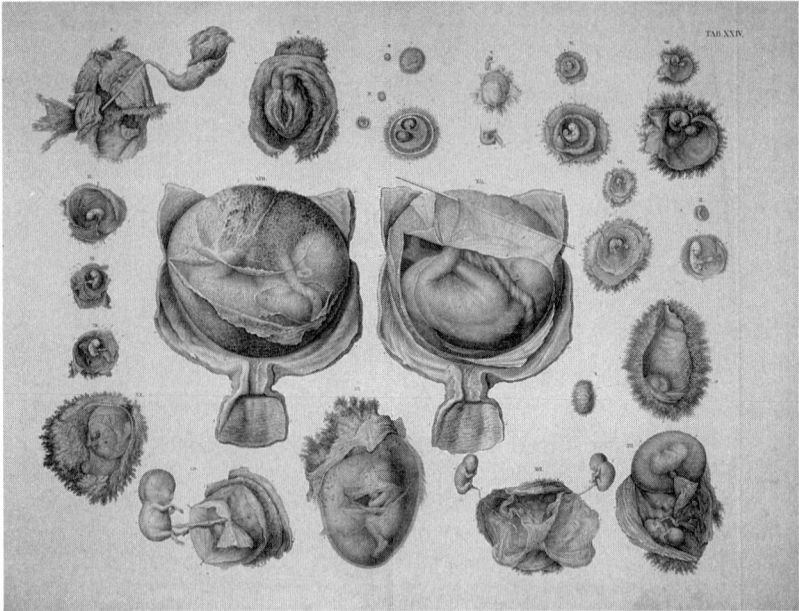

Fig. 1. Development of the fetus; images from the mid-nineteenth century.
In Herman Friedrich Killan, *Geburtshülflicher Atlas in 48 Tafeln und erklärendem
Texte* (Dusseldorf: Verlag vor Arnz, 1835–44). Wellcome Images.

malformations—is relatively new. It stems from an encounter between two
distinct domains of study: embryology and hereditary diseases. In the nineteenth century, zoologists and physicians began the intensive study of ani-
mal and human embryos (fig. 1). At the same time, physicians expanded
their earlier interest in "morbid heredity" and often paired it with investigations of the origins of intellectual disability (known at that time as "inborn mental deficiency").[4] Since "hereditary defects" were transmitted in
families, the main way to prevent the birth of children with such defects was
to discourage—and, in some cases, to actively prevent—the reproduction
of carriers of "hereditary taints," a trend that culminated in the twentieth-
century eugenic movement.[5]

From late nineteenth century onward, a different strand of medical activity focused on promoting the birth of healthy children through subjecting
the bodies of pregnant women to surveillance. Such surveillance, named
antenatal or prenatal care, was intensified in the mid-twentieth century and
became the dominant trend in obstetrics.[6] The assumption behind guided
prenatal care was that to promote the birth of healthy babies, it was important to pay attention to the health of pregnant women: to supervise their

blood pressure, blood sugar, and weight; prevent and treat infections; and provide advice on appropriate diet and lifestyle. Women, and not their unborn children, were the main focus of medical attention. The expected well-being of pregnant women's offspring was a statistical outcome of the proper management of their bodies.

Ann Oakley's well-known critique of the overmedicalization of pregnancy was entitled *The Captured Womb* (1986), not *The Captured Fruit of the Womb*.[7] In it, Oakley mentions the usual methods for checking on fetal health—palpation of the woman's abdomen and listening to the fetal heartbeat—and one technique for scrutinizing the fetus: X-rays. She also could have mentioned a test developed in the 1950s that in cases of suspected destruction of fetal red blood cells by maternal antibodies caused by maternal-fetal Rhesus factor (Rh) incompatibility detects signs of such destruction in the amniotic fluid.[8] However, these approaches had a limited capacity for visualizing fetal problems. Listening to fetal heartbeat and palpating the pregnant woman's abdomen were very crude measures of fetal health; taking X-rays was mainly employed to identify potential sources of difficulties during childbirth; and testing for destruction of fetal red blood cells by maternal antibodies addressed only one among many causes of fetal problems. But perhaps the main difference between all these approaches and prenatal diagnosis, such as it developed in the 1960s, was that the unique goal of the early attempts to assess fetal well-being was to improve the outcome of a pregnancy.

Prenatal diagnosis in its present form—scrutiny of the fetus coupled with the option to terminate pregnancy—came into being in the late 1960s and early 1970s. It was the result of a partly contingent coming together of four distinct developments in biomedicine—the perfection of amniocentesis, the rise of cytogenetics (the genetic study of cells), the application of new biochemical approaches to the study of amniotic fluid and pregnant women's serum, and the development of obstetrical ultrasound—along with a social innovation, the decriminalization of abortion. In the late 1960s, thanks to the newly developed ability to study the chromosomes and enzymes of cultured fetal cells derived from amniotic fluid, researchers acquired the capacity to directly investigate the genetic makeup of the fetus, an approach named antenatal or prenatal diagnosis.[9] This book follows the material, social, and cultural consequences of the introduction of this important biomedical innovation.

A Difficult Subject

I became interested in prenatal diagnosis (hereafter PND) when looking for a research topic that would be stress free, or at least less taxing than the previous one—malignant tumors. For a decade I had studied women's cancers, focusing on the history of the concepts of precancer and hereditary cancer risk and their use in the diagnosis, screening, and prevention of malignant growths. During that period, my research frequently resonated with distressing personal experiences. Several family members and friends developed cancer, and a few died of this disease. After years of discussions with cancer specialists, observation of clinical practices, and immersion in materials about malignancies—scientific publications about cancer, patient files, "pathographies" (autobiographical narratives about a disease), and images of pathological specimens—I strove to move away from the cancer "war zone," often described as a place where patients and physicians valiantly fight a cruel and treacherous enemy. When a group of colleagues from my department invited me to join a collective study of PND and the prevention of disability, I was pleased to accept this invitation.[10]

Research on the history and present practices of PND, I initially thought, would be an excellent choice for my next project. Such a study, I believed, would combine my interest in women's reproductive health, the rise of new diagnostic approaches, and the links between the laboratory and the clinics. It would also allow me to move away from the bleak environment of suffering and dying patients to an optimistic universe of expectant mothers and newborn babies. Pregnancy and childbirth are universal symbols of revitalization and a new start. In many languages, childbirth is called "a happy event," a term that, thanks to women's increased control over their fertility and a sharp decrease in maternal and infant mortality, is more accurate today than it was when it was coined. Medical practices are frequently associated with suffering, impairment, and risk of death. Those practitioners who deal with pregnancy and childbirth are associated with anticipation of the wonders of a new life. Hospital wards often smell of disinfectant and decay, while maternity wards smell of milk and flowers. To an outside observer, interventions promoting the birth of healthy children may appear to be a hope-infused domain of medical activity.

My ignorance did not last long. When I started to interview physicians involved with PND—gynecologists, pediatricians, clinical geneticists, and fetal medicine experts—I was often asked why I elected to study this topic. I told my "ten years in the shadow of cancer" story, and my wish to switch to a more uplifting subject. Raised eyebrows and ironic smiles followed, and my interlocutors gently informed me that compared with their domain of expertise, oncology is a hope-laden field. In the twenty-first century, a significant proportion of cancers can be either cured or kept in check for years, while there is no treatment for the great majority of severe fetal im-

shift from selective abortion
to selecting embryos.

pairments diagnosed before birth. In the first half century of its existence, PND was transformed from a method for detecting a small number of conditions in "high-risk" women into a routine screening technology for detecting a great number of fetal anomalies, and offered to the majority of pregnant women in industrialized countries. However, the impressive extension of the scope and scale of PND was not followed by a parallel extension of the number of curable conditions. The main clinical contribution of PND has been the promotion of the selective abortion of affected fetuses. Despite many of its advocates' insistence on PND's role in effecting cures, now as in the 1960s, a woman who learns that her unborn child is severely impaired often has only two options: to continue the pregnancy or to terminate it. As the historian of science Susan Lindee explains, "The selective abortion of affected fetuses was and remains the primary intervention of genomic medicine."[11] This statement can be extended to the visualization of fetal anomalies by ultrasound, another important way to identify impairments.

Experts' pessimistic statements are strictly off the record, however. The official discourse of fetal medicine specialists focuses on the conditions in which detection of a fetal impairment leads to an effective curative intervention—mostly surgical—either during the pregnancy or immediately after birth.[12] Such cases are relatively rare, but many professionals are convinced that their proportion will dramatically expand soon. With the rise of new genomic technologies, they affirm, PND will become a truly therapeutic approach. As a review article on PND concluded in 2012,

> For the past 30 years the goal of prenatal diagnosis has been to provide an informed choice to prospective parents. That paradigm is now shifting. We are now at the point where that goal can and should be expanded to incorporate genetic and genomic data to pave the way for a personalized approach to fetal treatment.[13]

The hope that, thanks to the development of new genetic and genomic technologies, PND is at the point of producing effective cures is undoubtedly sincere, but the claim that PND's original aim was to allow prospective parents to choose whether to continue the pregnancy, and not to treat the future child, is inaccurate. From PND's earliest days, its promoters foregrounded its curative potential.[14] In the early 1970s, scientists, probably inspired by the success of treating an inborn condition—phenylketonuria—with a specific diet, believed that many other birth defects, including Down syndrome, might be linked to a metabolic imbalance and could be cured through the correction of such an imbalance.[15] When those hopes failed to materialize, the focus switched to hopes of intrauterine therapies.[16] Early proponents of PND believed that further advances in biochemical and genetic studies would enable the rapid development of therapies for the majority of fetal anomalies.[17] The bioethicist John Fletcher argued in 1982 that ethical dilemmas connected with the rise of PND would soon disappear:

> Unless present trends change, in the next few years the public will see earlier and safer methods of prenatal diagnosis combined with more efficacious methods of fetal therapy. For some time to come, the power to diagnose will far outstrip the power to treat, but with sufficient research and resources the two activities of fetal medicine will assume more balance. The effects of conjoining therapy to diagnosis will bring about a more ethical balance between the immediate risks and benefits of prenatal diagnosis. . . . The ability to offer treatment to more affected fetuses will create more assurance that progress in diagnosis and treatment will not be halted due to legal and constitutional efforts to protect the life of the fetus.[18]

PND has made important advances since 1982, but the possibility of "conjoining therapy to diagnosis" has not been among them. My previous research subject—the diagnosis of hereditary susceptibility to cancer—and prenatal diagnosis have in common a significant gap between the refinement of diagnostic approaches and the crudeness of solutions proposed to people receiving a positive diagnosis. In the twenty-first century, oncologists employ highly sophisticated tools in molecular biology to detect mutations that increase cancer risk, but the main solution they offer to mutation carriers is to get rid of potentially dangerous body parts: breasts, ovaries, and sections of intestine.[19] Similarly, fetal medicine specialists have developed increasingly powerful diagnostic tools, but in the great majority of cases the only "solution" proposed to women diagnosed as having a fetal anomaly is to terminate the pregnancy.

Historians of medicine know that a significant lag between the advent of accurate diagnosis and the development of an effective treatment of a given condition is not unusual. In the late nineteenth century, scientists firmly believed that identifying the role of pathogenic microorganisms in producing transmissible diseases would be immediately followed by developing effective cures for these diseases. This was an optimistic view. The new science of microbiology rapidly made important contributions to the improvement of human health: aseptic surgery, vaccination, isolation of the sick, inspection of food and water, and improved sanitation.[20] Yet for nearly a half century, bacteriology made only a modest contribution to the cure of transmissible diseases: a handful of serum therapies (the best known among them being the anti-diphtheria serum) and a few drugs, such as the anti-syphilitic drug Salvarsan (arsphenamine). Effective treatment of bacterial infections came into being with the development of sulfa drugs in the 1930s, and then antibiotics in the 1940s—a half century, that is, after the "bacteriological revolution." PND has reached its half-century mark—which occurred in 2010, if one is looking for an isolated "first," or in 2018, if one is looking for the first widespread use of this technology.[21] PND could very well be on the cusp of becoming a truly curative approach. Should that happen, this book will be of interest only historically, and that would be an excellent thing. But we are not there yet.

In the meantime, a focus on the hope that PND will contribute to a cure for inborn impairments supports (at the time of this writing, 2018) the presentation of a sociological issue—the great majority of fetal malformations detected during pregnancy cannot be cured now—as an epistemological one: in principle, PND can lead to the development of effective cures for the majority of fetal malformations.[22] This hope helps to neutralize professionals' discomfort about the main use of PND—selective abortion of impaired fetuses. Such discomfort can extend to researchers who investigate PND. In my own studies, I frequently wished that I was conducting "real" historical research that investigates people who died long ago, uses exclusively historical and archival materials, and is interested in topics that are no longer relevant. Such research is the exact opposite of the study of PND, which is deeply intertwined with problematic and untidy issues, some of which strike a very painful chord because of their resonance with difficult situations among family members and friends. A scholar who investigates a messy biomedical technology is frequently thrown into a zone of discomfort. The anthropologist Didier Fassin explains:

> A critical ethnography certainly does not "solve" the mystery: it proposes an interpretation of it. This interpretation is not a mere conjecture, though, as superficial reading of interpretive theory would infer. It is an ethnography because it associates the empirical evidence of fieldwork (and, for the historian, critical reading of written sources), and the reflexive concern of writing. And it is critical because it rejects the taken for granted and articulates the objective and subjective dimensions of life in society. Hence the propensity of critical ethnography to "unsettle." But the uneasiness it provokes is not an obstacle to knowledge or action: it is their condition.[23]

My unease in studying PND is rooted in my own mixed feelings. I am distressed by many of the claims of stakeholders in present-day debates about PND: physicians who maintain that only "backward" or "irresponsible" women reject PND, and who paint a uniformly bleak picture of the lives of impaired children; feminists who criticize expectant mothers undergoing PND on the grounds that these women wish to produce a perfect child; people who blame women giving birth to a child with a "preventable" inborn defect and thereby harming their family and society; observers who argue that women choosing to abort an impaired fetus reject human diversity and impoverish our shared humanity; women who had an abortion following a diagnosis of a fetal anomaly and strongly dissociate themselves from "selfish" women who terminated their pregnancy for "social" reasons; epidemiologists who affirm that their goal is to "eradicate" specific inborn impairments; disability rights activists who assert that a "eugenic" termination of pregnancy for a fetal malformation is a statement that the lives of disabled people are not worth living. I'm also bothered by the tendency of many of these PND debate participants to extrapolate from selected exemplary situations to all others, to avoid difficult questions, and to use emotions to achieve specific political goals.[24] In gathering materials for this book, I often have had the impression of walking on a very thin rope, unable to protect myself from hurtful falls. My informants were right: PND is decidedly not a stress-free research topic.

PND and Management of Birth Defect Risk

PND can be defined as a *dispositif*—a heterogeneous array of techniques and approaches that, combined, make the scrutiny of living fetuses possible.[25] Before the late 1960s, physicians and midwives had only limited access to the live fetus through palpation, listening to the fetal heartbeat, and using X-rays. X-rays enabled the visualization of selected fetal anomalies such as dwarfism, hydrocephalus, or Rhesus factor (Rh) incompatibility, any of which may induce changes of the skeleton (fig. 2). However, X-rays detected only a small number of fetal malformations. Moreover, use of this imaging was linked to significant risks to the fetus.

The initial stimulus for the development of PND was to help women from families with a known hereditary condition to have healthy children. Later, however, with the widespread application of PND, its use as a test destined for a small minority of women who had a high probability of giving birth to a child with a specific hereditary disease was switched to its use to screen the fetuses of most pregnant women. This change was favored in

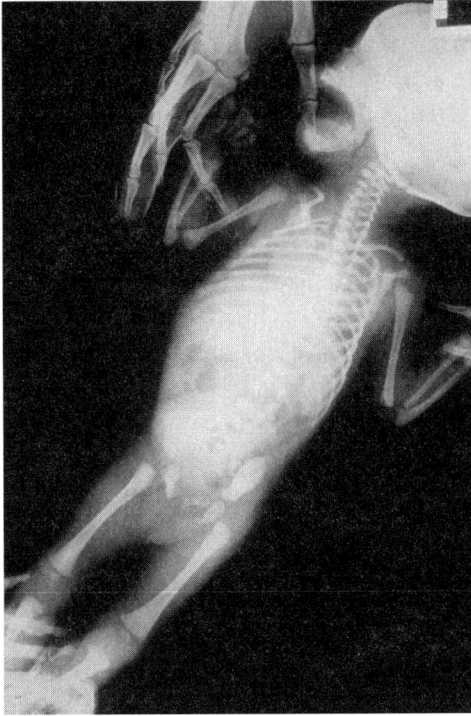

Fig. 2. Erythroblastosis fetalis (Rh factor incompatibility). X-ray of an infant's trunk and upper limbs. Wellcome Images.

many industrialized countries for the prevention of Down syndrome, a condition redefined in the 1970s as a public health problem. At the same time, the dissemination of PND was presented as both an expansion of individual reproductive choice and a public health measure destined to reduce the burden of inborn disability in societies: a private solution to a public health problem.[26]

In the 1950s, physicians, aware of the link between infection with the rubella virus early in pregnancy and birth defects, responded (especially in the United Kingdom and France) to the demands for PND from female patients who had rubella while pregnant and were fearing the consequences for their unborn child.[27] In such cases, abortion for a risk of birth defects was associated not with antenatal diagnosis of a fetal impairment but with the statistical probability of such impairment. The same was true for women who in the early 1960s used the drug Thalidomide early in pregnancy, learned about the link between this drug and fetal anomalies, and wished to terminate their pregnancy.[28] The first "official" PND, a direct study of fetal cells obtained through the sampling of amniotic fluid (amniocentesis), was the diagnosis of a risk.

In 1960, two Danish researchers, Povl Riis and Fritz Fuchs, determined the sex of the fetus of a woman known to carry the hemophilia gene. The fetus was female, and so the woman continued her pregnancy. If the fetus had been male, and consequently would have had a 50% chance of having hemophilia (but also a 50% chance of being healthy), she would have aborted the fetus, within the framework of the Danish eugenics law that legalized abortion for carriers of hereditary diseases.[29] In 1959, several groups of researchers discovered, nearly simultaneously, the link between an abnormal number of chromosomes (aneuploidy) and inborn anomalies. As a result, mongolism (now known as Down syndrome) was redefined as the presence of three copies of chromosome 21 (trisomy 21), Turner syndrome as the absence of one X chromosome (45, X0), and Klinefelter syndrome as the presence of a supplementary X chromosome (47, XXY). In 1960, researchers identified two additional trisomies: Edwards syndrome (trisomy 18) and Patau syndrome (trisomy 13).[30] These studies opened up the new possibility of a cytological diagnosis of chromosomal disorders, at first in adults, then in newborns and aborted fetuses.[31] In the late 1960s, researchers showed that it is possible to display anomalies in cultured fetal cells (fibroblasts) derived from the amniotic fluid (fig. 3).[32]

The new technology enabled the diagnosis of aneuploidies. The presence of an abnormal number of chromosomes in fetal cells is nearly always a random accident in the production of egg and sperm cells, unrelated to family history. At that time, however, amniocentesis was linked to a significant (up to 5%) probability of miscarriage. Scientists therefore focused initially on the PND of hereditary disorders, especially metabolic diseases such as Tay-Sachs disease, glycogen storage diseases, or maple syrup urine disease. In these cases, women were at very high risk (25% for recessive hereditary conditions and 50% for dominant ones) of having an affected fetus.[33] In the mid-1970s, risk of miscarriage following amniocentesis diminished thanks to the use of obstetrical ultrasound to guide the collection of amniotic fluid. Increasingly, amniocentesis was perceived as a relatively safe diagnostic approach, expanding the indications for its use.

Obstetricians and geneticists who promoted PND argued that they were responding to patient demand. This was indeed true for many women aware of the presence of a severe hereditary disorder in their family, and willing or even eager to undergo a risky diagnostic procedure to ascertain that they could give birth to a healthy child.[34] The "women's demand" argument was

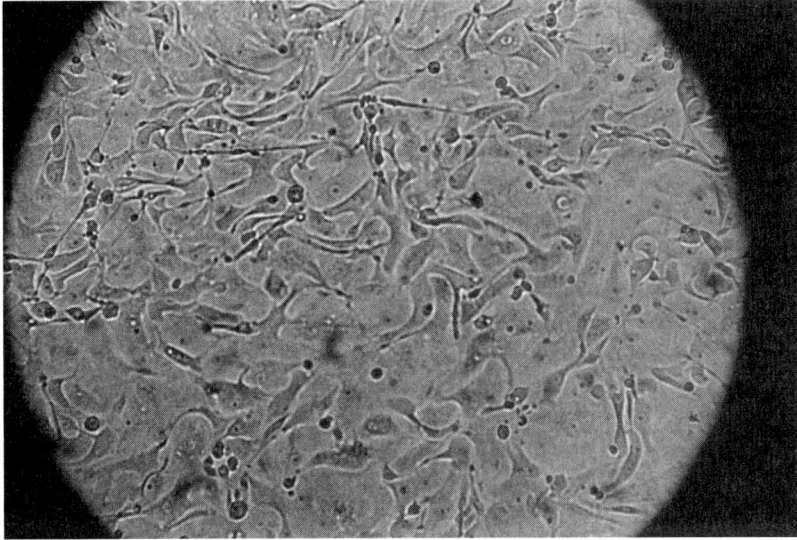

Fig. 3. Abnormal fetal cells in amniotic fluid. Amniocentesis culture of fetal cells from amniotic fluid, low power, light microscope. The contorted shape of these cells indicates that the fetus is abnormal. Wellcome Images.

less true for risks related to advanced maternal age, above all the risk of Down syndrome.[35] In the 1970s, when obstetricians started to use amniocentesis to detect chromosomal anomalies of the fetus, few pregnant women over the age of thirty-five knew that their age put them at greater risk of having a child with this condition. When amniocentesis became more widespread, women learned from their physicians and from the media to recognize age as a risk factor, and older pregnant women began requesting this test.[36] Yet the widespread use of PND was especially related to gynecologists' and obstetricians' desire to open new domains of professional intervention.[37] The professionals' push, rather than the users' pull, explains the rapid spread of amniocentesis during the 1970s and 1980s for determining risk of Down.[38] It also explains the equally rapid spread of the two methods of screening for risk of fetal impairment: serum markers and diagnostic ultrasound.

The history of serum markers of risk of a fetal anomaly is closely intertwined with the history of amniocentesis. Biochemists systematically studied amniotic fluid, looking for possible correlations of the chemical composition of this fluid with inborn defects. In the early 1970s, they noticed that the amniotic fluid of women who carried fetuses with major neural tube defects (anencephaly, spina bifida) contained higher-than-average levels of alpha-fetoprotein (AFP, one of the proteins secreted by the fetus). They found that these women also had higher-than-average levels of this protein in their blood. Testing for the levels of AFP in maternal serum provided a simple and risk-free way of indicating whether the fetus was at risk of neural tube defect.[39] The wide distribution of screening for AFP led to the incidental finding that women who carry fetuses with Down syndrome have an unusually low level of AFP in their blood, prompting the possibility for screening for this condition, too.[40] In early 1993, the sensitivity of a biochemical detection of higher Down syndrome risk was increased through the addition of two additional serum markers. This led to the development of the "tri-test," which aimed above all to detect a higher risk of trisomy 21.[41]

In the 1970s, obstetrical ultrasound, initially used mainly to detect potential gynecological problems and obstacles to childbirth, became increasingly employed to establish the age of the pregnancy and to scrutinize the living fetus.[42] Determining the accurate age of pregnancy became especially important when physicians started to measure pregnancy-related serum markers, since the normal level of such markers is strictly related to the pregnancy's age. In contrast, the usual method of calculating pregnancy age—the time of the woman's last period—is not very reliable, especially in women experiencing irregular menstrual cycles.[43] At the same time, obstetricians began using ultrasound to detect fetal malformations, a development made possible by the rapid increase of the resolution power of ultrasonography.[44] The next important step was the observation, made in the late 1980s, that Down syndrome fetuses often had increased nuchal translucency (a larger-than-average fluid-filled cavity behind the fetus's neck) in the first trimester of pregnancy. Ultrasonographic observations were then combined with the results of biochemical tests to produce an evaluation of the woman's risk of carrying a Down syndrome fetus.[45]

Before the development of screening for the risk of Down, the only way to rule out such a risk had been for the woman to undergo amniocentesis. The procedure is, however, linked to a risk of spontaneous abortion (miscarriage).[46] Physicians proposed this test only to women over thirty-five years of age, at a higher risk of giving birth to a trisomic child—an imperfect solution, since most children with Down syndrome are born to mothers younger than thirty-five.[47] The combination of serum and ultrasonographic markers for Down facilitated the development of more extensive screening for this condition. Such screening was adopted in the late twentieth and early twenty-first centuries by the majority of western European countries and some US states. Pregnant women who undergo the screening receive an estimate of their risk of giving birth to a Down syndrome child. If such a risk exceeds the risk of miscarriage following amniocentesis, health professionals propose that they take that test.[48]

In many settings, screening for Down has become a routine part of the supervision of pregnancy.[49] Moreover, pregnant women who reject such screening may have at least one, and often several, diagnostic ultrasounds during their pregnancy. These examinations detect many fetal anomalies, including ultrasonographic markers of Down syndrome.[50] In this way, a routine ultrasound was gradually transformed into an additional screening tool that detects structural anomalies of the fetus.[51] Nearly all pregnant women in industrialized countries therefore face some form of scrutiny of the fetus. Such scrutiny implicitly leads to the perception of each fetus as malformed until the contrary is proved, and, some critics of PND argue, transforms each pregnant woman into a rational manager of fetal risks.[52]

Methods employed to scrutinize the fetus are the same in all countries, but their use is shaped by local medical cultures. There are important differences in uptake, performance, and regulation of prenatal tests, even among countries with corresponding levels of development and wealth, and similar structures of health care delivery. This is especially true for screening for Down syndrome. Health professionals in France and Denmark strongly promote such screening, which is accepted (or at least not rejected) by most pregnant women in these countries. In the Netherlands, professionals (mainly midwives) do not encourage prenatal screening, and many women—of all social strata—do not seek it. The United Kingdom occupies an intermediary position. Screening is offered to all pregnant women, but its use is uneven, and there are important differences in its uptake among regions and among women from various social classes.[53] Consequently, disparities in uptake of prenatal screening for Down, and in frequency and quality of diagnostic ultrasound, produce parallel disparities in the probability that a pregnant woman will learn about the risk of fetal malformation.[54] In 2017, a Dutch fetus was submitted to less intensive scrutiny than a Danish one, and had a lower probability of being classified as flawed.[55]

More extensive prenatal screening and PND in the late twentieth and early twenty-first centuries produced a new technoscientific entity, "the scrutinized fetus," endowed with entirely novel properties.[56] The present properties of this entity may soon change, with the highly probable widespread application of a new diagnostic technology grounded in the study of circulating fetal DNA (cfDNA) in maternal blood. This approach, called noninvasive prenatal testing (NIPT), is risk free and detects genetic anomalies of the fetus (or, rather, a high probability of the presence of such anomalies) as early as the ninth or tenth week of pregnancy. At first, expert NIPT was presented as an intermediary test, proposed only to women found to be at a higher-than-average risk of carrying a Down syndrome fetus by older screening methods. If NIPT indicated that the fetus was indeed trisomic, the woman was invited to undergo amniocentesis to confirm the presence of an abnormal number of chromosomes.[57] NIPT is, however, predicted to soon replace serum markers as a "first screen" for Down syndrome, facilitating a further extension of such screening.[58]

The main consequence of a widespread application of NIPT will be a drastic decrease in the number of amniocenteses.[59] Another consequence will be earlier detection of fetal genetic anomalies, making earlier termination of pregnancy possible for such anomalies.[60] Pregnancy is a process, and for many women the moral status of the fetus changes during the pregnancy, as does the level of their attachment to their future child. Shifting the timing of PND of fetal genetic anomalies to the first trimester, when miscarriages are frequent and an abortion for societal reasons is seen as more acceptable by people who do not oppose abortion in principle, can modify the status of termination of pregnancies for such anomalies.[61] A further potential consequence of the introduction of NIPT is an increase in the number of genetic anomalies detected before birth. NIPT was initially devised to detect the main autosomal trisomies: 21, 13, and 18. It was then expanded to search for sex chromosome anomalies and, since spring 2014, selected chromosomal deletions.[62] In parallel, the introduction of a new cytogenetic approach—comparative genomic hybridization (CGH)—facilitated the detection of genetic anomalies in fetal cells.[63] Extending the possibilities of detecting anomalies in fetal DNA increases the number of fetal impairments that can be detected during the pregnancy. In some cases, it also increases uncertainty about the potential health problems of the future child.[64]

PND is frequently presented, especially in heated debates about the topic, as a monolithic entity: a technology that detects "flawed" fetuses and enables decisions about their future. Such a view is inaccurate. Far from being monolithic, the PND *dispositif* is a highly heterogeneous assemblage, with many situated applications. Moreover, PND has had important transformations in the half century of its existence: from a diagnostic to a screening device, and from a detector of well-defined, usually hereditary conditions to a detector of a risk, not infrequently of unknown magnitude. The widespread application of NIPT is expected to further reinforce this trend. Yet PND has seldom been investigated as a situated risk management approach. This book attempts to fill that gap.

[handwritten marginalia: diagnosis implies search for evidence of disease - not]

The Structure of the Book

As I propose in this study, PND is a gendered risk management technology. *[handwritten: gender]* Its goal is to prevent the birth of impaired newborns, ideally through the development of effective therapies yet in practice mainly through preventing the birth of children with inborn defects in the first place.[65] It is a risk management technology because a positive PND is frequently the disclosure of risks: risks to the health and well-being of the future child, and risks to the child's family, as it is difficult to predict how a given mother/family will react to problems generated by the child's special needs. Another form of risk is faced by health professionals who fail to accurately diagnose a fetal problem—in some cases a direct risk of being sued for mistaken diagnosis, or facing disciplinary measures and career setbacks; in many other cases, an indirect risk of harming their reputation and self-perception as competent specialists. Risks are also experienced by the community and society at large—economic risks of additional expenses for the care of impaired children and adults, especially in times of restricted health and education budgets, and societal risks of pressures to "normalize" human beings through preventing the birth of those whose bodies are "deviant."

[handwritten marginalia: double bind]

PND is a gendered technology, not only because it is women who become pregnant and face decisions about the pregnancy's outcome, but also because situated applications of PND are shaped by views on the role of women as mothers and caretakers, as well as situated definitions of responsible motherhood. Nonetheless, biomedical technology, risk management, and gender are rarely part of public debates about uses of PND. In this study, I aim to take them to the center of such debates.

In the pages that follow, I investigate different aspects of present-day uses of PND, such as the management of diagnostic uncertainty in clinical settings, the shifting understanding of responsible motherhood/parenthood, and the consequences of introducing PND for attitudes toward living and dead fetuses, and children and adults with disabilities in different settings.[66] The book is grounded in my close reading of archival materials, scientific articles and books, and sociological and anthropological investigations and interviews. It also based on my direct observation of the work of two groups of professionals who play an especially important role in the production of knowledge about anomalies of fetal development: fetopathologists and clinical geneticists. These observations were conducted in France and Brazil, but it is important to remember that each country represents only itself. The study by Dorothy Wertz and John Fletcher, who during the mid-1980s interviewed medical geneticists in nineteen and later in thirty-seven nations, displayed an impressive variability of both beliefs among genetic counselors and ways of implementing PND.[67] Later studies confirmed and extended Wertz and Fletcher's findings. Even countries that may seem quite alike have promoted dissimilar approaches to prenatal screening and testing: the Netherlands is very different from Belgium, and Norway from Denmark.[68]

The two countries I examine in chapters 3 and 4 of this book may be perceived as representing, respectively, western European countries (rich, industrialized, nonreligious, with a well-endowed national health care system) and Latin American countries (less affluent, more religious and traditional, partly industrialized, with uneven access to health care). Such a view is inaccurate. France, with its active promotion of PND, probably does not mirror developments in other countries in western Europe, while Brazil—with its unique combination of world-class hospitals and laboratories, restrictive laws criminalizing abortion for a fetal indication, and highly stratified access to health care despite the declaration of a "right to health" in its constitution— is probably not representative of Latin America. On the other hand, the contrast between France and Brazil underscores the great variability of PND practices and the need to study other countries and other sites.

This study covers new ground, but I could not have written it—or even conceived it—without the existence of many historical, sociological, and anthropological studies about the history of pregnancy, abortion and miscarriage, maternal and child health, the politics of motherhood, clinical genetics, genetic counseling, disability and disability rights, and PND itself. I hope that it sufficiently acknowledges the extent of my debt to scholars who had investigated these issues before I did.[69]

Prenatal detection of fetal impairments is inseparably entangled with selective abortion, and therefore with the status of the entity that develops in the pregnant woman's body. The first chapter, "The Liminal Fetus," studies the indeterminate and fluid status of the human fetus, especially the "biomedical fetus" shaped by new medical technologies. The oscillating status of the fetus/unborn child is at the very center of debates that link PND to the selective termination of pregnancy. Professionals who deal with second- and third-trimester terminations of pregnancy for fetal indications are often divided and undecided as to whether they are dealing with a "fetus," a "product of pregnancy," or a "baby."

The same may be true for some women who undergo this procedure. In the 1970s, when abortion for a fetal malformation became an accepted clinical practice, there were few formal rules about how to treat fetuses outside women's bodies. A variety of factors—the distress of women who underwent an abortion, the uneasiness of health professionals who treated these women, scandals arising from a disrespectful treatment of fetal bodies in hospitals, a greater recognition of the suffering caused by pregnancy loss, and the consolidation of the status of the fetus (partly under the pressure of anti-abortion activists)—led to a codification of rules for the management of fetal bodies. Such codification was also connected with the rise of semi-prescriptive views of the "correct" rites of mourning for dead fetuses and the "appropriate" expression of maternal love after a pregnancy loss. First re-served for miscarriages, such norms were extended to women who elected to terminate their pregnancy for a fetal indication. This chapter focuses on the fluid status of the fetus, the changing trajectories of fetuses inside and outside women's bodies, and their shifting role as symbolic and cultural objects.

Scrutiny of the fetus does not always provide a firm diagnosis of an anom-aly or a firm prognosis of its effect on the future child. As a result, many women are obliged to make difficult decisions about the future of their pregnancy based on partial knowledge. Chapter 2, "Genetics, Morphology, and Dif-ficult Diagnoses," examines efforts to limit the uncertainties of PND through the juxtaposition of results obtained to three approaches: obstetrical ultra-sound, analysis of the fetal genome, and fetopathology. Higher resolution of sonograms, improved operators' skills, and studying fetuses in three di-mensions have greatly extended the capability of scrutinizing the unborn. On the other hand, changes observed on the ultrasound screen are often insufficient to provide a firm diagnosis of a fetal impairment. Women diag-nosed as having structural fetal anomalies during an ultrasound are advised to receive genetic testing. The results of such tests, including observation of fetal chromosomes (karyotype) and a whole-genome-based approach such as comparative genomic hybridization, can clarify the nature of the observed structural anomalies. They can also produce more diagnostic and prognos-tic uncertainty, either because of the variability of the detected mutations' expression or because they detect changes in fetal DNA with unknown clinical meaning. Similarly, ultrasonographic observations can provide a firm diagnosis, but also findings with unknown consequences for the fu-ture child. Dissection data help to clarify the meaning of "suspicious" so-nograms and abnormal genetic findings. This chapter examines dilemmas produced by the observation of dysmorphic changes in fetuses (deviations from the standard form) and atypical genetic findings, and the key role of fetopathology in efforts to both reduce the prognostic uncertainty of PND and produce knowledge about fetal anomalies.

The often invisible labor of fetopathologists is further investigated in the third and fourth chapters. Chapter 3, "Diagnostic Puzzles," grounded in my observations made in the fetopathology department of a major French research and teaching hospital, examines the risk management of birth defects in France. French women can freely decide to have an abortion only until the fourteenth week of pregnancy. However, after fourteen weeks abortion for a fetal or maternal indication is possible only with the approval of an interdisciplinary ethics committee, and is thus fully controlled by medical experts. French fetopathologists dissect miscarried fetuses and those that died in the womb, but many among the fetuses they study were aborted for an anomaly. They collaborate closely with clinical geneticists and have multiple exchanges with ultrasound experts, gynecologists, and pediatricians. These French experts work together in a joint effort to uncover the cause of fetal problems. Such a high degree of collaboration in investigating fetal anomalies is—at least partially—motivated by a shared desire to limit diagnostic and prognostic uncertainty, and to reduce the risk of diagnostic errors which can lead to either the abortion of a healthy fetus or the unanticipated birth of a severely impaired child. This chapter describes French fetopathologists' practices, their interactions with other specialists, and the ways they solve—or fail to solve—difficult diagnostic and prognostic puzzles.

PND is controversial because it is linked to selective abortion. What happens when such a link does not exist, at least not officially? The fourth chapter, "Visible Disasters," is grounded in my observation of a fetopathology department in a major research and teaching hospital in Brazil. It is focused on the management (or absence of management) of the risk of birth defects absent the possibility of (legally) proposing abortion for a fetal indication. The impossibility of legally terminating a pregnancy with a malformed fetus defines and shapes PND in Brazil. Middle-class women treated in private obstetrical clinics have access to PND, and usually can arrange for an abortion of a "flawed" fetus. On the other hand, because the termination of pregnancy is illegal, pathologists study almost exclusively miscarried fetuses, stillborn babies, and dead newborns. Fetal medicine experts who work in public hospitals often practice in private clinics as well, and circulate between two very different social worlds. In private practice they can (unofficially) help women who learn of severe fetal malformations, but cannot verify the accuracy of their diagnosis through autopsy of the aborted fetus. In a public hospital they can study naturally aborted fetuses, but nearly always the only help they can offer women diagnosed as having a severe fetal impairment is to explain the consequences of the detected anomaly. Since nothing can be legally done until the child is born, fetal medicine experts' diagnostic errors in Brazil are not linked to the same professional risks as are similar errors in France. Consequently, collaborations among professionals are less developed, and fetopathologists often work in relative isolation. This chapter focuses on the difficult tasks of Brazilian fetopathologists, who strive to produce clinically useful knowledge with modest technical means. At the same time, they are aware of the limits of their intervention, and know that they are often dealing with the fallout of a system that is unjust to many women.

The argument at the center of this study—that PND is a gendered risk management technology—assumes that the birth of an impaired child can be regarded as a risk to the child, the family, and the community. Many disability rights activists reject this view. Chapter 5, "Balancing Risks," discusses the activists' opposition to selective abortion of impaired fetuses— which they equate with the physical elimination of impaired people—as the extreme expression of a normalizing, health-obsessed society in which disability is described as a tragedy and a situation to be avoided at any cost. Activists assume that a woman's decision to terminate a pregnancy following a positive PND (refusal of *this* child) is radically different from abortion because a woman is unwilling to become a mother at a given moment of her life (refusal of *a* child). Moreover, they often focus on the problems of permanently impaired "healthy disabled people," who with appropriate support can lead an autonomous or quasi-autonomous life, such as members of the Deaf community, visually impaired people, or people with reduced mobility. PND was developed to diagnose different kinds of disabilities: hereditary diseases, severe multilevel impairments, and intellectual deficiency. Many people with such disabilities need a heavy investment in their care. Care tasks may be especially challenging when dealing with people who have substantial intellectual disabilities, who may need lifelong care—often provided by their mothers, who may predecease them. This chapter investigates the multiple meanings of the terms *disability* and *birth defect*, and problems associated with the care of people with inborn impairments, especially those with intellectual limitations and neuropsychiatric complications. It also examines the challenging decisions of women who learn that their future child may be affected by such difficulties, but cannot be told what their magnitude will be.

PND was originally used in the effort to prevent the transmission of familial hereditary conditions. Yet in the twenty-first century, only a small fraction of pregnancy terminations for a fetal indication (5–10%) are connected with such conditions. Nonetheless, the management of hereditary disorders continues to play an important role in reflections on PND. In addition, the number of prenatally detected hereditary conditions may increase with the extensive use of new genomic tests that diagnose such conditions in fetuses and adults. Chapter 6, "PND, Reproductive Choices, and Care," focuses on the reproductive decisions of women aware of a hereditary condition in their family. It examines how women and their partners balance the multiple risks they may face—to their future children, themselves, and their families—when confronted with the choice of whether to undergo PND and consider a selective termination of pregnancy. Such a decision is often presented by participants in debates about prenatal diagnosis and selective abortion in absolute terms, either entirely positive—promoting the birth of healthy children and putting an end to a familial "malediction"—or entirely negative—disavowing the value of the lives of family members with an inborn condition. In practice, however, one can observe a great variety of situations, patterns of perception of specific hereditary disorders, and attitudes toward PND. Women's reproductive decisions are intertwined with the more general issue of situated and contradictory views of women's duties and obligations as mothers of children with special needs. This chapter follows debates about PND and hereditary/inborn conditions as well as their emotional underpinnings, especially the multidimensional politics of maternal and paternal love.

Fetuses have become increasingly accessible to the experts' gaze, but as the concluding chapter, "A Nonscrutinized Diagnosis," argues, PND itself has eluded critical scrutiny. The focalization of debates about PND on the abortion of "flawed" fetuses and these debates' transformation into "total social conflicts" have deflected attention from situated dilemmas and the concrete problems produced by this biomedical technology. Some of these problems—such as false-positive and false-negative results, the risk of embarking on a taxing diagnostic odyssey, the stress produced by prognostic uncertainty, and the consequences of incidental findings—are shared with other screening and diagnostic procedures. Others—such as the material aspects of interruption of pregnancy, the liminal status of the fetus, and women's ambivalence about their reproductive decisions—are specific to PND. But concrete, situated problems generated by the widespread use of PND are often absent from debates about this technology, which are frequently centered on abstract questions and broad moral issues. The conclusion ends with a subjective list of topics, especially those I consider problematic and unsettling, which are missing from present deliberations about PND.

A positive PND—that is, the diagnosis of a serious anomaly of the fetus—is always an indication of a forthcoming loss: either loss of pregnancy or loss of hope for the birth of a healthy child. Academic research has a limited capacity for dealing with emotionally charged issues. Art, by contrast, provides a very different means for reflecting on uncertainty and sorrow. The book's coda, "Minerva's Owl and Apollo's Lyre," juxtaposes two artworks—one relatively unknown and the other famous—which deal with pregnancy loss, reproductive ambivalence, and the liminal status of the fetus: a 2010 collection of short stories, *Obsoletki* (from the Latin term for pregnancy loss, *gravidas obsoleta*), by the Polish poet Justyna Bargielska, and a 1932 painting, *Henry Ford Hospital*, by the Mexican artist Frida Kahlo. Although very dissimilar in intention and execution, both works combine raw sentiment, empathy, and irony, and both firmly place failed pregnancies within the fragile, glorious, messy, tragic, banal, ambivalent, risky, and hopeful stream of life.

Finally, I would like to note my inconsistent use of the terms *pregnant woman/mother* and *fetus/child*. Pregnant women who accept their pregnancy talk, from the very beginning of the pregnancy, about their future baby/child. Health professionals frequently call a pregnant woman "mother," even in situations where it is likely that she will decide to have an abortion. The usual term for fetal DNA found in the blood of pregnant women is "free fetal DNA in maternal circulation," despite the fact that the goal of studying the free fetal DNA is to provide information about fetal malformations, which can lead to the termination of the pregnancy. Similarly, the entity in the woman's womb is usually called the embryo until between the eighth and tenth week of pregnancy, when it is called the fetus, but physicians often talk about the woman's baby/child. If the woman has a miscarriage or an induced abortion, rules on the disposal of the expulsed body always use the term *child*, while fetopathologists who dissect the body always speak of the fetus. My inconsistency in the use of these terms follows the stakeholders' inconsistency. Outside the rarified atmosphere of philosophical and ethical debates, human reproduction is rarely simple or free from ambivalence. The uncertain and shifting status of the pregnant woman and the entity that is developing in her body, and the difficulty of stabilizing them, are at the center of this study.

Debates about PND are dominated by the thorny question of selective abortion.[70] However, an exclusive focus on this issue masks all the other aspects of PND as a gendered risk management technology: material, professional, organizational, institutional, juridical, and economic. It also masks the heterogeneity of PND's uses and the irreducible complexity of specific situations. In studying a contested biomedical approach, God (in French) or the Devil (in English) is in the details, or rather in the juxtaposition of details, which produces a thick, multilayered description. In this book, I aim to examine closely the contextualized uses of the PND *dispositif*. I follow the advice of the British sociologist Anne Kerr: avoid dwelling on thought experiments about choices that may become possible in the future, and concentrate instead on the messiness and complexity of the present in a time of uncertainty, as well as the things that really matter to people in the business of reproduction.[71] With the advent of new genomic technologies, the real-life business of reproduction is changing rapidly. Earlier radical transformations of PND, such as widespread screening for Down syndrome and the use of obstetrical ultrasound as a powerful diagnostic tool, were adopted without public scrutiny of their potential consequences. The advent of non-invasive prenatal testing and the predicted arrival of other diagnostic approaches grounded in recent genomic technologies are expected to radically modify scrutiny of the fetus once more. In the face of such developments, it may be helpful for the public to shift its gaze from sweeping generalizations, abstract ethical principles, and futuristic speculations to the tangled and untidy real-life effects of an extensively used biomedical innovation that makes it possible "to see what is about to be born."

ONE

The Liminal Fetus

non selective abortion [handwritten annotation]

PND and the Aborted Fetus

PND has become a risk management technology thanks to its use in determining impaired fetuses for the possibility of selective abortion.[1] Selective abortion became and remains the primary intervention of genetic and fetal medicine.[2] The latter development was facilitated by the liminal status of the fetus, a material entity endowed with a fluid and unstable status. It was also facilitated by the long history of nonselective abortion, used as a method to control female fertility.[3] Abortion opponents firmly believe that life begins with conception, and that the status of the embryo/fetus is equivalent to that of the newborn child. People who do not unconditionally oppose abortion have more fluid and indeterminate perceptions of the unborn. The same can be said of women who undergo an abortion; moreover, these perceptions mirror their attitude toward their pregnancy and may change with time.[4] Contrasting images of PND reflect such perceptions.

The fetus, as the anthropologist Lynn Morgan notes, has become an icon. Images of fetuses are everywhere: in textbooks and popular books, in health publications, in the media, and on billboards. These omnipresent images are the result of the impressive labor involved in making many other things invisible:

> The embryo visualizations portrayed in a contemporary coffee-table book about gestational development is to be a remarkable political achievement predicated, in part, on keeping hidden the unsavory details of anatomical technique that transform dead specimens into icons of life.[5]

Unretouched photographs of dead fetuses and images of their dissections, Morgan argues, are nearly universally considered revolting.[6] She remembers seeing by chance the photograph of a severed fetal head: the image haunted her for a long time. It is nevertheless important to remember how fetuses really look:

> It is undoubtedly easier and more pleasant for feminist scholars to study beautiful artistic renditions of fetal development or those produced through sanitizing reproductive imaging such as ultrasound, than it is to contemplate the gory techniques and practices described above. Yet by focusing so exclusively on the visual practices through which embryos and fetuses are materialized, feminists risk colluding in the cultural practices that produce and perpetuate such beautified fetal images.[7]

The invisibility of the "real-life" fetus is inseparable from the invisibility of products of abortion. Scholars who study abortion seldom investigate the material aspects of this procedure. It is difficult to discuss the harsh physiological aspects of termination of pregnancy for fetal malformation, especially in the second and third trimesters of pregnancy. The invisibility of material aspects of abortion contributes to the stigma associated with this procedure. Women who decide to terminate a pregnancy for a fetal anomaly are caught in the contradictions between the expectation that women love their "baby" from the moment they learn they are pregnant, the belief that it is better not to give birth to a severely impaired child, and the difficulty of the procedure itself. Those who terminated a pregnancy following a PND of a fetal anomaly have reported intense feelings of grief and despair, which sometimes began up to a few weeks after the abortion, and which refused to go away, despite their health care providers' earlier reassurance that everything would be all right. The intensity and length of the grief following their supposedly rational decision frequently caught these women unawares.[8]

Women who decided to terminate a pregnancy for a fetal indication may be harshly judged for that choice by opponents of abortion, while pro-choice activists who rigidly adhere to the view that a fetus is not a baby may fail to empathize with their very real grief. As a woman who aborted following a positive prenatal diagnosis explains, "It's hard to use people like us as political propaganda when we freely admit that we aborted our babies. When so much of the debate is based on when life does begin and when we proclaim that our babies were alive and are our children. That makes it messy."[9]

In the early days of abortion for fetal indications, the confusion and difficulty of this situation were already acknowledged by the physicians involved in the diagnosis of fetal disorders. Experts who counseled pregnant women saw themselves as directly responsible for the consequences of the diagnoses they had made. The British geneticist Bernadette Modell, one of the pioneers of PND for thalassemia (a severe hereditary blood disorder), remembers:

> When the news got about, people started getting on planes in Cyprus, Italy and Greece to come for the service. We were under no illusion whatsoever that this was a service that was needed. But the experience was awful because we were only seeing patients at 25 per cent risk, which meant that we had 25 per cent affected fetuses, and 25 per cent of those women terminated that pregnancy, and they came at 18 weeks and they terminated at 20 weeks. As part of our clinical responsibility, we accompanied them; sat with them through the termination; showed them the baby and discussed the baby with them at the end. . . . The experience of genetics for these women at that time was with techniques for prenatal diagnosis which were at the absolute borderline of acceptability.[10]

The sense of responsibility was also felt by genetic counselors. In the early days of PND, many counselors were involved with the women's health movement and had a strong commitment to women's reproductive rights. They, too, felt the consequences of their interventions. June Peters, a US genetic counselor who started practicing in the early 1970s, recalls:

> You couldn't do amniocentesis until 16 weeks, we didn't have results to disclose until 18, maybe 19 or 20 weeks. It was late. If the baby had to be delivered, the mother was admitted to the hospital and labor was induced. So I'd go sit with the moms and wait, and wait, and wait for hours and sometimes days. And they'd talk about their mixed feelings as the pregnancy was ending about how they really wanted this baby . . . and it was tragic for them to have to terminate. Then I was there when the baby was delivered. Most of the fetuses were still-born. They were the size of your hand; we would wrap them up in miniature blankets or towels and bring them to the parents and help them say goodbye. And afterwards, we would go to the autopsy.[11]

From the 1980s onward, the geneticists' and genetic counselors' task became dissociated from the material aspects of termination of pregnancy and from the contentious issue of handling the products of such a termination.[12] Yet today the fate of the aborted fetuses remains a very difficult topic.[13] Discussions about PND tend to steer away from the minefield of the practical modalities of abortion and from the fate of the fetal bodies outside the womb. Researchers who study PND usually focus on the decision to terminate a pregnancy, and stop short of asking what happened next to the woman and to the fetal remains.

Choosing—or Not—How to End a Pregnancy

In some cases, the termination of pregnancy for fetal indications, especially genetic anomalies, takes place at the end of the first trimester. In other cases, it occurs during the third trimester, because some fetal anomalies, such as malformations of the brain, are visible only late in pregnancy. Yet most abortions for a fetal indication are conducted during the second trimester (thirteen to twenty-seven weeks). In France, abortion for a fetal indication is performed in a maternity ward by the same physicians and midwives who attend a regular childbirth.[14] The organization of medical labor, the spatial arrangement in the hospital, and the legal and administrative decisions blur the boundaries between a natural miscarriage, fetal death in the womb, and a termination of pregnancy following a positive PND.[15] The second- or third-trimester termination of pregnancy is always a medical interruption: induction of expulsion of the fetus by drugs, then a birth-like process. Only those performed during the first trimester of pregnancy, 10.3% of all the abortions for fetal indications in France, are surgical abortions by aspiration.[16]

Information leaflets distributed to French women undergoing an interruption of pregnancy explain that past the first trimester, the only alternative to a medical termination is a C-section, described as a bad choice because it weakens the uterus and may compromise the woman's reproductive future.[17] These leaflets also explain that while some women scheduled for this procedure are initially shocked to learn that they will be going through labor, this is by far the better solution for them: "Parents have taught us, that, far from being traumatic, a well-accompanied birth process helps them to cope with this difficult event."[18] Many French midwives are convinced that this is indeed the case. A normal childbirth is a physiologically extreme event, but its intensity is attenuated by an expected happy ending: the presence of a baby. The unavoidable physiological violence of an abortion for a fetal malformation does not have such a silver lining; nevertheless, midwives believe that the process of actively giving birth facilitates the grieving process. A minority among them also think that women who elect to terminate the pregnancy should bear the physical consequences of their choice.[19]

The explanation given to French women that when a woman is more than fourteen weeks' pregnant her only choices are natural birth or a C-section is, to put it mildly, inaccurate. When an interruption of pregnancy takes place before twenty-four weeks, it is also possible to propose a surgical procedure: dilation and evacuation (D&E), performed under general anesthesia. In D&E, the fetus is dismembered before the elimination.[20] There is

birthing death

no fetal body and no dilemmas about its management. For some women, this is an important advantage: they are spared the experience of "birthing death." For other women, this is an aggravating element: they want to be able to see and touch the fetus and perform mourning rites.[21]

D&E was developed in the United States as an attempt to improve second-trimester abortions.[22] The first method employed by physicians to induce second-trimester abortions was the injection of saline into the amniotic sac, a slow and painful process. It was later replaced by the use of prostaglandins, hormones promoting the expulsion of the fetus. Variants of this method continue to be employed today, using more effective products: mifepristone (RU486), misoprostol, or a combination of both medications. Paralleling the development of medically induced abortions, US surgeons in the 1970s refined the D&E technique.[23] They described this method as an important improvement over the medical induction of abortion, because it causes less discomfort for the patient, is much faster, and yields a more certain result.[24] More than 90% of second-trimester terminations of pregnancy in the United States continue to be surgical, although increasing limitations on the provision of abortion may seriously restrict women's access to this method and to second-trimester abortions in general.[25] D&E is the most frequently employed method of second-trimester terminations worldwide, although in some European countries such as France, Finland, and Sweden, practically all the second-trimester abortions are medical, while in others such as the United Kingdom, it is used much less frequently than a medical termination.[26]

Comparative studies indicate that in the second trimester, D&E may be seen as a good choice because, among other reasons, this intervention produces lower rates of post-abortion complications than a medical abortion.[27] In addition, it is experienced by many women as a less painful and less traumatic procedure.[28] Attempts to conduct randomized trials of the two abortion methods in the United States (where D&E is widely available) failed, because the majority of women preferred a D&E and refused to be randomized to the medical abortion branch.[29] All the women interviewed wanted to be able to choose their abortion method, and those able to decide how they would abort reported satisfaction with the method they had chosen.[30]

French women are not given such a choice, and it is reasonable to assume that only some of them know that such a choice exists. They are obliged to confront the expelled fetus and to decide how they should deal with it. French specialists claim that D&E presents a greater health risk to the woman.[31] They also argue that D&E deprives the woman/couple of the possibility of seeing and holding the fetus and "properly" mourning the pregnancy loss.

Another disadvantage of D&E, they say, is that this method destroys the fetus's body, a serious obstacle to an investigation of the causes of fetal malformation.[32]

Nonetheless, the main reason for the unavailability of D&E in France and other countries may be the higher cost of this procedure, the need to train physicians and nurses to perform it, and the emotional difficulty for health professionals to carry it out.[33] The statistician Sarah Lewit, a leading US expert on contraception and abortion, explained in 1982 that with the D&E procedure, the psychological burden was lifted from the patient and shifted to the physician, "who henceforth must rely on a strong sense of social conscience which focused on the health and desires of the women."[34] In countries that reject the option of D&E for second-trimester abortions, physicians may be unwilling to shoulder this burden.

Extreme Cases: Late-Term Abortions and Nonviable Fetuses

If a fetal malformation is diagnosed after twenty-four weeks, the only choice for a woman who wishes to terminate the pregnancy is to undergo a medical termination. Such a late abortion is, as a rule, accompanied by feticide—an active killing of the fetus by injection of a lethal product before the induction of its expulsion. Most health professionals consider feticide as an indispensable element of an abortion after the viability limit. They find the alternative—birth of a living, breathing, and occasionally crying child destined to die—insupportable. Feticide, an especially traumatic experience, is rarely discussed in the context of induced abortion. Health professionals are reluctant to discuss it, while women who never heard about this procedure until they were expected to undergo it wonder whether the questionable morality of this act is the main reason for the quasi impossibility of discussing their experience. The Israeli sociologist Ronit Leichtentritt, one of the few social scientists who studied feticide, reports that women she interviewed vividly remembered the moment of fetal demise:[35]

It was the worst moment in the feticide procedure—the experience of your baby dying inside you.—Sara

It took forever until he finally managed to inject that stuff into her. Then he needed to inject some more. And again some more. He kept on saying she is moving—it's like he was surprised she wasn't cooperating.—Tami

This is the worst part, when they [*pause*] when they inject [*pause*] they gave me Valium by injection at that point because I was in a terrible shape. I couldn't stop shaking, my whole body was shaking.—Hadas

I saw the last breaths he took. I saw the change, from being alive to [*crying*]. I saw the indications of his last breaths.—Ruth

You feel that these are his last movements, they're slower. . . . It's horrible. It haunts you later, the sensation of a body changing from something alive and kicking to something dead. . . . I could feel him dying slowly, resisting death, his last movements.—Noa[36]

Prognostic uncertainty may aggravate the women's distress: as one of the women who underwent feticide explained, "Maybe I took his life by mistake. Maybe I did the most horrible thing and maybe not."[37]

The violence of a late-term abortion coupled for some women with an unwillingness to feel responsible for the death of a well-formed child may explain why women who learn late in pregnancy that the fetus has a lethal malformation sometimes elect a natural birth and the demise of their child.[38] As one woman put it, "She was alive and kicking. Termination was not an option to me. It was not the right thing to do it, morally."[39] The Israeli anthropologist Noga Weiner describes the case of a woman who had one living child with a genetic disease. Her second child, born severely impaired, died in the hospital's intensive care unit after the child's physicians decided to withdraw care. In her next pregnancy, this woman underwent PND, found out that the fetus was affected, and elected to terminate the pregnancy. In her fourth pregnancy, she refused PND, partly because her physicians implicitly reassured her that they would again withdraw intensive care if the newborn child was severely disabled. This woman had found the process of separation from a live child less traumatic than a late-term abortion. She was haunted by her recollection of the aborted fetus looking healthy, whole, and intact. By contrast, she was able to see the struggle and the suffering of her newborn child and was willing to let him go.[40]

Some women decide to continue a pregnancy after the diagnosis of a lethal condition of the fetus. When the refusal to abort a fetus destined to die is justified by the woman's religious faith or her "cultural background" (often shorthand for non-Western origins), physicians usually support her decision.[41] Religious women can see the birth and the short life of a nonviable child as a positive event rather than a senseless tragedy. This was the case for a Brazilian adept of spiritism (a religion that strongly opposes abortion), who perceived the birth and short life of her daughter, who had a severe chromosomal anomaly, as edifying and transformative events.[42] Physicians may be much less understanding when the woman's/couple's motivation is not religious and they are not recent migrants. A Norwegian couple, anesthesiologists who learned that their baby girl had trisomy 18, described in 2012 the lack of support for their decision to continue the pregnancy and enjoy their time with their daughter, however limited. According to this couple, nearly all the physicians they met spoke about a "lethal baby" and hinted that they dealt with a different species, not a human being. They were also distressed to discover that in Norway, trisomy 18 babies were not monitored during the delivery and did not receive medical care after birth.[43]

In the United States, associations such as Be Not Afraid provide advice and support to women pregnant with a fetus diagnosed as having a condition usually incompatible with life.[44] Such support can also be provided by perinatal hospices. The first project of such hospices, writes the freelance journalist Amy Kuebelbeck in association with the US National Hospice and Palliative Care Association, is to focus on the need to support pregnant women who decide to continue a pregnancy when the child is expected to die during birth or shortly afterward. Kuebelbeck explains that choosing to continue a pregnancy after a diagnosis of a lethal malformation of the fetus is not desperate but rational and imbued with hope. The parents "may hope the baby is born alive. They may hope that the baby be treated with dignity. They may hope that the baby be remembered. Those are profound kinds of hope." Kuebelbeck, who did not write from a religious perspective, links palliative prenatal care with palliative postnatal care for newborns with severe conditions whose parents decide to forgo aggressive treatment and let the child die peacefully. In both cases, she notes, the goal is to promote peaceful death, facilitate grieving, and support the inscription of the child's life in the family's history.[45] Later, however, the project of perinatal hospices was taken over by anti-abortion organizations such as Americans United for Life. In 2014, that organization proposed the Perinatal Hospice Information Act, which aggressively attempts to dissuade women from seeking abortion for lethal fetal malformations, and argues that women who decide to continue the pregnancy achieve sublimation and peace through the uplifting experience of the short presence of their dying/dead child.[46]

An additional element in the debate about perinatal hospices is the variability of survival among children with some of the conditions labeled as lethal. This term covers malformations, such as anencephaly or the absence of kidneys, which invariably lead to the child's rapid death, often several hours after birth, but also other conditions, such as trisomies 13 and 18, which in selected cases allow for a much longer survival: months, and sometimes years. Children who live with these trisomies are severely disabled, physically and mentally. Nevertheless, families that elected to give birth to these children and are active in specialized support groups claim that the child was happy and that her/his brief life enriched the family's experience.[47] By contrast, the survival of these children was often achieved thanks to a significant investment of medical resources.[48] Nowadays, there seems to be a general agreement that women should not be pressured to terminate the pregnancy following a positive PND. Such an agreement may be eroding, however—as seems to be the case already in Vietnam—whenever health professionals are convinced that their duty is to persuade women to make the "right" choice for their future child, their family, and society and abort an impaired fetus.[49]

Handling the Products of Abortion

In France, abortions for fetal indications are supervised by midwives. In the United Kingdom, late termination of pregnancy is delegated to gynecological nurses. In that country, the spatial organization of the hospital and its internal division of labor radically dissociate childbirth that leads to a live birth from an identical physiological act that ends with a lifeless body.[50] D&E is rarely used, and the gynecological nurse manages the long process of the expulsion of the fetus. Her work is seen by other health professionals as important and skilled, yet also as a "dirty" job of handling dead bodies. Gynecological nurses have been described as bothered by criticism of their involvement in late-term abortions, yet proud to be helping women in a difficult moment of their lives. As one of these nurses explained,

> Midwives get all the praise for delivering healthy babies and they take all the credit. But ask them to take a late miscarriage and they will do everything in their power not to have to deal with it, even when it would be in the woman's best interests to be on the labor ward. That's too much like hard work for them. You can actually see them wrinkle their noses sometimes at the thought of it. They don't know how to deal with the trauma of it; they have no idea really. If there isn't a baby you can coo over and happy smiley parents then forget it as far as obstetrics are concerned.[51]

In Canada, too, a late termination of pregnancy is accompanied by gynecological nurses. Their task is viewed by their coworkers as essential and highly skilled, yet "unclean." In addition, it often has a low visibility, because some hospitals fear interventions from anti-abortion activists. Canadian gynecological nurses frequently feel isolated and insufficiently recognized. They have complained about their lack of experience in guiding women through what essentially was a birthing process, and the difficulty of dealing with the product of this process, sometimes described as an "expulsed fetus" and sometimes as a "stillborn child." These nurses report that some women who undergo a termination of pregnancy do not understand that they will be having a full birth, making the situation even more stressful. Another source of stress is a near-total absence of physicians on the ward. Although they are readily involved in many aspects of "normal" births, physicians are reluctant to participate in the care of women undergoing an abortion, another sign of the inferior/stigmatized aspect of this procedure.[52]

In the 1970s and 1980s, feticide was not a routine part of an induced abortion procedure in Canada, so sometimes the fetus/baby was born breathing. One nurse recalled that when first faced with such a situation, "I was shocked! I didn't know what to do so I called the doctor and asked him what to do. He simply said, 'Drop it in the saline solution.' There was no way I was going to do that. These babies are human beings that deserve to be treated with dignity and caring." She ended up wrapping the baby in a blanket, then waited until it had stopped breathing before taking it to the mother. Another practical difficulty was the lack of dignified transport and storage of dead fetuses. Gynecological nurses commented on the types of containers that were filled with saline solution and used in transporting the aborted fetuses, especially smaller ones, to the pathology department. One such container was a peanut butter jar; another was a twenty-four-hour urine bottle; still another a Tupperware jar.

The absence of standardized procedures for dealing with fetal bodies—at once potential human beings to be treated with dignity, surgical waste, and specimens to be conserved—along with a lack of standardized recipients for these bodies, also underscored the difficulty of classifying and containing dead fetuses. The unstable status of the fetus was especially visible when the parents, encouraged by gynecological nurses, spent some time with their dead baby, then gave the body to nurses, who placed it in a container of saline solution in order to transport it to the pathology department. Not infrequently, the parents would change their mind and want to see their baby again. Most nurses did what they could to retrieve the body. This meant removing the body from the saline solution, rewashing and re-dressing it, and also warming it before it could be presented to the parents.[53]

French midwives, too, have been distressed by their major role in the termination of pregnancy for fetal indications.[54] In interviews about the most stressful aspects of the procedure, they pointed to the strain of providing psychological support to the couple, the loneliness of their task, and the difficulty of handling the dead fetus/baby. Nearly all agreed that the wide-spread use of feticide in late-term abortions—a procedure accomplished by a physician—makes their work easier. One of their greatest fears is a patient's giving birth to a live child after an elective termination of pregnancy.[55] Correspondingly, physicians assisting in induced abortions say that feticide is the most difficult part of their job, even when they are certain that the child cannot survive outside the womb. Despite their fears, midwives view this act, which clearly differentiates termination of pregnancy from birth and an aborted fetus from a living child, as an indispensable intervention, a view that attenuates the stress it produces.[56]

French midwives gathered in a focus group to discuss the difficulty of dealing with situations in which birth coincides with death, and said they often did not know the right words for speaking about such situations:

> I say childbirth [accouchement].
> At 15 weeks—I wish we had a word to speak about it.
> I try to say nothing . . . to avoid the terms "baby" or "fetus" . . . I say "he," "she."
> It's their baby, non?
> An interruption at 21 weeks, you speak about a baby?
> I don't think so. I say nothing.
> For me it's their child.
> He or she, something, but not a baby.
> For me it's obvious that one should personalize.
> I'm surprised by this question, I never thought about it.
> For an early abortion, one does not speak about a child. For termination at 18–20 weeks, one speaks about childbirth.
> When you pronounce the word birth.
> As if one could not speak about birth. But it is a birth of a dead child! He needs to go through [passer] this child! Birth is also a transition.
> Yes it is a transition.
> And the official papers say: she gave birth to a dead child.
> Born without life [né sans vie]; this is the official term.
> Therefore a child is born![57]

Several midwives were aware that they always use gloves when they handle the bodies of fetuses/babies who were born dead, yet never do so when touching a living baby: "Dead babies is not something I can touch with my hands . . . and yet they are not dirty . . . still, the contact with death. Perhaps to put a barrier between me and the body, to keep a distance, to protect myself. I put gloves on every time, I just realized this while talking about it."[58]

The Fetus's Multiple Identities

Debates about PND and selective termination of pregnancy focus on the fate of the fetus. For some scholars, such as the German philosopher Barbara Duden, the fetus—or at least the public fetus—is a very recent invention, a combined consequence of the development of obstetrical ultrasound and the abortion controversy.[59] Yet fetuses are far from being newly invented entities. They have existed as objects of medical and scientific curiosity since at least the Renaissance, and the prevention of miscarriage, stillbirth, and newborn deaths had been regarded as important issues long before the advent of new approaches to prenatal care.[60] In the nineteenth and early twentieth centuries, fetuses were also objects of public curiosity.[61] Emerging scientific developments and medical technologies nevertheless modified the status of the unborn through a shift of the medical gaze from already formed fetuses to embryos. In the past, women had been sure that they were pregnant only when they felt fetal movements, which occur in the early second trimester. From the 1930s onward, a positive pregnancy test became an inconvertible proof of pregnancy. Then, from the 1970s onward, following the widespread availability of home pregnancy tests, many women considered themselves to be pregnant in the first month after conception.[62] Before the advent of these tests, a chemical pregnancy, which ends with a very early miscarriage, would have been seen as delayed menstruation. After the use of such tests became common, a chemical pregnancy became a real event, interpreted by some women as the loss of a child.[63]

In vitro fertilization technologies similarly reinforced the perception of the embryo as a "baby."[64] Women desperately trying to conceive have a high level of emotional investment not only in their pregnancy (if they are successful in becoming pregnant) but also in the embryos produced in the test tube with their eggs and their partner's or donor's sperm. These embryos, microscopic agglomerates of cells, are seen by many women as their potential offspring, and by some even as their already existing "children." Many women are willing to give or sell their ovocytes to another woman.[65] By contrast, women/couples are much more reluctant to donate their supplementary frozen embryos to another couple ("embryo adoption") or for research, and elect to have these destroyed instead.[66]

zarodzkyos

The Polish anthropologist Magdalena Radkowska-Walkowicz has observed the attachment of Polish women undergoing infertility treatment to their frozen embryos, which are kept in the fertility clinics' liquid nitrogen tank.[67] Women she interviewed and those who participated in specialized Internet forums spoke about the "tiny embryos" (using an endearing diminutive, *zarodeczki*). Typical statements were as follows:

> My embryos are my children . . . from the moment of fertilization each tiny embryo is a human being. . . . I keep thinking how my little ones are doing. . . . I love them from the first moment and keep thinking about them. . . . I have two tiny ones frozen in the clinics; I cannot stop thinking about them. . . . I want to convey to all my five embryos the message that they are wanted. . . . I'm waiting for each one of them and each one will receive his or her teddy bear. . . . My husband promised our embryos that he will buy them plenty of Lego building blocks and asked them to develop well, because the building blocks will be waiting for them.[68]

One of the Polish fertility clinics in Radkowska-Walkowicz's study advertised its services as "love from the first cell." Clinics she researched gave their in vitro fertilization patients photos of "their" embryos, and occasionally films that recorded development. One woman posted that she could not stop looking at the photos of her tiny embryos. Another shared that after she was shown photographs of her embryos, "you just look at them and you think: this is your child, at the age of five days, it's absolutely extraordinary." A woman who received a film of the early development of her embryos said that she was surprised by how dynamic this process is: "One can see that before the division the embryo is trembling; it's like struggling to undergo a division!" Women who "love from the first cell" and then face a failed embryo transfer spoke about their grief when the embryo—or, perhaps, their first child—died.[69]

In some circumstances at least, embryos and fetuses in the West can become children from the moment of fertilization. And in the twenty-first century, they can have a long post-miscarriage life: the sociologist Mary Ebling discovered that her dead fetus continues to live in cyberspace.[70] Non-Western fetuses have a different status. In rural India, some observers have noted, miscarriage is culturally insignificant, as are stillborn babies. Women speak about loss of pregnancy without emotion, while they openly acknowledge the pain of a child's death.[71] In Ecuador, Lynn Morgan has explained, abortion is seen as a transgression, not because the fetus has a "personality," but because its elimination is seen as self-mutilation, and as guilty opposition to God's will. The entity in the pregnant woman's belly is not called a fetus—even by educated women—but rather *criatura* (creature) or *venidero* (the one to come). Pregnant women view themselves as fused with the creatures they gestate. On the other hand, the precise nature of this "creature" is neither fixed nor consensual. Ecuadorian women often contradict themselves when asked whether the fetus is formed, and how such formation might affect the legitimacy of abortion. They also disagree on ways to properly dispose of miscarried or stillborn children, and on their status: should they be considered as a woman's children or as potentially dangerous entities, apt to be transformed into frightening ghosts, *aucas*? The coexistence of such contradictory opinions on the nature of the *criatura* is not seen as problematic. Ecuadorian women do not think it necessary to have a single view or coherent position when questioning one of nature's profoundest mysteries.[72]

Cameroonian women, too, are often ambivalent about abortion. They regard it as illegitimate; tolerated, especially in early pregnancy (up to five months), when the fetus is perceived as an "unformed" mass of water and blood; and as justified by the woman's social suffering. And also like women in Ecuador, they maintain a deliberately ambivalent view concerning pregnancy loss, which may be spontaneous, induced, or, when a woman who attempted to get rid of a pregnancy in its early stages miscarries later, both provoked and spontaneous.[73]

Many societies, Lynn Morgan and Claudia Browner argue, do not recognize embryos, fetuses, and even young children as persons, or even as humans, until these social designations are bestowed on them by the social group. Persons are created through the active agency of others, and some anthropologists argue that an intuitive, maternal, and metaphysical respect for embryos and fetuses is no more universal than belief in the tooth fairy.[74] Not only can the status of the fetus vary, but in many cultures birth is not seen as the crucial moment of acceptance of a newborn into society: additional rites are needed for such an acceptance. The killing of a newborn is not a murder, and sometimes even not a serious transgression, if it occurred before these rites are performed. Society, not biology, produces identity and personhood.[75]

The Western fetus is usually considered to be a stable biological entity. Its anchoring in biology is, to an important degree, a consequence of biomedical innovations: pregnancy tests; prenatal diagnosis and screening; improved survival rates of premature children; in vitro fertilization and manipulation of human embryos in a test tube; stem cell and fetal tissue research;

and fetal surgery. This may also reflect a growing identification of the fetus as a baby, not only by anti-abortion movements but in medical information routinely distributed to pregnant women that exhorts them to avoid behavior which will harm "their baby." Educational materials aimed at persuading pregnant women to quit smoking systematically construct the fetus as a child, and a pregnant woman who smokes as a child abuser. The strong language of these publications can be contrasted with the paucity of reliable knowledge about the effects of smoking, especially in moderation, on the fetus.[76] In several US states, it is possible to jail a pregnant woman who through her behavior (such as taking illegal drugs) is perceived as putting the health of her unborn child at risk. In a widely reported case, a pregnant woman, Alicia Bertan of Wisconsin, was threatened with jail for refusing to enroll in a drug rehabilitation program despite her affirmation, bolstered by a negative urine test, that she had stopped using drugs. She ended up being forcibly enrolled in such a program for nearly three months.[77]

For many scholars, the advent of the "modern"/"scientific" fetus is linked to the rise of omnipresent fetal images. This phenomenon began in 1965 with Lennart Nilsson's famous photo essay in *Life* magazine, "Drama of Life before Birth," and his book, *A Child Is Born*, published the same year. Both became huge international best sellers.[78] Yet Nilsson's book was not the first to present fetal development to the lay public. In the 1940s, 1950s, and early 1960s, several books aimed at pregnant women included drawings and photographs of fetal development. Probably the best known among them was Geraldine Lux Flanagan's *The First Nine Months of Life*, published in 1962 and heavily illustrated with black-and-white photographs, obtained mainly from leading US embryologists.[79] Yet Nilsson was a well-known professional photographer who dedicated several years to the photography of fetuses, and produced aesthetically compelling—and heavily retouched—color photographs. The outstanding visual qualities of these photographs facilitated their worldwide exposure and their transformation into codified images of fetal development.

Many scholars have discussed the role of Nilsson's photographs in the rise of the "public fetus."[80] The historian Solveig Jülich examined a much less known role of these photographs: their contribution to Swedish debates about abortion. In the early 1950s, Nilsson was invited by the head of a leading obstetrics and gynecology department in Stockholm to photograph miscarried and aborted embryos and fetuses at its hospital. At that time, Swedish law permitted abortion on medical, humanitarian, and eugenic grounds, but abortion remained nevertheless an issue. In 1952, Nilsson's photograph of a fetus aborted at five months was published by a popular magazine under the heading "Why Must the Fetus Be Killed?" Twelve years later, publication of another aborted fetus as photographed by him was criticized as anti-abortion propaganda. His 1965 photo-essay in *Life* was criticized on similar grounds. Critics argued that Nilsson's effective visual manipulation of very tiny fetuses to present them deliberately in his photographs as "virtual babies" led future mothers to believe that a ten-millimeter-long fetus is already a fully formed child. Such images might have distressed women who decided to have an abortion. Nilsson himself did not mention abortion, insisting that his photographs displayed the "miracle of life." He and his editors carefully avoided mentioning that all the photographed fetuses were dead and that the great majority had been aborted. The staff of the publishing house that produced Nilsson's book was explicitly instructed to answer evasively or vaguely when customers asked about the origin of the pictures. Any other response, they were told, was impossible—"it would have destroyed the market."[81] The transformation of dead specimens into "icons of life" would not have been conceivable without a careful masking of the material realities behind the stylized photographs.[82]

Nilsson's images of dead fetuses were made "alive" through a subterfuge. Ultrasound enables the direct observation of a living fetus in the womb. Images generated by obstetrical ultrasound were at first blurred and imprecise. They became increasingly sharp and well defined, and recently became available in three and sometimes four dimensions (short films of the fetus in the womb).[83] The possibility to see on an ultrasound screen that the fetus, even when very small, looks like a miniature baby is often presented as a key element in the identification of a fetus as a child. In an oft-told story, a woman/couple see for the first time a sonogram of the fetus, suddenly realize that she/they are really going to be parents, and immediately become viscerally attached to the future child. In a "pro-life" version of this story, a woman considering an abortion changes her mind when she sees "her child" on an ultrasound screen. There is, however, one problem with this narrative: it exists only in a specific time and place. In other places and periods, seeing a tiny, baby-like form might have elicited very different feelings.

Lynn Morgan's study of the collection of fetuses by early twentieth-century US anatomists indicates that at that time people who, one may reasonably assume, were no less able than those alive today to perceive similarities between aborted fetuses and newborn babies had a very different attitude toward fetal bodies.[84] Emilie Wilson's study of physiological experiments conducted between the 1930s and the 1960s by the Pittsburgh anatomist Davenport Hooker on aborted nonviable fetuses (under twenty-four weeks old) tells an even more striking story. Scientific films depicting the stimulation of such fetuses with soft hairs to elicit reflex movements were described in positive terms in articles in *Time* magazine and other popular publications. Experiments on dying fetuses were seen as a legitimate investigation of human development and an important contribution to embryological knowledge. Entities that looked very much like small babies were treated like experimental material. At the same time, some fetuses employed by Hooker in his experiments were also treated as babies: a few among them were baptized, sometimes during an interval between two experiments. Baptism did not change their designation as "exteriorized fetuses," or the distribution of films that recorded their movements.[85]

Dead fetuses were occasionally a source of public entertainment. At the 1933 Century of Progress Exposition in Chicago, two enterprising showmen, Lew Dufour and Joe Rogers, produced a highly profitable "Life's Exhibit": a display of normal and impaired fetuses. The fetuses had been bought on the black market: they were reported to have been smuggled out of a hospital by technicians or impecunious interns. At that time, hospitals had a rule that such specimens were to be destroyed, but it was seldom rigidly enforced.[86] The main use of embryos and fetuses was, however, as research material. In the United States, embryologists had forged an alliance with civil authorities and legislators to ensure a steady supply of embryos for research. Accordingly, state policies were tailored to meet the embryologists' needs. Early stage embryos were especially rare: they could be obtained only when a woman who had undergone a surgical ablation of the uterus was accidentally found to be pregnant.[87]

The history of the "premature babies shows" in the late nineteenth and early twentieth centuries shows that not only dead or dying fetuses but also premature babies born alive did not automatically acquire the status of human beings. The showman and inventor Dr. Martin Couney, specialist in the care of premature infants, produced highly popular shows of the feeding and care of these infants. The shows combined a demonstration of scientific activity—standardized care of infants by professional nurses, and their maintenance in incubators—with the spectacle of very tiny babies who nevertheless survived, at least for a while. Couney had no problem acquiring "material" for his exhibits. Premature babies were "loaned" by parents who believed that the infant would soon die. In one case, Couney did not have enough premature babies for a London show, but was able to import several baskets of them from France. More surprisingly, until the 1920s, parents who loaned their babies to Couney seldom visited them, and were not very interested to learn how they were doing. Often, when a baby survived (the initial survival rate was 30–40%; it increased with the perfection of methods of care for premature newborns), the "loan" became a "gift," because nobody claimed the child.[88] In the early twentieth century, a living and breathing humanlike form was not always perceived as a "child." This still can be the case in the early twenty-first century, but only in well-defined and well-sheltered spaces such as a neonatal resuscitation unit. In these spaces, specialists who decide, or help parents to decide, whether to revive very premature and/or severely disabled newborns continue to test the boundary between a fetus and a child.[89]

For many women, the experience of pregnancy is a shifting state that may include oscillation between several conflicting perceptions of the fetus. Such perceptions may vary not only between a woman's different pregnancies, but also within a given pregnancy. Her experience and the context in which she finds herself when pregnant, the sociologist Deborah Lupton explains, shape her view of the entity she is carrying. The fetus can be conceptualized as mine/not mine, part of me/separate from me, companion/antagonist, baby/ parasite, self/other.[90] At the same time, biomedical technologies contribute to both the stabilization and the destabilization of the fetus being perceived as the woman's child. They produce images of fetuses labeled "baby's first photos," but these photos also make pregnancy more tentative: if a fetal malformation is found, the images can transform a wanted future child, already accepted by the family, into an entity with an uncertain status.[91] Such fluctuating and fluid definitions of what a fetus is, and the irreducible ambivalence of many decisions about the fate of pregnancy, spill into the management of fetal remains outside the woman's body.[92]

Fetal Trajectories outside the Womb

The period of the rapid spread of PND was also a period of changing attitudes toward pregnancy loss, be it natural (miscarriage) or provoked (abortion for a fetal malformation). The latter instance—a termination of an initially wanted pregnancy following a positive PND—is seen by some women who underwent this intervention as a variant of miscarriage: nature had produced severe fetal flaws, and their decision to terminate the pregnancy was a quasi-inescapable result of this accidental development.[93]

In the eighteenth century, miscarriage, especially before quickening—the perception of fetal movements—was considered by women and their physicians to be not the loss of a pregnancy but rather the expulsion of a "growth," an entity not destined to become a child. The expulsion of such "false conception" was not a mistake made by the body but just the opposite: a sign that the womb does its work well.[94] In the nineteenth century, miscarriage might have been regarded by some women as a relatively benign event, or even a positive one. Letters written by Victorian women attest that some among them, exhausted by repetitive pregnancies and apprehensive of the dangers of childbirth, were pleased that a miscarriage would increase the interval between two full-term births.[95] Women's willingness to contribute fetuses for research in the nineteenth century was explained by their perception of miscarriage as a relief, among other things. Some even employed expressions such as "bliss" and "rapture" to describe pregnancy loss.[96] Sorrow after such an event probably became more frequent in the twentieth century, when contraception gave women better control over their fertility. Yet women did not speak about their miscarriages, and were expected to get over them as quickly as possible. Their distress and grief, especially when lasting beyond the aftermath of the loss, were perceived as a pathological failure of adjustment.[97]

Social scientists have described dramatic stories about the insensitive treatment of women who miscarried late in pregnancy or gave birth to a dead child. These women were not allowed to choose whether to treat the "product" of miscarriage as a child, while some physicians believed that they should be protected from the need to sign a birth certificate. Occasionally, dead fetuses were mishandled by hospital staff, and even landed accidentally in a laundry machine. In some cases, physicians did everything they could, including the abandonment of a newborn baby without any care, in order to present the birth as stillbirth rather than a live birth followed by the baby's demise. A stillbirth label allowed them to avoid the trouble of producing a birth certificate followed by a death certificate, and saved the hospital the costs of treating a "condemned" baby in an intensive care unit.[98] A woman who gave birth to a severely malformed baby recalled that the physician exclaimed, "My God, it is alive," before crushing its skull; another woman remembered reading in her stillborn child's chart that he was described as a "hairy creature with extra teeth."[99] Such extremely distressing accounts were a powerful stimulus for changes in the treatment of pregnancy loss.

In an oft-quoted passage, the anthropologist Rayna Rapp tells of a pregnant woman and her husband who were crying upon hearing a diagnosis of a fetal malformation. When her physician asked her why she was in tears, she answered because of the baby, the baby is going to die. The physician then said firmly, "That isn't a baby. It's a collection of cells that made a mistake."[100] This statement is usually perceived as shocking, because few women who endure a miscarriage, especially a late one, see this event as a loss of "a collections of cells." The same is true for women who decide to put an end to a wanted pregnancy following a positive PND.[101] Rapp also described the distress of women who decided to terminate a pregnancy after a diagnosis of a fetal anomaly and were unable to bury the dead fetus. After they were denied access to fetal remains, some families chose to bury a sonogram of the fetus instead.[102]

In the late 1970s, the labels *baby* and *fetus* as used in a US hospital practice reflected organizational rather than medical, biological, or legal divisions. The category of life was imposed on certain objects that were handled by medical practitioners and health care organizations in specific ways. At that time, only obstetrical patients could deliver, only patients who delivered became mothers, and only mothers produced babies. Other women were treated as gynecological patients—they aborted, and the product of an abortion was always a fetus. The line dividing the two was fetal viability. "Products" of less than twenty-eight weeks' gestation were declared to be incompatible with life and were described as "fetuses." "Products" of more than twenty-eight weeks were usually labeled "stillborn babies." Babies and fetuses were handled by different areas of the hospital: the pathology department versus the morgue. They were also treated differently by the hospital's administration. "Products of abortion," that is, fetuses, could be disposed of without a signed permit; dead babies always required either a disposal form signed by the parents or a direct transport to the morgue.[103]

In the 1980s and 1990s, pregnancy loss became an acceptable topic of public debate. The recognition of grief associated with miscarriage promoted the awareness of the physical and emotional challenges arising from pregnancy termination for a fetal indication. It also promoted the blurring of boundaries between fetuses and babies. Since women who had lost a wanted pregnancy nearly always conceptualized their experience as the loss of their baby, health professionals gradually adopted the same language. Regulations were adapted to these new sensibilities. Increasingly, dead fetuses were being regarded not as hospital waste but as human remains, a trend accelerated by a growing pressure on hospitals to give couples the option of burying or cremating their lost baby.[104] Scandals such as the storing of more than four hundred dead fetuses at Alder Hey Hospital in Liverpool, the dumping of fetal remains in a field in Chino Hills, California, or the discovery of dead fetuses stored in the Saint Vincent-de-Paul Hospital in Paris illustrate the radical shift in the perception of a "pregnancy product" in the late twentieth and early twenty-first centuries.[105]

In the hospital, pregnancy loss has become a codified event. In the twenty-first century, with the growing popularity of medicalized abortions, which in the first trimester usually take place at home for financial and other reasons, women are obliged to deal with abortion products themselves. After eight to ten weeks of pregnancy, these women expel a discernible embryonic sac, and sometimes a semi-formed fetus. Some have no problems with disposal of an aborted fetus: "I don't think I thought anything. I just wrapped it up, put it in a nappy [diaper] bag and put it in the bin." Others are distressed by being obliged to get rid of the tiny fetus. The usual disposal methods, flushing it in the toilet or putting it in the trash bin, feel somewhat wrong, but the women are not sure what would be a better alternative: "I said to my mum, 'Do I put it down the toilet?' Because I thought will it flush, because it's quite big? Where will it go? We just wrapped it up and put it down the toilet. We didn't know what to do with it. You don't know [how] to dispose of it if it comes out like that." Women who decide to abort are still unsure of their feelings toward the material fetus: "Now I want to know where it is, which is really strange because obviously I'm never going to know"; "You feel protective, even though I knew I won't be keeping it."[106]

Pregnancy Termination and Ritualized Grief

Before the 1980s, many institutions had failed to recognize women's heartache after a pregnancy loss. The Canadian sociologist Deborah Davidson describes her desolation after being denied the opportunity to connect with her stillborn children:

> In 1975 and 1977 I gave birth to premature babies who were whisked away before I could see them, or likely so I would not see them either alive or dead. They were phantasms of another sort, or so it would seem. Even though my son, born in 1975 at 29 weeks gestation, lived for ten hours, I did not see him. I learned of his death when I was moved to a room where I would not make yet-to-deliver women uncomfortable. Even at that point, they did not tell me, but *I knew.* My daughter, born in 1977 at 27 weeks gestation, died in the delivery room shortly after her birth. It would have been so easy for them to give me at least a glimpse, perhaps even let me hold her. But they did not. Official forms regarded their birth, death, and disposal. Yes, "disposal," in the parlance of the hospital. The forms were given to my children's father to be filled out with the help, the "guidance," of the hospital staff—Baby Whatley was sufficient—they said. I had no opportunity to name my babies for public record. They were cremated in the hospital and disposed of without invitation or ceremony. Hushed and hidden—ghost babies—phantasms.[107]

In the 1980s and 1990s, denial of the emotions surrounding pregnancy loss was replaced in many Western countries (but less often in non-Western ones) with the new orthodoxy of appropriate grief for this traumatic experience.[108] Women who had a late miscarriage, those whose fetus died in the womb, and those who underwent an abortion were increasingly encouraged to perform specific grief rites. Women and their partners were invited, and sometimes urged, to view their child, hold it, keep mementos such as photographs and objects that came in contact with the child's body, and, when applicable, to receive psychological help and/or join a support group. At first, such mourning rites were reserved for spontaneous pregnancy loss, but later were extended to induced abortions. Nurses or midwives responsible for the management of second- and third-trimester medical terminations of pregnancy increasingly encouraged parents to hold the fetus/baby, and regarded the preparation of the fetus for the ceremony of "saying goodbye to the baby" as an essential part of their task. A typical argument: "It helps to see the baby. Maybe not today, but weeks down the road when you are struggling you will know that you made the right decision."[109]

In the United Kingdom, the new view of the right ways to grieve for pregnancy loss was grounded in ideas promoted by the psychologist Emanuel Lewis in the 1970s. Lewis strongly denounced the conspiracy of silence surrounding stillbirth and, inspired by observations from his practice, recommended physical contact between women and their stillborn/aborted children. He was especially impressed by a woman who insisted on holding her dead and partly macerated stillborn child, and another who went into a frenzy of kissing the stillborn child's body. When there are no shared memories of life with the dead person, Lewis argued, a grief process cannot be resolved without seeing and preferably touching the body: "Our hospital culture tends to impede the normal healing process of mourning. We do this even by asking the mother whether she wishes to see her dead baby. Really, what could be more natural?"[110]

Lewis's ideas resonated with those of health professionals. The genetic counselor June Peters, who assisted in abortions for fetal indications in the early 1970s, recalled that nurses swaddled the tiny fetuses and brought them to the parents to help them say goodbye.[111] In the 1980s, Lewis's views became increasingly popular. Nurses and midwives were especially active in initiation and codification of grief protocols, perhaps partly because of their unease when handling fetal remains after expulsion.[112] Guidelines published in 2002 by gynecologists in the United Kingdom recommend that "staff . . . create an atmosphere which encourages parents to see and hold their baby. . . . Parents may need to be informed that if they do not see their baby they may regret it as it could make mourning more difficult."[113]

The UK geneticist and pioneer of PND Bernadette Modell recalls how the rules for mourning the miscarried fetus were extended to abortion for a fetal indication:

> We agreed . . . that late termination of a wanted pregnancy fell into the same category, discussed it with the midwives, and used the same approach from the beginning of our prenatal diagnosis service. I was the person principally involved. It was, of course, very emotional, but from the beginning we understood it was the right thing to do. The mothers really appreciated the opportunity to greet and say goodbye to the baby. We were privileged to share these deep moments of truth with our patients—and it also helped to give us confidence in the service we were providing. This was a new concept at the end of the 1970s: it is standard practice now. Some mothers want photos. Some Church of England chaplains (and perhaps others) also offer a small service for the baby. I remember particularly one Cypriot uncle, resident in the UK, who accompanied his young niece who had come alone from Cyprus for PND. When we discussed this, he was shocked at first, but then said, "Oh, I see, at least she will know she was a mother."[114]

In France, too, the sociologist Dominique Memmi explains, health professionals were behind the widespread establishment of grief observances for dead fetuses/babies. Before the 1990s, only a small minority of women had expressed a desire to have physical contact with the fetus/stillborn child. However, health professionals had been deeply moved by stories of those women who wanted to treat a miscarried/aborted fetus as their child, and started systematically proposing to women that they view and touch it. At first, many of these patients were positively shocked by this invitation, but between 1996 and 2007 the percentage of those choosing to hold the fetus greatly increased—mainly, Memmi proposes, because of persuasive medical staff members.[115] These professionals acted as moral entrepreneurs who incited couples to humanize the fetus and transform it into their child.[116] In the twenty-first century, French midwives assisting at a miscarriage or a termination of pregnancy for a fetal indication are expected to wash and dress the fetus's body, then take commemorative photographs. If the woman does not want to have these photographs, they are saved in her medical file for retrieval if she changes her mind.[117]

Most (94%) of the French midwives interviewed believe that women who undergo medical termination of pregnancy should see and hold the fetus, and 56% of them actively encourage these patients to do so.[118] When talking to such a patient, they speak about birth and baby, not expulsion and fetus, even when the fetus/baby is very small. Midwives also pay special attention to the presentation of this fetus/baby in ways that maximize the humanity of its body and minimize the visibility of abnormal traits. Clothing the fetus—sometimes in specially prepared tiny clothes provided by French support organizations for pregnancy loss—is seen as an important step in transforming the dead fetus into the family's child. For many midwives, this allows each woman to begin the "right" kind of mourning process and helps her avoid long-term psychological harm.[119] Similarly, Canadian nurses and midwives urge women to see and hold the dead fetus, and when the fetus is malformed, wrap it in a way that hides the malformation. They also systematically emphasize its "cute" features. For example, when the body is severely misshapen, they draw the woman's attention to perfectly formed tiny hands and feet.[120]

Some scholars claim that clinical studies have validated the initial intuition that physical contact with the dead fetus/child helps grieving women and their partners to get over the trauma of pregnancy loss.[121] Other scholars have argued that pressuring parents to conform to specific rites of mourning can add to the woman's trauma.[122] Patricia Hughes and her collaborators, who studied posttraumatic stress disorder in women after miscarriage or stillbirth, were surprised to find out that contrary to their expectations, holding the prematurely born baby aggravated rather than attenuated the women's stress, probably because some had been left with images that haunted them afterward. The profound change in clinical practice as regards seeing and holding the dead infant, Hughes and colleagues have argued, was introduced in maternity units based on very limited and unsystematic clinical observations.[123] Some women strongly preferred seeing and holding their dead child, taking photographs, and producing mementos, while others categorically refused to do so. However, most women, in shock over their loss, had no clear plan of how to manage the situation and simply went along with whatever was expected of them. They could be persuaded to do whatever the medical staff recommended: "I didn't really want to hold him but they said it would be better for me."[124] In addition, women who chose termination for a fetal indication may have feared accusations of being selfish and uncaring, and so were especially vulnerable to pressures to show their commitment to the ideal of "good motherhood" by observing fetal bereavement rites.[125]

The sociologist Lisa Mitchell's qualitative study questions the universality of the assumption that failing to hold and see the fetus would always prolong the woman's grieving process and lead to emotional problems.

She interviewed nineteen women who had undergone termination of pregnancy for a fetal indication. Among them, six had been certain before the abortion that they wanted to see and hold the fetus, and afterward felt that contact with the dead fetus/stillborn child had attenuated their grief. Some added that viewing the malformed fetus helped them accept their decision to terminate the pregnancy. Of the ten women who initially did not want to see the fetus, five did not view it and did not regret their decision. The other five, and three women who were unsure about what they wanted to do, could not avoid seeing the dead fetus. Several described this experience as traumatic. Mitchell explains that the "new normal" of seeing and holding the fetus makes other preferences difficult. Women who are told that "most women choose to see their baby" are essentially informed what they should be seeing, namely, "their baby."[126]

Despite a few critical voices, pregnancy loss has become a site of ritualized grief in the twenty-first century. Today, it may be difficult to imagine a woman from an earlier generation stating, as the British editor and writer Diana Athill (born in 1917) had, that she did not grieve at all over the loss of an unplanned but strongly desired pregnancy.[127] Testimonies about stillbirth published in the *New York Times* in 2015—a self-selected sample of parents willing to share their experiences—record the extensive use of mourning rites for dead/stillborn children. These rites include photographing the child, collecting mementos, conducting burial ceremonies, naming the child, and including him or her in the family through symbolic acts such as celebration of birthdays.[128] Grieving for an unborn child can also be seen as signifying a "responsible motherhood." A panel reviewing fetal deaths in Florida and investigating whether the pregnant woman—often poor and/or of color—had contributed to the pregnancy loss interpreted the woman's holding the fetus and taking home memorabilia prepared with the help of the hospital staff as positive signs, thereby discounting suspicions that she had hastened the fetal demise.[129]

Some feminist scholars propose that the recent universal acceptance of grieving rites for unborn children, while undoubtedly helpful for some women, can be linked to the reinforcement of conservative views of motherhood and an anti-abortion agenda. An Indiana bill of April 2016, HB 1337, requires hospitals to bury or cremate miscarried fetuses no matter their gestational stage, and denies women who miscarried the previously existing option of treating the remains as medical waste, thereby cementing the connection between "abortion products" and "babies." Indiana's governor at the time, Mike Pence (as of January 2017, vice president of the United States), declared that he "is signing this legislation with a prayer that God would continue to bless these precious children, mothers and families."[130] The feminist historian Leslie Reagan, who had miscarried in the first trimester of pregnancy, before the fetus was formed, was shocked to discover that the "support material" offered by her hospital contained images of miniature footprints of the lost "baby": "this material was not just helpful medical material; it was sympathy with vested political interests. Baby footprints are one of the symbols—along with roses and fetuses in jars—of the anti-abortion movement."[131] Lisa Mitchell explains that the omnipresent discourse on the benefits of mourning rites for the unborn transforms fetuses, including malformed ones, into desirable babies; fetishizes the fetus as naturally possessing the qualities of babyhood; and implicitly labels women who refuse these increasingly popular mourning rites as failing to behave as a "good mother" should.[132] For Leslie Reagan, too, public recognition of pregnancy loss comforts the woman, but at the same time performs ideological work by reinforcing conventional gender and family norms, the primacy of motherhood for women, and individual, especially maternal, rather than social responsibility for children.[133]

Sonograms, PNDs, and Memorial Rites for the Unborn

In her pioneering study of pregnancy loss, the US sociologist Linda Layne argues that feminists feel uneasy about the personification of the fetus, because it is perceived as threatening a woman's right to terminate an unwanted pregnancy. However, women who face pregnancy loss often wish for recognition of their maternal role despite the absence of a baby. Normal pregnancy, Layne argues, is a process during which the materiality of the future child and the woman's status as a mother are reaffirmed through a series of milestones: a positive pregnancy test, hearing the fetal heartbeat, seeing a sonogram of the fetus, learning the fetus's sex, buying things for the future child. When the pregnancy ends without a baby, the woman who wishes to confirm her identity as a mother replaces memories of a living child with objects related to the material "would-be child": items that had come in contact with the dead baby's body such as clothes and blankets, handprints and footprints, official documents such as birth and death certificates, and sonograms. The latter can play an especially important role when no body exists because the fetus was too small or severely misshapen. In such cases, sonograms are the only material evidence of the existence of a "child."[134] Material traces of the lost child's life can be collected in a memory box.[135] Some hospitals give women scheduled to undergo a termination of pregnancy a prepared box for the collection of memorabilia, complete with information about the memorabilia other families cherish. Such a box may also contain an inexpensive disposable camera for taking commemorative pictures of the fetus/dead child.[136] A collection of material traces of a lost pregnancy confirms that, as one woman put it, "we have/had a son, we didn't give birth to an 'it.' Collin was real; he existed. If we have/had a son, then we must be parents. If not, what are we?"[137]

The charity Now I Lay Me Down to Sleep specializes in providing artistic black-and-white photographs to families who have lost a baby. The photographs, produced by volunteers, often depict fully developed dead babies/stillborn children, who indeed often look like sleeping newborns.[138] In contrast, photographs made by bereaved parents and other family members and posted on specialized Internet sites do not attempt to achieve an artistic quality, and frequently represent very small fetuses.[139] Posting fetal images on the web can play a role akin to that of the memory box: commemoration of the child as a member of the family. Photographs of stillborn babies are unusual, however. Our culture, unlike the Victorian one, does not cherish photographs of the dead, death masks, and death mementos such as lockets containing a lock of hair from the deceased. In the twenty-first century, commemorative ceremonies frequently integrate photographs of deceased persons at key moments of their life, often when they look full of vitality and energy, and as a rule do not present photographs taken after their death. Yet in the case of stillborn babies, who had died before they were alive, photographs of the dead are the best available proof of their existence.

Memorial sites for stillborn and miscarried babies communicate the anguish of grieving parents and at the same time mirror beliefs that became popular in a given time and place. Public expressions of grief over pregnancy loss, the anthropologist Helen Keane proposes, are shaped by the rise of conventional ways of constructing the "realness" of lost babies, and a parallel need of the grieving woman to show that she is/would have been a truly loving and caring mother. Thus, they produce socially intelligible norms of womanhood and motherhood. It is implausible that a woman would openly acknowledge in such a memorial that she was, for instance, relieved because miscarrying a malformed fetus saved her the stress of deciding whether to terminate the pregnancy.[140] Memorial sites sometimes contain letters to the unborn child that enumerate the things parents had hoped to do with this child, often through a conventional description of a generic girl or boy. While real-life children may not fulfill their parents' gender-stereotyped visions, imagined children always conform to such hopes and aspirations.[141]

Children dead before they were born are very frequently described as "angels." Some French parents active on specialized Internet forums have even invented the term *parange*—from *parent* and "ange" (angel)—and speak about themselves as "mamange" and "papange" (angelmom and angel-dad).[142] The image of the lost child as an angel, omnipresent in the Internet memorial sites, strongly resonates with nineteenth-century commemorations of dead children, but with one important difference. The nineteenth century's lost angel was a real child, not infrequently one that had lived long enough to acquire individual traits and leave precise memories. The unborn angel exists only in his mother's/parents' thoughts and dreams.[143]

In her study, Margaret Godel analyzes in detail two Internet memorial sites (by permission of the women who created them), both for stillborn babies who suffered from lethal malformations but whose mothers refused to terminate the pregnancy because of their strong religious beliefs. Ruth, stillborn at twenty-eight weeks, had anencephaly and spina bifida. Anna, stillborn at twenty-six weeks, had trisomy 18 and Turner syndrome. Anti-abortion propaganda often focuses on the similarities between fetuses and babies: an aborted fetus looks like a baby and therefore is a baby. Ruth and Anna, one may assume, did not look like typical babies, but were nevertheless ideal babies for their parents.[144]

Ruth's website includes photographs that had been taken just before her funeral, along with her mother's poems. These poems express the mother's feelings about her loss and describe Ruth's imagined life. Ruth's parents note that she will always be a very important member of the family.[145] Anna's website includes many pictures and videos taken during the twelve hours her family kept her body. Her mother explains that when a couple has a living child, they are able to show this child to everybody. When there is no child at the end of a pregnancy, a memorial website is the only way to show the world that she was beautiful and perfect: "To us, [Anna] is as special as any child who lives, and we're as proud of her as any parent could be of their own child."[146] Anna's father says that the time they spent with her was "the most amazing 12 hours of my life. I held her, talked to her for so long. I brought her to the window and showed her the night sky, something I'd always imagined me doing with her, teaching her the constellations. I told her of all the things I'd imagined doing with her. . . . [We have] given her all the love and cuddles we hoped would last a lifetime."[147]

A material entity, a dead baby that can be held and told stories, provided a bridge between the real Anna—a stillborn baby with trisomy 18, a condition usually leading to rapid death or, in the rare cases when the child survives for more than a few months, causes a severe intellectual disability—and the imaginary Anna, who can be taught about constellations. Ruth, born without a brain, might have been similarly cherished as a perfect member of her family. The contrast between the idealized image of a lost child and the imperfect body of a stillborn baby may be disturbing, but an outsider's attempt to compare them may miss the point of these memorials: they celebrate not a flawless dead child but a perfect and unconditional parental love.

Just as some Internet memorial sites include unretouched photographs of the stillborn/miscarried child, an article about miscarriage published by the London *Daily Mail* in January 2014 was illustrated with a dramatic photograph of an aborted fetus/baby and his mother. The photograph depicts a visibly exhausted woman dressed in a blue hospital gown, her hair caught in a plastic cap.[148] She is resting, eyes closed, on a hospital bed; on her belly lies a tiny, bloodied fetus enveloped in a white frilled cloth. The woman, Alexis Fretz of Pennsylvania, named the nineteen-week-old male fetus Walter Joshua.[149] This photograph recalls the iconic image of an exhausted but happy woman with a newborn baby in her arms, once printed on paper and mailed to family and friends and today tweeted, e-mailed, and posted on Facebook. The contrast between the typical photographs of a new mother and her baby and an image in which a woman cradles a tiny blood-covered fetus may be especially upsetting, as it contains the simultaneous presence of maternity and non-maternity, and of a being that is a minuscule red fetus and at the same time the Fretzes' baby.

The blog in memory of Walter Joshua Fretz contains only photos of a dead fetus. In contrast, a video on a website commemorating the miscarried twins Cate and Cole Van Nyhuis focuses on sonograms. Nearly half of the two-minute-long video is dedicated to these images. The second half comprises photographs of tiny aborted fetuses. The fuzzy gray captures of the ultrasound screen are the main proof of the life of future babies before they became small dead fetuses. Cate and Cole, the sonograms attest, were really alive.[150]

A video on the memorial site of Jeffry Drake, miscarried at fifteen weeks, has a similar structure. It starts with a photo of a pregnancy test stick displaying positive results. The video then shows many sonograms of the fetus; we also learn that the mother was diagnosed as having low-placed placenta—a sign of serious risk of miscarriage—which indeed took place three weeks later. The video ends with photographs of the miscarried fetus/baby in his crib, and the arms of his mother. The story of Jeffry Drake started with a blue line on a pregnancy test stick, continued with gray sonograms, and ended with color pictures of a minuscule fetus. Diagnostic technologies transformed an early second-trimester miscarriage into the condensed narrative of a life.[151]

In cultures such as Ecuador's in the late twentieth century or that of the United States in the first half of that century, it was possible to simultaneously maintain several, sometimes contradictory, views about the nature of the fetus/unborn child. The indeterminacy of the fetus's status might have helped pregnant women—especially those whose pregnancy failed to produce a live child—to accept a pregnancy loss. It also facilitated the task of scientists and physicians who studied life before birth. In the twenty-first century, technological innovations coupled with the "abortion wars" led to the identification of the fetus as an already-existing baby. This identity was reinforced by the development of mourning rites for the unborn. Internet memorial sites for miscarried/lost babies participate in equating a fetus with a baby. They also reinforce the role of PND technologies in making the pregnancy real, and consolidate the woman's status as a mother. The presence of concrete data on the sonograms (date, time, name, and number of weeks of gestation) further reinforces the "real child" aspect of a pregnancy, because it attests to the process of a specific child's biological and social production. Sites such as those in memory of Anna and Ruth also include information about the medical condition that led to the child's demise. Anna's anencephaly might have been recognized in the nineteenth century, too, although only after her birth; but Ruth's diagnosis, Turner syndrome and trisomy 18, could not have been possible before the rise of cytogenetics circa 1960.[152] Memorial sites dedicated to lost pregnancies incorporate parental emotions and lost dreams, but also knowledge developed by gynecologists, geneticists, fetal medicine experts, and fetal pathologists. The lost child comprises elements borrowed from Christian and New Age visual language, the Victorian cult of the dead, and images and data produced by twenty-first-century biomedical technologies. The virtual fetus of the Internet memorial, like the material fetus outside a woman's body, is a complex mixture of facts and feelings, data and dreams.

TWO

Genetics, Morphology, and Difficult Diagnoses

Atypical Morphology and Abnormal Heredity

Prenatal diagnosis (PND) is strongly identified with genetics. While this statement is historically accurate, it is factually misleading. At first, PND was indeed a diagnosis of a well-defined hereditary pathology. PND detected fetuses with diseases transmitted in families, such as Tay-Sachs disease, Niemann-Pick disease, or maple syrup urine disease, and the presence of a genetic (but usually nonhereditary) anomaly, Down syndrome (trisomy 21).[1] In the United States and the United Kingdom, a new professional group, genetic counselors, became specialized in advising women who underwent amniocentesis. In turn, the rise of genetic counseling strengthened the equation of PND with the search for changes in the genetic material of the cell.[2] An abortion for a fetal indication was also called "genetic abortion," and this name often encompassed as well the termination of pregnancy for a structural anomaly of the fetus detected during an ultrasound.[3] The history of PND is, to an important extent, a history of the investigation of morphological anomalies—more precisely, a combination of studies of atypical form and abnormal heredity. These two distinct lines of study meet within a relatively little-known medical subspecialty: dysmorphology.

Dysmorphology, the UK sociologist Joanna Latimer and her colleagues argue, is an important site of a recent "rebirthing of the clinics."[4] The notion of rebirthing assumes a previous death. Sociologists and anthropologists who studied new developments in biology and medicine occasionally spoke about the "death of the clinics." They evoked a switch from a "molar" to "molecular" approach and its consequence, the (presumed) decline of the older clinical understanding of healthy and sick bodies.[5] Historians of science and medicine are often more interested in the multiple juxtapositions and recombinations of older and newer approaches.[6] Dysmorphology is an important site of such blending of the old and the new knowledge and practices, the coexistence of molar and molecular knowledge, and the integration of traditional clinical knowledge and recent developments in genetics and genomics.[7] One of the reasons dysmorphology can play this role is the privileged access of specialists in this domain to rare genetic anomalies. Such access opens new possibilities for the study of human heredity: the abnormal facilitates the investigation of the normal.

Dysmorphologists specialize in the study of recognizable human malformations: those visible to the naked eye. This specialty started as a study of children with atypical facial traits and bodies (the "funny-looking child") presented to specialists in order to find out what may be wrong with them.[8] The name dysmorphology was coined in a journal article published in 1966 by the US pediatrician David Smith to replace the term *teratology*—literally, the science of monsters—which is difficult to use when talking to parents of an affected child.[9] In 1970, Smith published the first textbook of dysmorphology for use by pediatricians and general practitioners, *Recognizable Patterns of Human Malformation*.[10] This book is still widely used by pediatricians and fetal medicine experts today.[11]

The development of dysmorphology is inseparably associated with the application of genetic knowledge to the clinics. This specialty can be described as a key component of the clinical part of clinical genetics. The first journal dedicated to this specialty, *Clinical Dysmorphology*, was founded in 1992 by the UK clinical geneticists Dian Donnai, Robin Winter, and Michael Baraitser. Winter and Baraitser also developed the Internet-based *London Dysmorphology Database*, one of the main resources for pediatricians and geneticists who study inborn impairments.[12] The US journal *Teratology*, founded in 1968 and renamed *Birth Defects* in 2003, also publishes many articles on dysmorphic traits and the links between such traits and genetic anomalies.[13]

Dysmorphologists usually do not treat children with well-studied inborn conditions, even when people with such a condition, be it hereditary (for example, mucopolysaccharoidosis), genetic (for example, Down syndrome), or produced by environmental factors (for example, fetal alcohol syndrome), have typical "dysmorphic" traits. Dysmorphologists specialize in rare genetic anomalies and those which cannot be easily recognized.[14] "Every doctor," the British dysmorphologists William Reardon and Dian Donnai explain, "likes to make a good diagnosis. If the condition in question is rare, so much the better—at least in terms of professional satisfaction. No branch of medicine affords more opportunity for the diagnosis of rare disorders than clinical genetics."[15]

Dysmorphology is above all a traditional clinical discipline. The recognition of dysmorphic syndromes depends more often on careful observation and experience than on sophisticated laboratory investigations. Dysmorphologists may be proud of their capacity to detect a genetic condition solely by looking at a child, an especially impressive variant of the clinician's traditional embodied expertise.[16] A diagnosis at first glance may, however, be problematic. If it is not well grounded or, alternatively, when it is correct but given without an appropriate dialogue with the family, it may worsen the patient's experience. Dysmorphologists insist therefore that a rapid diagnosis always be tentative and reserved strictly to an intraprofessional communication. It should also be, as far as possible, grounded in a knowledge that can be transmitted to colleagues and taught to students.[17] Nevertheless, no matter how standardized dysmorphology training becomes, some individuals stand out through their ability to make a diagnosis intuitively, a kind of a "sixth sense," which, combined with relentless reading and enviable powers of recall, enables them to succeed diagnostically where their peers fail.[18] The role of such internalized and quasi-instinctive diagnostic skill may be greater in dysmorphology than in many other branches of medicine.[19]

Inborn syndromes are infrequent, but they are not very rare in the pediatric wards of major teaching hospitals.[20] Typically, parents and/or pediatricians observe, either at birth or during an early period of the child's life, anomalies and developmental delays that may be the result of a specific pattern of perturbation of fetal development. The child is then directed to a clinical geneticist or a dysmorphology clinic. In some cases, the experts are able to rapidly diagnose the child's condition, although at the time of writing (2017), only in a small number of genetic disorders, frequently advanced by clinical geneticists, does a correct diagnosis lead to an effective treatment. Often, the parents learn that physicians cannot cure their child's condition, and can only attempt to slow the disease's progression or attenuate its symptoms. Even such a limited intervention is not always possible. In some especially difficult cases, such as Tay-Sachs disease, the physicians can only tell the parents that their child will inexorably deteriorate with time.[21] When the experts are unable to provide a diagnosis rapidly, they first attempt to find out if the child's condition is hereditary. The dysmorphologist draws a pedigree, collects biological samples from parents and siblings, attempts to get as much information as possible about the medical history of the family, and asks for family photographs to trace the presence of "dysmorphic" traits. If the child looks "syndromic" (that is, may be affected by an anomaly of genetic material at the cellular level) and the specialists are unable to find abnormal traits or pathological manifestations in other family members, the child's symptoms are tentatively attributed to a new mutation.[22]

A genetic diagnosis of the child's problem produces an explanatory framework that allows parents and health professionals to discuss the child's condition, prognosis, and, if applicable, treatment.[23] Children seen by dysmorphologists are often severely impaired. Parents want to understand what has happened to their child and why. Providing a name for the child's condition and linking it with specific changes in genetic material help the parents to create meaning in a very difficult situation, and can help them to alleviate their potential guilt through learning that the child's impairment was an unavoidable accident and not, for example, a consequence of the mother's behavior during the pregnancy.[24] The diagnosis of a genetic condition has practical consequences as well. It diminishes clinical uncertainty, guides the child's treatment, offers the parents a possibility of joining a syndrome-centered organization, and allows the parents to know their chances of having another child with an inborn impairment. Dysmorphology and clinical genetics also deal with reproductive futures.[25]

Dysmorphology and the Fetus

In the late twentieth century, the rapid increase in the resolution power of obstetrical ultrasound enabled the study of structural malformations of the fetus, and the articulation of morphological and genetic investigations. In some cases, such as spina bifida (incomplete closure of the spinal cord), fetal medicine specialists can make a definitive diagnosis of a structural inborn defect based on sonograms alone (fig. 4).

However, in many other cases, a sonogram is not enough to make a diagnosis. Change in fetal DNA (a mutation) is but one possible explanation of a structural problem detected during an ultrasound. The same fetal anomaly can be the result of a genetic defect or a mechanical problem during fetal development. Cleft palate is associated with several genetic syndromes, but also with a mechanical distortion of the formation of facial bones. Polymicrogyria—a brain anomaly characterized by an excessive number of small convolutions (gyri) on the surface of the brain—is frequently the consequence of a genetic anomaly, but in some cases it is connected with infection or mechanical causes of insufficient supply of oxygen to the fetal brain. Microcephaly—an insufficient development of the fetal brain—is linked with several genetic syndromes, but can also be produced by a viral infection or a toxic substance: it is one of the main manifestations of fetal alcohol syndrome.[26]

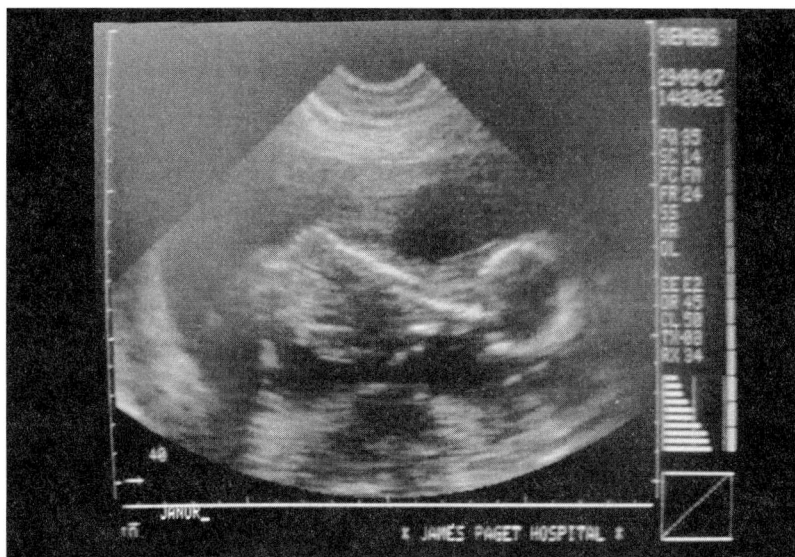

Fig. 4. Lumbosacral spina bifida, fetal scan at sixteen weeks. The head is to the right of the picture; the spine is in the center. Wellcome Images.

If an observed structural anomaly can have several causes, experts usually recommend sampling fetal cells through amniocentesis or chorionic villus sampling, and studying the DNA of these cells. Often, such a study does not detect genetic anomalies. When an analysis of fetal DNA does lead to the detection of a mutation, fetal medicine experts and genetic counselors can provide the pregnant woman with more precise information. However, "more precise" needs to be qualified. The range of inborn impairments associated with a given genetic anomaly may be relatively narrow; it may also be very broad. All the children with Tay-Sachs disease will die young, and all those with hemophilia will have a severe blood-clotting disorder. But while all the children with Down syndrome will have an intellectual disability, a diagnosis of trisomy 21 does not predict what the IQ of the future child will be—it can be as low as 10 or as high as 70—and it does not tell whether the child will have health problems.[27] Many genetic anomalies cause an even broader spectrum of pathological manifestations, from very mild to very severe.[28]

One of the most studied hereditary syndromes, Smith-Lemli-Opitz syndrome, first described in 1964, illustrates the dissociation between the certainty of a molecular diagnosis and the uncertainty of its clinical meaning. Smith-Lemli-Opitz syndrome is a recessive hereditary disorder produced by a mutation in the DHCR7 gene, which causes the deficiency of an enzyme involved in the metabolism of cholesterol (7-dehydrocholesterol reductase). This mutation produces a generalized cholesterol deficiency that leads to developmental delays, severe behavioral problems such as autism, and "dysmorphic" facial traits. The intensity of symptoms in mutation carriers is, however, variable: some people are severely impaired, while others have only a mild manifestation of this condition. The presence of "typical" facial traits, found in some but not all people with Smith-Lemli-Opitz syndrome, is not an indication of the degree of functional impairment. Some people with this condition look "abnormal" but do not have a significant intellectual disability, while others look perfectly "normal" but have learning difficulties.[29] When a mutation in the DHCR7 gene is detected in a fetus, this information, combined with ultrasonographic data, is not sufficient to eliminate uncertainty about the child's future.

Another example of a mutation with variable expressions is DiGeorge syndrome. It is produced by the absence (deletion) of a segment of a long arm of chromosome 22 (22q11.2 del). A relatively frequent genetic anomaly (its incidence is estimated at 1 in 2,000 live births), DiGeorge syndrome is associated with a long list of health problems: heart defects, "dysmorphic" features, cleft palate, anomalies of calcium metabolism, insufficient thyroid activity, immune deficiency and autoimmune disorders, digestive system anomalies, hearing and dental problems, seizures, and skeletal anomalies.[30] It is also linked with mild to moderate intellectual impairment and a high risk of mental illness, especially schizophrenia, in young adults; the estimated risk of the latter condition is 25–30%. Most adolescents with DiGeorge syndrome need some kind of educational and psychiatric support, while many adults are under psychiatric care. On the other hand, people with milder variants of this condition are often unaware of its presence. They see themselves as "normal" people, or "normal" people with a few health issues.[31] In 70% of the cases, the 22q11.2 deletion is the consequence of a new mutation, and in 30% it is inherited. Half the children of a person with DiGeorge syndrome will inherit the parental mutation.[32] People who learn that they have DiGeorge syndrome only after it is diagnosed in their child nearly always have a mild form of this condition.[33] It is, however, impossible to predict the severity of pathological manifestations produced by an inherited 22q11.2 deletion. When DiGeorge syndrome is diagnosed in a fetus, the prognostic uncertainty makes decisions about the pregnancy's fate difficult, especially for a parent who is a mutation carrier.[34] In the twentieth century, genetic conditions such as Smith-Lemli-Opitz syndrome or DiGeorge syndrome had been rarely detected before birth. In the twenty-first century, the application of genomic technologies such as comparative genomic hybridization to PND increases the number of mutations detected during pregnancy, and therefore the number of women who learn that they are carrying a fetus with an uncertain future.

Diagnostic Puzzles and Genomic Approaches

In the twentieth century, fetal genetics was limited to the study of stained chromosomes taken from fetal cells.[35] Cytogeneticists counted the number of chromosomes in fetal cells, looking for aneuploidies (an abnormal number of chromosomes): Down syndrome (trisomy 21), Edwards syndrome (trisomy 18), Patau syndrome (trisomy 13), and sex chromosome aneuploidies (the most frequent among them being Klinefelter syndrome [47XXY], Turner syndrome [45X0], and triple X [47XXX]). In the early 1970s, cytogeneticists developed a new staining technique, banding, which made possible the identification of specific regions on the chromosomes, deletions (part of a chromosome is missing), duplications (part of a chromosome is duplicated), or translocations (the transposition of a segment of one chromosome on another chromosome).[36] Such chromosomal anomalies are associated with inborn impairments and developmental delays.[37] Banding could have led to PND of chromosomal anomalies in women who underwent amniocentesis. In practice, however, PNDs of deletions, duplications, and translocations were very rare. Visual observation of chromosomal anomalies in a routine setting is difficult. Such anomalies have been diagnosed during pregnancy only when cytogeneticists were already looking for a specific anomaly because they were aware of its presence in the family, typically following the birth of an affected child or a miscarriage of a fetus diagnosed as having an identified mutation. Thus, in the mid-1970s the French cytogeneticist Joelle Boué identified a specific chromosomal anomaly in cells of a child who had died of multiple inborn malformations. When the child's mother was pregnant again, Boué was able to show that the fetal cells from her amniotic fluid carried the same chromosomal anomaly as the cells of the deceased child. The woman decided to terminate the pregnancy.[38]

In the 1980s, a new technology, fluorescence in situ hybridization (FISH), made it possible to get a more detailed analysis of chromosomal anomalies. FISH uses molecular "probes" (segments of DNA) marked with a fluorescent stain. The probe is then hybridized (mixed) with fixed (in situ), denatured chromosomes (that is, with an "open" double-helix structure), and is allowed to attach to its complementary sequence. The presence of a hybridized fluorescent probe—that is, the existence of a complementary

DNA sequence—is then revealed under a light that excites the fluorescent dye. Fluorescence is visible to the naked eye or, more often, is measured with specialized instruments. At first, FISH was mainly employed for prenatal detection of hereditary diseases produced by a change in a single gene, such as cystic fibrosis or hemophilia. Later, this method was also used when ultrasonographic findings led to a strong suspicion of the presence of a specific genetic condition, for example, achondroplasia (dwarfism). Its introduction improved the understanding of some hereditary syndromes and the expansion of the definition of others, as well as, in some cases, contributed to the delineation of new pathological conditions.[39] FISH "fishes" for specific mutations. This and related approaches such as multiple ligation probe amplification (MPLA) are effective when the physicians have a precise idea of the mutations they are looking for.[40] When fetal medicine experts observe a fetal anomaly but are not sure what it may be, they frequently apply a different genomic approach, comparative genomic hybridization (CGH). This approach can uncover a much wider range of chromosomal anomalies.[41]

CGH (also called chromosomal microarrays) is an extension of FISH technology. With this approach, DNA from a test sample is labeled with a red fluorescent dye, while DNA from a reference sample is labeled with green. The two samples are mixed, and observers measure the ratio of red to green fluorescence at each chromosome, focusing on deviations from the expected 1:1 ratio. The use of fragments of DNA printed on a chip (microarrays) makes it possible to rapidly compare two DNA sequences. The resolution power of the technique depends on the size of the microarrays used. A smaller size increases the method's sensitivity, because it is possible to observe more deviations from the norm. Specificity, however, decreases: a greater percentage of the observed changes will be variants of unknown/uncertain clinical significance—VUS. VUS may reflect normal variation among individuals; they may also be related to yet-unidentified pathological conditions. Recent introduction of diagnostic approaches grounded in analysis of free fetal DNA in maternal circulation (non-invasive prenatal testing) further enhanced the uses of FISH and similar technologies, since women who receive a diagnosis of higher probability of rare chromosomal anomaly of the fetus often undergo sampling of fetal cells by chorionic villus sampling or amniocentesis. The cells are then analyzed by a genomic technology that looks specifically for the suspected fetal anomaly.

CGH was introduced in the 1990s to study malignant cells. In the early twenty-first century, the perfection of the methods of genomic analysis and the construction of increasingly effective sequencing machines made a much wider use of this technology possible.[42] The extensive use of CGH in the early twenty-first century facilitated the identification of genetic anomalies linked to inborn syndromes, especially developmental delays.[43] In 2005, the American College of Medical Genetics stated that approximately one-third of the people with severe developmental delays were diagnosed as having chromosomal anomalies, and recommended the inclusion of genetic and genomic studies in each evaluation of an intellectual impairment in a child.[44] Such systematic evaluations accelerated the "geneticization" of birth defects. In the first edition of Smith's *Recognizable Patterns of Human Malformation* (1970), only a handful of inborn malformations were associated with genetic anomalies. In the 2013 edition of this book, 80% of the described inborn malformations are connected to changes in the genetic material of the cell; many of these changes can be detected during pregnancy.[45]

The possibility of detecting many genetic anomalies before birth radically alters the meaning of the diagnosis of such anomalies. The main aim of a diagnosis of a rare mutation in a child being treated in a dysmorphology clinic is to help this child and her/his parents. A diagnosis points to the cause of the child's problems and guides medical and educational interventions.[46] The aim of a prenatal detection of the same mutation is very different: to allow the woman to decide whether she wants to continue her pregnancy. Such duality of aims of genetic diagnosis complicates the task of dysmorphologists and clinical geneticists. The same experts who follow children with severe inborn impairments, and frequently become deeply attached to these children and their parents, may provide genetic counseling to women who wish to have an abortion in order to prevent the birth of a child with a similar condition.[47]

Sonograms and Uncertainty

High-resolution obstetrical ultrasound, like the new genetic technologies, coproduces accurate diagnostic and prognostic dilemmas. In some cases, ultrasonographic findings coupled with a precise genetic diagnosis can provide additional elements that can help the pregnant woman decide whether to continue the pregnancy. A fetus with Turner syndrome (45X0) and a significant hygroma (fluid-filled cavity behind the fetal neck) is classified as "severe Turner," and is seen as being at a greater risk of health problems after birth. Women whose fetus is diagnosed as having a "severe Turner" are therefore more inclined to have an abortion than those who learn that the fetus has Turner syndrome, but appears normal on the ultrasound screen.[48] Similarly, a woman who learns that she is carrying a Down syndrome fetus that also has been diagnosed as having a severe cardiac anomaly may be more inclined to interrupt the pregnancy than a woman whose trisomic fetus does not have visible malformations and is predicted to be healthy.[49] On the other hand, physical and intellectual impairments of children with Down syndrome are independent variables. Some children born with severe heart defects and successfully treated become high-functioning individuals with Down, while some Down syndrome children free of physical impairments have very limited intellectual capacities. The same is true for Turner syndrome. Some women with this condition have significant health problems such as kidney anomaly or the narrowing of the aorta but no intellectual impairment; other women with this condition have only minor physical health issues but significant learning difficulties.[50]

In other cases, ultrasonographic findings produce diagnostic dilemmas that cannot be resolved through studying the genetic material of fetal cells. Such dilemmas can be illustrated by the ultrasonographic diagnosis of three conditions: absence (agenesis) of corpus callosum, non-immune hydrops fetalis, and increased nuchal translucency with a normal karyotype. Agenesis of corpus callosum is a major anomaly of the neural system: the absence of the white band that connects the two hemispheres of the brain. This anomaly appears at the end of the first trimester of pregnancy but is usually detected in the second or third trimester. The ultrasonographic diagnosis of this condition is frequently confirmed by magnetic resonance imaging. The consequences of the absence of corpus callosum vary from minor cognitive and neurological problems to severe motor and intellectual impairments. Before the development of advanced methods of visualization of the brain such as high-resolution magnetic resonance (fig. 5), the absence of corpus callosum had been detected nearly always in children/adults with severe neurological problems or developmental delays. More recently, this condition was increasingly observed—often by chance—in people who live normal lives. With the possibility of diagnosing this condition before birth, ultrasound experts are not always sure of what they should say to pregnant women and their partners.

When the absence of corpus callosum is associated with other brain anomalies or with the presence of an abnormal number of chromosomes, specialists usually inform the pregnant woman that the outlook for her child does not look good, indirectly encouraging her to interrupt the pregnancy.[51] When the only observed defect is the absence of corpus callosum, several small-scale studies have indicated that 65–75% of children were free of major neurological complications. However, in 15% of cases classified before birth as an isolated absence of corpus callosum, physicians

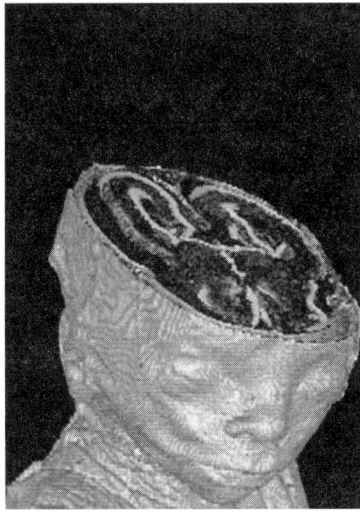

Fig. 5. A high-resolution magnetic resonance imaging (HRMRI) scan of a miscarried twenty-week-old fetus, showing the head and a cross section of the brain. Wellcome Images.

found later that this anomaly was associated with other inborn defects. As a consequence, the child's prognosis was worse than initially believed. Some experts argued that even children diagnosed as having an absence of corpus callosum and classified as "normal" had a lower IQ than their siblings, and some displayed impairment of their cognitive, executive, and social skills later in life. Experts also noted a higher frequency of psychiatric disorders in people with this condition. On the other hand, these data, too, were grounded in the study of a small number of cases, and therefore are not seen as very reliable.[52]

Non-immune hydrops fetalis (NIH) is a generalized swelling of the fetus. In the past, such a swelling had nearly always been produced by Rhesus factor (Rh) incompatibility between the pregnant woman and the fetus. When the pregnant woman is Rh negative (her red blood cells do not have an Rh marker on their surface) and the fetus is Rh positive, the woman, especially if sensitized by a first Rh-incompatible pregnancy, develops antibodies against fetal red blood cells. Such antibodies produce a swelling (hydrops) of the fetus. In the 1970s, physicians developed a treatment that prevents the production of antibodies against an Rh-negative fetus by an Rh-positive pregnant woman.[53] Today, the majority of cases of hydrops fetalis (76–87%) observed in industrialized countries is non-immune, that is, not produced by Rhesus factor incompatibility. One important cause of a non-immune hydrops fetalis is an abnormal number of chromosomes (aneuploidy). Women diagnosed as having NIH are often advised to undergo amniocentesis, and if the fetus is diagnosed as having aneuploidy, they are informed that the combination of NIH and chromosomal anomaly is often fatal. As a consequence, many women elect abortion.

When the fetus has a normal number of chromosomes, the three main (and unrelated) causes of NIH are respiratory disorders, heart and circulation disorders, and infections such as toxoplasmosis, cytomegalovirus, and parvovirus B19. In approximately 20% of cases, NIH physicians are unable to find a cause for this anomaly.[54] NIH is associated with a poor survival. Of babies born with this condition, 50 to 80% die either immediately after birth or during the first months of life. Approximately half the children who survive are healthy; the other half suffer from moderate to severe developmental delays. Combined uncertainty about the child's survival and the prognosis of the surviving children makes the counseling after a PND of NIH especially challenging.[55]

Increased nuchal translucency (accumulation of fluid behind the fetus's neck), observed usually between the eleventh- and the thirteenth-week ultrasound scan, is often a sign of aneuploidy, especially Down syndrome and Turner syndrome (figs. 6a, 6b). In both cases, the combination of chromosomal anomaly and significant nuchal translucency aggravates the prognosis for the future child.[56]

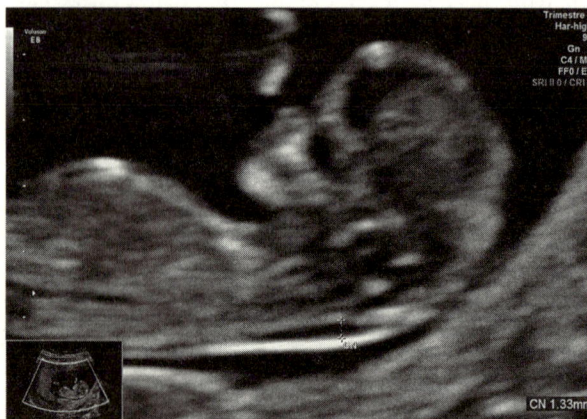

Fig. 6a. Normal nuchal translucency in a twelve-week-old fetus, ultrasound scan. Photograph courtesy Dr. Veronique Mirlesse.

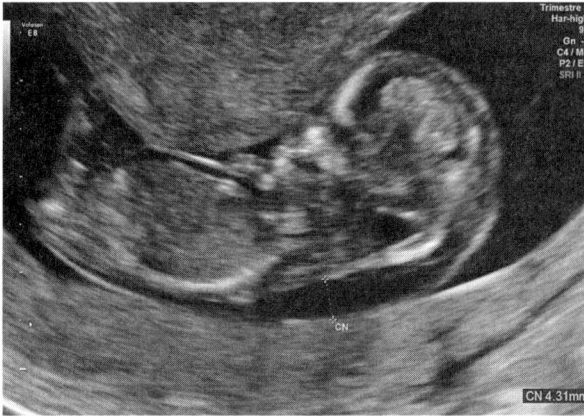

Fig. 6b. Enlarged nuchal translucency in a twelve-week-old fetus,
ultrasound scan. Photograph courtesy Dr. Veronique Mirlesse.

When ultrasound experts observe an increased nuchal translucency without aneuploidy, specialists often recommend further genetic tests, looking especially for Noonan syndrome, a genetic anomaly that also induces an increase in nuchal translucency.[57] On the other hand, a diagnosis of Noonan syndrome often does not put an end to prognostic dilemmas, because it is linked to multiple physical and intellectual impairments with a variable and unpredictable degree of severity.[58] When the fetus displays an increased nuchal translucency and all the genetic tests are normal, the main prognostic indication is the size of the translucent zone. If it is greater than 3.5 millimeters at eleven to thirteen weeks of pregnancy, the experts believe that the woman is at an increased risk of miscarriage and fetal death in the womb; if she gives birth to a living child, the child has a higher risk of birth defects. Such a pessimistic evaluation is often interpreted as indirect advice to elect an abortion.[59]

If the nuchal translucency is smaller than 3.5 millimeters, experts recommend an ultrasonographic monitoring of the fetus during the rest of the pregnancy. Until recently, they agreed that when such monitoring does not detect structural anomalies or retardation of fetal growth, the child will be healthy.[60] A 2014 study of a large series of fetuses with an increased nuchal translucency revises this optimistic conclusion. The authors of the study argue that even when all the ultrasounds were normal, a diagnosis of increased nuchal translucency at eleven to thirteen weeks of pregnancy increased by three times (from 1.7 to 5.2%) the probability of a major birth defect.[61] For some specialists, these results are reassuring; 95% of the women will still have a healthy child.[62] On the other hand, some women are risk averse, and may find a 5% chance of a severe impairment in their future child unacceptable (for a comparison, a forty-year-old woman, defined as being at "high risk" of Down syndrome, has an approximately 1% probability of giving birth to a child with this condition). They may also apprehend embarking on a long and stressful medicalized path. Some women who learn about an increased nuchal translucency of the fetus at the eleventh or twelfth week of pregnancy may elect an abortion.[63]

Dissections as "Quality Control" of Ultrasonographic Diagnoses

When a fetus is diagnosed as having agenesis of the corpus callosum, non-immune fetal hydrops, or increased nuchal translucency, the main difficulty is not identification of a structural anomaly, which is usually accurate, but the uncertain meaning of this anomaly for the future child. In other cases, ultrasound experts are not entirely sure of the exact significance of the images they observe, or whether the detected anomaly (missing fingers, cleft palate) is isolated or part of a syndrome and associated with other, invisible problems. One of the most effective ways to improve the accuracy of the correlation of sonograms with pregnancy outcomes and genetic data is to compare these images with the bodies of aborted or miscarried fetuses. In fetal medicine, probably more than in many other medical specialties, dissection plays a key role in improving the experts' diagnostic and prognostic skills, especially when dealing with rare or little-studied pathological conditions.

Fetopathologists usually do not dissect fetuses aborted for well-studied genetic anomalies, such as Down, Edwards, and Patau syndromes (respectively, three copies of chromosome 21, 18, and 13). The diagnosis of these conditions is made by counting the number of chromosomes in fetal cells, a reliable test. Studies of aborted fetuses and of the children of women who elected to continue the pregnancy after a diagnosis of aneuploidy have shown a high degree of agreement between prenatal laboratory findings and observation of fetuses or newborn children. In one exceptional case, the autopsy of a fetus aborted after being diagnosed as having Edwards syndrome (trisomy 18) by chorionic villus sampling failed to confirm the initial diagnosis: the fetal cells had a normal number of chromosomes.[64] The authors of this study later found out that the original diagnosis of trisomy 18 had reflected an anomaly of the placenta. Chorionic villus sampling, a method that can be performed early in pregnancy (ten to twelve weeks), examines the fetal part of the placenta. In the exceptional case, the placenta was composed of a mixture of normal and trisomic cells (mosaicism), while the fetus had only normal cells. This unusual observation led to a recommendation to always use an ultrasonographic study of the fetus to confirm a diagnosis of trisomy 18 made by a chorionic villus sampling, since trisomy 18 is often associated with visible fetal malformations. If the ultrasound does not show such malformations, a diagnosis of Edwards syndrome has to be confirmed by amniocentesis, a method which examines cells produced by the fetus and shed into the amniotic fluid.[65] Discrepancies between the diagnosis of an abnormal number of chromosomes and postmortem findings are, however, very rare. As a consequence, many hospitals do not perform autopsies on fetuses aborted after a diagnosis of aneuploidy.[66]

Diagnosing fetal anomalies by ultrasound is less definitive. Images are more open to interpretation than is the counting of fetal chromosomes. When a woman's decision to terminate a pregnancy is grounded exclusively in ultrasonographic data, fetal medicine experts want to verify the accuracy of the findings that led to her choice. This is especially true when the reason for the abortion is not a well-studied condition that can be diagnosed based on the sonograms alone (anencephaly, spina bifida, severe skeletal malformations, absence or abnormal form of limbs) but a less understood morphological anomaly or a polymalformative syndrome, meaning that fetal development went awry for an unknown reason. In such cases, the images observed on the ultrasound screen may be insufficient to define the extent and severity of the observed anomalies. A systematic juxtaposition of ultrasound and postmortem data helps improve the predictive power of obstetrical ultrasound and other visualization technologies such as magnetic resonance imaging.[67] According to a survey made in the 1980s, that is, in the early days of the diagnostic application of obstetrical ultrasound, a postmortem study of 133 fetuses aborted because of abnormal results of a scan led to a change in the original diagnosis in 53 cases.[68] More recent surveys have demonstrated a much better correlation between ultrasonographic diagnoses and postmortem studies of fetuses aborted for fetal anomalies. The level of agreement between prenatal and postmortem findings has steadily improved with time. Such an improvement was especially dramatic when data from the 1980s were compared with those from the 1990s.[69] The detected discrepancies mainly concerned minor structural anomalies. Moreover, the reported discrepancies between ultrasonographic and dissection findings went nearly always in the "right" direction: anomalies found postmortem were more prevalent and/or more severe than those uncovered during an ultrasound. Postmortem data thus provided retroactively a stronger justification for an abortion for a fetal indication.[70]

Occasionally, anomalies detected by ultrasound or other prenatal tests are not confirmed postmortem, but such cases seem to be rare. For example, a Norwegian group studying the results of 274 fetal autopsies describes a single case in which a postmortem did not confirm the initial ultrasonographic observations—bone anomalies and abnormal facial features—which led to the decision to terminate the pregnancy. In all the other cases, while they note that discrepancies between ultrasonographic findings and autopsies were frequent (approximately 40% of the cases), these were mostly minor and did not affect the original diagnosis.[71] One exception involved Dandy-Walker syndrome, a congenital malformation of the brain linked to neurological impairment. Postmortem observation in fetuses aborted after a diagnosis of this syndrome often diverged from ultrasonographic findings. Fetal medicine specialists found a way to overcome this difficulty by recommending that each ultrasonographic diagnosis of the Dandy-Walker anomaly be confirmed by magnetic resonance imaging (MRI). The use of the latter approach has reduced the disagreement between prenatal and postmortem data, but does not eliminate it entirely.[72]

Discrepancies between medical imaging data and those of the dissection of the fetus, fetal medicine specialists argue, do not always mean that ultrasonographic or MRI data were erroneous and that the autopsy revealed an "ultimate truth" about fetal impairment. Ultrasound and MRI observe the living fetus in real time. Images on a screen can reveal changes in physiological functions that may not be visible during an autopsy, such as the accumulation of fluid around the heart, inversion of the direction of blood flow in the aorta, or defective functioning of heart valves. In other cases, the discrepancy between ultrasonographic and postmortem findings can be attributed to changes that took place after fetal demise. For example, in some cases brain anomalies observed during an ultrasound cannot be found during an autopsy, because brain tissue is very fragile and deteriorates rapidly, especially if the dead fetus had remained for some time in the womb and underwent a partial self-destruction (autolysis) of tissues.[73]

Despite its limitations, a detailed postmortem of the fetus continues (in 2018) to be perceived as the gold standard in evaluating the causes of fetal malformation or fetal demise. In the past, the great majority of women facing a second- or third-trimester pregnancy loss had agreed to an autopsy of the fetus because they wished to know what went wrong and what the risk of another reproductive disaster was. Usually, an autopsy aims to better understand the cause of the patient's death and to contribute to medical knowledge. An autopsy of a fetus has additional goals: to advise a couple about their reproductive options, reassure the woman that she is not responsible for the unfortunate outcome of the pregnancy, and help her project herself into a future in which she will give birth to a healthy child.[74] It is not surprising that many women/parents agreed to a fetal autopsy. Recently, however, the publicity of cases in which hospitals treated dead fetuses or their tissues with disrespect; suspicions, fanned by the US anti-abortion movement, that fetal tissues are sold for profit; and the rapid development of mourning rites for the fetus have led to an increase in parental resistance to autopsies of fetuses and stillborn children and a decrease in parental authorizations for an autopsy.[75] An additional problem may be cost. Fetal autopsies are time consuming and have to be done by specialized pathologists assisted by skilled technicians.

To compensate for the decrease in fetal autopsies, some fetopathologists are developing a new approach: minimally invasive postmortem, also called virtual autopsy: the analysis of an intact fetal body. This method combines visualization technologies such as an MRI scan—and, if not available, a sonogram—with biopsies made by needle punctures, and advanced genetic studies. Advocates of virtual autopsy claim that it provides reliable results which are only slightly inferior to those obtained through the dissection of the fetus, especially when combined with detailed genetic investigations. In addition, virtual autopsy can overcome the main problems of conducting fetal autopsies today: parental opposition and high costs. It can also improve the study of fetuses in intermediate and developing countries. Tissue samples and sonograms collected in provincial hospitals can be sent to a small number of well-equipped centers staffed with competent pathologists, geneticists, and fetal medicine experts.[76]

Minimally invasive autopsies illustrate the dynamics of continuity and change in fetopathology. Fetopathologists' practices were standardized and codified in the 1940s and 1950s, mainly through the efforts of the US pediatrician and pathologist Edith Potter.[77] In many ways, their specialty still follows the basic rules established by Potter in the mid-twentieth century. In the twenty-first century, fetopathologists continue to perform the same tasks, often in the same order. There are, nevertheless, two important changes. One is the role of visual evidence. Today, fetopathologists—at least those who work in well-equipped hospitals in industrialized countries— use sophisticated microscopes with incorporated cameras to take many photographs of each dissection. They stock these photographs in the hospital's computer, then compare them with photographs of inborn anomalies of the fetus in centralized databanks. Far from disappearing in the Internet era, the pathologist's embodied ability to recognize forms plays a key role in diagnosing fetal anomalies. The second and probably even more important difference is the fetopathologists' increasing reliance on genetic data. Such data bridge the divide between clinical and pathological observations. This is a two-way street: in some cases, genetic data explain the fetopathologists' findings, while the rare and unusual anomalies they describe are a precious resource for geneticists who strive to expand their knowledge.

Today, the fetopathologists' goal is to elucidate the reason(s) for a fetal anomaly. In an ideal case, they provide an accurate diagnosis (for example, an inborn genetic disease or syndrome), a justification for this diagnosis (macroscopic and microscopic observations that support it), and a precise cause of the fetal malformation (for example, display of a mutation linked to the diagnosed condition, demonstration of the presence of an infectious agent). The number of cases classified as solved by fetopathologists and fetal medicine experts increases steadily, as does the understanding of the links between genetic and structural anomalies of the fetus. Yet such an understanding has not translated into effective cures for the majority of inborn impairments, and has not put an end to diagnostic and prognostic dilemmas.

Most pregnant women who undergo a routine ultrasound are reassured that "the baby is all right," but others learn that something may be wrong with the fetus, then embark on a long and difficult diagnostic odyssey which may end without a resolution. The Dutch media expert José (Johanna) Van Dijck argues that decisions about the fate of a pregnancy did not become more transparent or rational when physicians had better diagnostic technologies at their disposal. Just the opposite happened: the availability of increasingly advanced prenatal tests has rendered choices more complex and pregnant women less autonomous.[78] A "suspicious" result from a prenatal test frequently increases the pregnant woman's stress. Such anxiety-laden uncertainty—a rarely discussed consequence of the upsurge in the number of fetal anomalies detected by PND—can last throughout the woman's pregnancy and sometimes beyond it.[79] The cost of PND is not limited to the tests' price.

THREE

Diagnostic Puzzles:
Fetopathology in France

Medical Interruption of Pregnancy

The aim of PND is "to see what is about to be born," that is, to predict what the future of a scrutinized fetus will be. A positive diagnosis of a fetal anomaly is frequently an indication of a wide range of possibilities, and pregnant women are expected to become rational managers of their future child's risks.[1] The increase in the number of fetal anomalies that can be detected before birth has also increased prognostic uncertainty.

The fetopathology department, as I found out, is a privileged site for observing how health professionals deal with diagnostic and prognostic uncertainty of prenatal diagnosis. As I began my research on PND, I was barely aware of the existence of fetopathology. This medical subspecialty has a low visibility, partly because of the relatively low status of pathology and dissection, and partly because of the liminal status of human fetuses. Sociologists, historians, and anthropologists who studied PND have been mainly interested in genetics. A few investigated obstetrical ultrasound, although their studies focused more often on the production of "baby's first photograph" than on its importance for the diagnosis of fetal anomalies.[2] As far as I know, social scientists have not examined the role of the dissections of dead fetuses in the production of knowledge about fetal development.[3] As my work progressed, I became increasingly convinced that fetopathology, sometimes perceived as a poor relative of more prestigious scientific and medical specialties, is an "obligatory passage point" for the development of new knowledge about fetal malformations. At the same time, it is an important site for managing the uncertainty in PND.[4] I decided to take a closer look at the fetopathologists' practice.

I elected to observe the work of two fetopathology departments, one in France and one in Brazil. Both are situated in major research and teaching hospitals, and both are staffed by highly competent professionals, but their similarities end there. The two departments have different capacities for performing specific acts, have distinct patterns of collaboration with other experts, and study dissimilar pathologies. France has a long tradition of pro- natalist policies and protecting mothers and children, which includes effec- tive treatment of pregnant women within the national public health care system. Most of these women are treated in public hospitals and clinics. Upper-middle-class women who choose to give birth in a private maternity clinic do so because such clinics often provide more material comfort and more personalized attention, not because they believe that they will receive more competent medical care. In Brazil, health care is shaped by dramatic economic inequalities which produce sharp differences in access to health care according to wealth, and a split between the private and the public health care system. The latter faces significant financial and organizational difficulties, which persist despite the government's striving to improve its functioning. Pregnant middle-class women are practically always treated in the private sector. The public hospital I observed provided services for the poor. Another important difference between the two fetopathology depart- ments I studied is that while abortion for a fetal indication is legal in France, it is criminalized in Brazil, with the sole exception of anencephaly—the absence of a brain. Criminalization of abortion shapes not only the feto- pathologists' practice but the entire domain of fetal medicine in Brazil.

Most of the fetuses dissected by the French fetopathologists I observed had been aborted after a diagnosis of a fetal malformation. This was partly a local particularity. The teaching and research hospital where I made my observations is a tertiary referral center treating complex cases sent from regional hospitals and clinics. A significant proportion of these cases ended with a diagnosis of severe fetal impairment, followed by a decision to termi- nate the pregnancy; often, a dissection of the aborted fetus was undertaken. In hospitals that mainly provide routine care for low-risk pregnant women, a greater proportion of dissected fetuses either were miscarried or died in the womb, not infrequently near the pregnancy's end. In many of the latter cases, pathologists were unable to determine the cause of fetal demise.[5] In these cases, the autopsy's main role is to rule out the existence of a heredi- tary condition and to reassure the pregnant woman and her partner that the pregnancy's loss was an unpredictable accident that should not happen again. When a pregnancy ends with an induced abortion, the need to find out why things went wrong is even greater. Here the experts are more often able to uncover the cause of fetal problems and, if applicable, to provide genetic counseling. The determination of a cause retrospectively legitimates the decision to terminate the pregnancy, or, to quote the French philosopher Anne Fagot Largeaut, "takes moral decisions from the frail hands of humans, without burdening God with them."[6]

French gynecologists and fetal medicine experts frequently express satisfaction with their working conditions. They believe that PND in France is well organized and carefully regulated. Biochemical and genetic studies of fetal cells were submitted to strict quality-control measures in the late 1980s. Ultrasonographic diagnosis was formally regulated much later, in 2004, when the new version of the so-called bioethics laws required standardization and supervision of ultrasonographic diagnosis of the fetus.[7] French experts are convinced that these regulatory measures were effective and promoted a nationwide acceptance of standards of good practice. Some give the United Kingdom as an example of a country whose health care system's problems were caused by a lack of effective quality-control measures. In that country, French specialists believe, insufficient standardization of diagnostic practices, especially obstetrical ultrasound, produced local and regional disparities in PND: it is excellent in leading teaching centers, but may be less effective in other settings, especially small provincial hospitals. French fetal medicine experts are pleased to work in a country devoid of a strong anti-abortion movement and endowed with a legal framework that usually protects medical decisions. A third reason for the professionals' pride is a good collaboration between medical specialties, an important element in solving complex diagnostic puzzles. The fetopathologists I had observed worked in a close alliance with obstetricians, pediatricians, gynecologists, ultrasound experts, and clinical geneticists.

Strong interprofessional alliances often had been forged in scrutinizing living fetuses during the early days of PND's development. At that time, French geneticists, pediatricians, and obstetricians interested in the new approach started to meet informally to discuss difficult cases. They were later joined by obstetrical ultrasound experts. Many hospitals that specialized in maternity care created an informal collaborative structure to discuss fetal malformations. Until 1999, such structures had only a consultative role, but a 1999 law imposed the creation of specialized interdisciplinary committees responsible for decisions about the interruption of pregnancy for fetal or maternal indications. These committees were named pluridisciplinary centers for prenatal diagnosis—Centres Pluridisciplinaires de Diagnostic Prénatal, or CPDPNs.

The French health care system is far from being fully egalitarian, but women of all social strata usually have access to good-quality maternity services.[8] The main exceptions are women at the margins of society: illegal migrants, those who live in extreme poverty, women with mental health problems, and those isolated by language or cultural barriers.[9] Other women's care is followed by competent health professionals, above all midwives, who are responsible for supervising most normal pregnancies. During my observations at a French tertiary referral center, I was impressed by the speed with which a pregnant woman with a suspected fetal impairment and unclear diagnosis could gain access to a leading fetal medicine or genetics expert—often without knowing that the person she was seeing is a world-class specialist.[10] The usual pathway to such a consultation was an urgent e-mail message or phone call to the expert from the woman's primary health care provider or, more often, a physician from a secondary referral center, explaining why this is a truly complex case.[11]

The organization of PND in France is framed by laws on induced termination of pregnancy. Abortion became legal in that country in 1975, later than in the United Kingdom (1968) and the United States (1973).[12] French law differentiates a voluntary interruption of pregnancy (*interruption volontaire de la grossesse*, or IVG) for a "social reason" from a medical interruption of pregnancy (*interruption medicale de la grossesse*, or IMG) for either a fetal or a maternal indication. IVG is permitted up to fourteen weeks of pregnancy. Theoretically, women who wished to undergo IVG were expected to prove that the pregnancy produced a "severe distress," but this rule was never implemented. A 1980 ruling confirmed the woman's unrestricted right to have an early abortion, and the "distress" rule was formally abolished in August 2014.[13] When introduced in 1975, legal abortion in France was not fully paid for by the national health service. From 1982 until 2013, the service reimbursed 70% of abortion-related costs; for the majority of French women, their complementary health insurance picked up the rest of the bill.[14] Since 2013, the national health service has reimbursed all abortion-related expenses.

Interruption of pregnancy for medical reasons was tolerated in France long before the legalization of abortion for social reasons. Since the mid-1950s, physicians were allowed to induce an abortion for a medical reason—at that time, mainly because of serious risks to the pregnant woman's health, including mental health. In some hospitals, a liberal interpretation of the term *maternal distress* made possible the termination of pregnancy for fetal indications. Many French physicians were performing abortions on women at risk for giving birth to an impaired fetus: those who had been infected with the German measles virus early in the pregnancy, and those who had taken thalidomide when pregnant.[15] Then the 1975 law legitimated the termination of pregnancy for medical reasons. A pregnancy can be interrupted until its very end if a competent medical body allows this procedure. In 2017, access to IVG is controlled exclusively by the pregnant woman, while access to IMG is controlled exclusively by medical experts. The separation between IVG and IMG produces a sharp distinction between an unconditional and a conditional refusal of maternity: the first depends only on the pregnant woman's decision, while the second has to be approved by health professionals. Pregnant women are often unaware of this difference—when told that they need to ask permission for interruption of pregnancy for a fetal indication, many were surprised, and some were angry: the latter believed that it goes without saying that deciding what is a "severe" impairment should be the woman's/couple's alone.[16]

According to the 1975 law, the permission document for performing an IMG had to be signed by two physicians who attest that continuing the pregnancy constitutes a serious risk to the woman's health, or that the child very likely will suffer from an "especially serious medical condition, recognized as incurable at the moment of diagnosis." The French bioethics law of 1994 changed this rule, proclaiming that decisions about IMG had to be made in the pluridisciplinary prenatal diagnostic center (CPDPN) of each hospital or region. These centers came into being in 1999 with the publication of "application decrees" of the bioethics law. A CPDPN decides whether the fetal impairment corresponds to a legal definition of "serious disease" and "incurability."[17] The upper limit for voluntary interruption of pregnancy in France, fourteen weeks, is more restrictive than in many other western European countries. French women who are more than fourteen weeks' pregnant and want to have an abortion often travel to nearby countries, mostly Belgium and the United Kingdom. Most of these women seek an interruption of pregnancy for social reasons. Some missed the fourteen-week limit because they had not recognized that they were pregnant (a relatively frequent occurrence, especially among women who have irregular periods or believe that they are infertile); others initially accepted the pregnancy, then changed their mind. However, in a small number of cases, women who had an abortion abroad failed to receive the CPDPN's permission for an abortion for a fetal indication.

In the 1970s, most IMGs in France were justified by a risk to the pregnant woman's health. With the development of PND, the percentage of abortions for maternal indications diminished rapidly. In the twenty-first century, such indications have been a small proportion of all the IMGs. Among the 7,141 IMGs performed in France in 2010, 38.5% were for chromosomal defects (mainly Down syndrome), 6.2% for hereditary diseases, 44.0% for structural malformations of the fetus, 7.8% for other fetal indications, 2.7%

for maternal indications, and 0.8% for maternal infections that put the fetus at risk. The overall number of IMGs in France and their redistribution among different indications remained similar between 2006 and 2010.[18] The number of registered medical interruptions of pregnancy in France is relatively high. For example, in 2009 the UK health authority reported 2,006 cases of terminations of pregnancy for fetal indications, while the French authorities reported 6,768 such terminations. The number of pregnancies in the two countries was similar.[19]

French experts attribute their country's high number of abortions for fetal indications to effective organization of prenatal care, good quality of prenatal screening for fetal anomalies, good diagnostic skills of ultrasound specialists, and full reimbursement of all the expenses related to PND and termination of pregnancy. They explain that in countries where women can have an abortion "on demand," up to twenty to twenty-four weeks of pregnancy, abortions for fetal indications are probably underreported, because in some cases, especially of "borderline" conditions, women and health professionals may choose to present them as social abortions to avoid criticism for refusing to give birth to a disabled child.[20] French physicians also believe that national statistics on IMG are more accurate than in many other Western countries, because France mainly escaped the conflicts around abortion for fetal indications, while such conflicts have been more frequent in English- and German-speaking countries.

French Professionals and Medical Interruption of Pregnancy

A French woman who wishes to terminate her pregnancy after a positive PND is legally obligated to receive a CPDPN's permission for an IMG, even if the fetal impairment was detected before the fourteen-week limit for an IVG. On the other hand, many French women today undergo first-trimester screening for Down at eleven to twelve weeks of pregnancy. If the results indicate that they are at a higher-than-average risk of having a child with this condition, they can undergo chorionic villus sampling and receive early confirmation of a chromosomal anomaly of the fetus. The predicted more systematic use of non-invasive prenatal testing, performed usually at ten weeks of pregnancy, may support extending the first-trimester diagnosis of selected fetal anomalies.[21] A woman who learns about a fetal impairment before she is fourteen weeks' pregnant and does not wish to be evaluated by a CPDPN—for example, because the fetus was diagnosed as having a "borderline" condition such as Klinefelter syndrome (47, XXY), and she is not sure whether the CPDPN will grant her permission to have an abortion for this indication—can go to a different medical center, refrain from mentioning her PND results, and undergo an IVG for social reasons. French experts believe (in 2018) that such cases are rare, but also that there are no statistics on "hidden" abortions for fetal indications.

Abortion in the second trimester is more strictly supervised in France than in several other European countries and the United States. By contrast, abortion at the very end of pregnancy is less restricted in France than in several western European countries and many US states, which prohibit abortion beyond the limit of fetal viability (usually twenty-two to twenty-four weeks of pregnancy).[22] In France, a woman can have an abortion until the child's birth; only at that moment does the child become a distinct juridical entity and cease to be a part of the maternal body.[23] The possibility of waiting as long as necessary to produce a reliable diagnosis of the level of fetal impairment, according to French fetal medicine experts, prevents unnecessary abortions conducted under pressure to terminate the pregnancy before the end of the legal limit. It also reduces pressures to perform fetal surgery, an intervention that, many French specialists believe, is not only dangerous for the fetus but linked to a serious health risk to the pregnant woman.

In countries that set a strict upper limit on performing an abortion for a fetal indication, the discovery of a severe fetal anomaly beyond this limit obliges the pregnant woman to accept the birth of an impaired child. In such cases, physicians may be inclined to propose a risky fetal surgery as a last-ditch effort to prevent/reduce the child's impairment. Attempts at a fetal surgery having a low success rate, and a high risk of fetal death may also be motivated by an unacknowledged belief that fetal demise may be preferable to inaction. Since in France (but also, for example, in the United Kingdom or Israel) it is possible to interrupt an advanced pregnancy, women who learn about a severe fetal anomaly late in pregnancy can choose not to give birth to a disabled child without putting their own health at risk. French physicians who praise the flexibility of French law because it does not define an upper limit for having an abortion for a fetal indication often fail to notice the rigidity of a law that does not allow women to decide the fate of their pregnancy after fourteen weeks—perhaps because in both cases, the law consolidates the power of the medical expert.

The law that a medical interruption of pregnancy is justified only when a child faces a serious risk of a severe, incurable condition can produce paradoxical results: better cures may lead to more interruptions of pregnancy. For example, in 2018 the success rate of the surgery for repairing a moderate diaphragmatic hernia (a malformation of the separation between the intestines and the chest cavity) is 50–70%. Parents are told that if the operation is a success, the child will be healthy; if it fails, the child will die.[24] When a choice is presented in these terms, experts can reason that a pregnant woman who receives a diagnosis of diaphragmatic hernia of the fetus does not have a legitimate reason to interrupt the pregnancy. Moreover, the improvement of surgical methods to treat an extended diaphragmatic hernia may increase the chances of the child's survival, but also the chance that some surviving children will remain permanently impaired. In such a configuration, a woman's decision to interrupt pregnancy may be justified by a significant risk of severe health problems for the future child.[25]

When asked about their role in termination of pregnancy, French midwives explain that they are frequently distressed by the act itself; but since the woman's decision to have an IMG had been validated by a CPDPN, they are not entitled to judge her. Their duty as responsible professionals is to suspend their personal feelings and help women/couples in a difficult situation.[26] In France, the termination of pregnancy for fetal indications is fully integrated with other tasks of a maternity ward. Leaflets and instructions for women scheduled to undergo an abortion explain that they will be hospitalized the evening before the procedure, and will be given drugs to induce the abortion early the next morning. An early hour for the intervention is chosen, the leaflets explain, because a planned interruption of pregnancy is difficult not only for a woman and her partner but for the maternity staff as well. Staff members elect therefore to start their day with this demanding task.[27] The act of giving birth to a nonviable child is usually monitored by midwives, as are the majority of normal births in French public hospitals. However, obstetricians perform specific tasks, such as feticide (an act which produces fetal death in the womb). Feticide is systematically performed in France if the fetus is over twenty-two weeks old. Instructions for midwives also explain how to present the fetus to the woman and her partner, since midwives are responsible for the handling of fetal bodies.[28]

The legal status of the fetus/stillborn child and of the woman who gave birth is defined in France by a complex network of laws and regulations.[29] These rules are identical for miscarriage, stillbirth, fetal death in utero, and termination of pregnancy for a maternal or fetal indication.[30] The main distinction is between the nonviable and the viable fetus/child. Birth after the twenty-second week of pregnancy and when the child weighs more than five hundred grams is classified as childbirth. The woman is therefore entitled to all the legal privileges of a new mother, such as a full maternity leave. The father is also entitled to a legal paternity leave.[31] Jurisprudence established that these privileges are not linked with the detection of signs of life upon birth. Children born before twenty-two weeks or those weighing less than five hundred grams can be declared as stillborn children (*enfant né sans vie*) by their parents, but are not entitled to the full legal status of a child.[32]

The declaration of a fetus as a stillborn child is possible from the fourteenth through the fifteenth week of pregnancy. The handling of a declared fetus is dramatically different from the handling of an undeclared one. Undeclared fetuses are treated as surgical waste and transported by "ordinary means," which may include a trash van. Declared fetuses are transported by official transporters of dead bodies, provided by the hospital or external funeral services. If an undeclared fetus undergoes an autopsy, it is not necessary to reconstitute the body after the dissection. The remains are put in a sealed bag and are disposed of as surgical remains. After an autopsy, a declared fetus is sewn back together and reshaped, dressed with clothes provided by the family or the hospital, and placed in the hospital's morgue. The parents are then informed that the body is prepared for burial, and they can arrange for the body's retrieval by their funeral director.[33]

French gynecologists, obstetricians, and fetal medicine experts view their country's system of pregnancy monitoring as competent, effective, and equally available to all women, a statement occasionally accompanied by stories about other countries' problematic management of diagnoses of fetal malformations.[34] French physicians also believe that decisions about medical interruptions of pregnancy issued by CPDPNs are fair, and correspond to the spirit of the law on interruption of pregnancy for fetal indications.[35] Arguably, the fairness of the French system is far from being absolute. Not all women have an equal capability of stating their case before the experts, and decisions made by a given CPDPN may be affected by local traditions and contingent elements. Nevertheless, the physicians' conviction that the French system works well reflects their faith in the efficacy of the internal regulatory mechanisms of their professional practices.

Obtaining Permission for an Abortion for a Fetal Indication

French physicians have a twofold power over women who receive a diagnosis of a fetal malformation: they interpret the meaning of such a diagnosis and, if the woman wants to interrupt the pregnancy, decide whether her decision is legitimate. With the advancement of diagnostic technologies and the increased detection rate of conditions with variable or unknown consequences for the future child, the interpretation of a positive PND has become an increasingly complex endeavor. Words chosen by practitioners to describe a given fetal problem often affect women's/parents' decisions about the future of the pregnancy. The same inborn condition can be described by one expert as frequently associated with multiple health and behavioral problems and by another as producing only minor difficulties that will not prevent the future child from having a normal life.[36]

The great majority of the CPDPNs' decisions is consensual. An analysis of the reasons provided for pregnancy termination for fetal malformation in France reveals that—according to the authors of that study—most terminations were indeed for "severe and incurable conditions." Only in a small proportion of the cases (around 6%, according to data up to 2010) were couples granted permission to terminate the pregnancy after a diagnosis of a condition that can be perceived as borderline or controversial: late-onset pathologies such as Huntington disease, a moderate impairment produced for example by an absence of a limb, or a chronic but treatable disease such as hemophilia or sickle-cell anemia.[37] Members of a CPDPN usually agree that a given fetal impairment justifies an abortion. This is true even when there is no complete agreement among the French specialists. For example, a majority of French physicians support termination of pregnancy for Down syndrome, but some physicians see it as unjustified and misguided.[38] Nevertheless, CPDPNs practically never refuse a woman's request to have an abortion for this indication. Similarly, experts frequently recognize that future parents' view of a chronic condition may be different from its perception by professionals: a family's experience with a failed treatment for sickle-cell anemia may lead them to have a pessimistic view of life with this condition.[39]

CPDPNs reject approximately 2% of women's/couples' requests to terminate pregnancy for a fetal indication—and approve 98% of these. For example, in 2009 CPDPNs delivered 6,993 permissions for IMG, and refused such permission in 109 cases; in 2010 there were 7,141 accepted requests and 119 refusals.[40] In some cases, the CPDPN members feel that a fetal anomaly such as cleft palate does not fit the description of a "particularly severe and incurable condition." Nevertheless, they grant the woman permission to interrupt the pregnancy because of "severe maternal distress," with the underlying assumption that her strong rejection of the pregnancy will compromise her and her future child's well-being. In such a case, the interruption of pregnancy is legitimated by a risk to the woman's mental health, and is classified under the heading "maternal causes."

CPDPNs' rejections of requests for termination of pregnancy do not obey fixed rules. Such rejections may reflect contingent elements such as the composition of a given committee or their situated views of their obligations and duties. Some specialists are convinced that the CPDPN's task is to define what a medically acceptable act is.[41] Others think that physicians should not have the right to decide what a pregnancy's fate will be. Physicians can attempt to persuade the parents to accept a fetal problem that they perceive as minor, but they cannot impose their views: the final decision should belong to the future parents.[42] In some cases, the opinion of a charismatic leading specialist, be it restrictive or permissive, influences the views of other members of a local CPDPN. Experts' decisions may also be affected by their familiarity with a given fetal anomaly. Hospitals specializing in monitoring problematic pregnancies have a much higher percentage of IMGs. Members of the CPDPN in these hospitals are often more knowledgeable about the expected outcome of diagnosed fetal anomalies than those from hospitals that mainly treat women with normal pregnancies, and may provide different evaluations of the "severity and incurability" of a given condition than their less experienced colleagues.[43] Finally, the CPDPNs' decisions may be affected by the rhetorical skills of the pregnant woman who presents her request.

In one instance, the same CPDPN accepted one request for an abortion for a missing hand (a paradigmatic—and relatively rare—case of a "borderline" fetal anomaly) and refused a second similar request. In the first case, the pregnant woman argued that the presence of a genetic disease in her family, and her being a special education teacher and familiar with the plight of impaired children, decreased her ability to cope with the education of a disabled child. The committee accepted her request, justifying its acceptance by "maternal distress." In the second case, the woman was unable to present convincing arguments for her rejection of the child's impairment, and her request was refused. One of the physicians present at the CPDPN session added that it is already difficult to accept one request for termination of pregnancy for the absence of a fetal hand; the CPDPN cannot accept two such requests in the same session. In another borderline case, a couple was at risk of transmission of Stickler syndrome, a condition linked with ocular anomalies and dysmorphic traits which may compromise the child's respiration and swallowing. The syndrome has a variable expression. It is classified as treatable, but people with more severe cases face serious difficulties in daily life. The couple already had one affected child with multiple health challenges. They were determined not to have a second child with this condition, and asked for a PND of Stickler syndrome early in the pregnancy. The CPDPN members could not agree whether Stickler syndrome corresponds to the definition of a "condition of a particular gravity." They also hesitated about whether to permit an early PND for this condition, because they knew that if the test was positive and they refused termination, the woman would still be within the fourteen-week limit of IVG, and would be able to end her pregnancy at another French center.[44]

The French system of granting permission for IMG works well, one may conclude, when an abortion for a fetal indication is seen as noncontroversial by the CPDPN's members—that is, in the great majority of cases—but more problematic when the fetus is diagnosed as having a borderline or minor impairment. Women of higher socioeconomic status are often more skilled in convincing health professionals that the continuation of a pregnancy will be a source of distress for them, their family, and the future child. They can more easily find out which CPDPNs are more liberal and which are more restrictive, and address the former. Finally, they have more options if their request is rejected: an abortion at a different center in France if they are under the fourteen-week limit of IVG, and abroad if they are over this limit. French women from upper socioeconomic strata are, one may assume, at lower risk of giving birth to an impaired child they do not accept than are poor, uneducated, or otherwise marginalized women.

Fetopathology in France

In France, fetopathology is a separate subspecialty, with a distinct training trajectory.[45] The French Society of Fetopathology (Societé Française de Fœto-pathology, or SOFFŒT) was founded in 1984. Approximately twenty of its members are full-time fetopathologists; approximately one hundred other members combine work as fetopathologists with that of a general pathologist. SOFFŒT's activities were and continue to be shaped by a strong interest in the role of genes in fetal development; its members see themselves as important contributors to the development of clinical genetics in France. The society's first official site was located in the embryology and genetics department of a prominent teaching hospital. SOFFŒT is affiliated with the French Federation of Human Genetics, and its members participate in the federation's meetings and other professional activities.[46]

Since 1999, SOFFŒT has had exclusive responsibility for providing instruction in the diploma of fetal pathology program, managed conjointly by several universities (*diplome inter universitaire*). The program is designed for all physicians interested in this specialty. Previous experience in pathology is an advantage, but is not a prerequisite for studying toward a fetopathology diploma. The teaching of fetopathology is practice-based. Students take six intensive theoretical courses, each four days long—some take place in Paris, and some in the provinces—and undergo intensive practical training. Such training, distributed over two academic years, is conducted part-time, enabling the students to have other clinical activities and thus a source of income. Physicians preparing for the fetopathology diploma need to perform at least fifty fetal autopsies, learn how to read prepared microscope slides of fetal tissues, pass a theoretical examination, and write a short research dissertation that should demonstrate the candidate's ability to conduct independent research. In practice, SOFFŒT's experts, keen to transmit their knowledge to a younger generation, often propose "fail-proof" research subjects to promising candidates. The modest tuition helps strengthen SOFFŒT's budget. Otherwise, the society is financed—according to its members, on a shoestring—through member contributions, teaching income, and sales of a fetopathology textbook it has edited.[47]

Fetopathologists study *birth defects*, a term with rapidly evolving boundaries. One of SOFFŒT's leading experts argued in 2014 that she would like to persuade her colleagues to eliminate this term altogether. When physicians coined *birth defects*, they were unable to scrutinize the living fetus, and had only vague notions of the role of mutations in producing fetal malformations. In the twenty-first century, many "birth defects" are detected long before birth, while other inborn conditions, such as late-onset genetic diseases, are present already in the fertilized egg but revealed only in adults. A rapid increase in the physicians' capacity to detect fetal anomalies reduced the importance of birth as a crucial moment of revelation of an individual's health status. There are no birth defects, only a specific subset of human pathologies that are produced by either mutations or events that took place during embryonic and fetal development.

Several French fetopathologists recently expressed a fear that their specialty will disappear soon. Its future is threatened, they argue, by the combination of growing resistance to the dissection of fetuses and stillborn children and an increase in the resolution power of genetic analyses of fetal tissues. It is probable that genetic studies of fetal tissues, an intervention that does not affect grieving parents' sensibilities, will soon allow the diagnosis of nearly all the fetal anomalies. Clinical geneticists who have elected to collaborate with fetopathologists hold the opposite view. The rapid development of new genomic technologies, they maintain, will increase the demand for fetopathologists' services. These powerful innovations increasingly reveal anomalies of fetal DNA such as microdeletions and microduplications, but fail to provide information about the clinical meaning of many of these anomalies. Fetopathology is a key site for the correlation of genetic, anatomic, and cytological data about normal and abnormal fetal development. Far from being threatened by "extinction," fetopathology's progression will be augmented by the need to provide clinical meaning to the rapidly accumulating genetic data.[48]

Observing Fetopathologists in a French Hospital

The hospital where I made my observations of its fetopathology department, Andral Hospital (AH), is a prominent French research and teaching facility. It is named after the nineteenth-century French physician Gabriel Andral, famous for his pioneering studies of pathology and blood physiology. AH's physicians are known for their cutting-edge research in biochemistry and genetics as applied to gynecology, obstetrics, and pediatrics, especially the latter. The hospital's architecture reflects its multilayered history: a complex mixture of buildings, some dating from the early twentieth century, many from the 1960s and 1970s, and a few constructed in the late twentieth and early twenty-first centuries. The fetopathology department is situated in one of AH's older buildings. It does not have a space of its own, but instead occupies several rooms within the histology, embryology, and cytogenetics department. The fetopathology department is usually overcrowded, especially on days when everybody on the staff is at work there.

Cupboards and equipment overflow the corridor, which is also where patients' files are stored (fig. 7). There is a tiny kitchen where one can make a cup of tea or coffee, and a space with two toilets and a shower. A small closet bears a sign, Radioactive Materials, on its door, perhaps to discourage its use by outsiders. The fetopathology department also has a small shared library/meeting room, which is occasionally used as additional work space.

The center of the department's activity is its only rigorously organized space: a midsized but well-equipped dissection room with a tightly regulated access.[49] One of the main features of the room is an array of sophisticated equipment for producing high-quality photographs of the dissected fetuses (fig. 8). The analysis of pathologists' findings and the circulation of their work among other experts depend to an important extent on the production of such photographs.

The fetopathology department's head, Joelle, is the only person entitled to an office. But this office is very small and tightly packed with papers and boxes of slides and files. She can, however, close her door and work in relative calm. Other fetopathologists, fetopathology technicians, students, and visitors share two common rooms. On busy days, it is a challenge to find work space or even a free chair in these two rooms. On less busy days, the rooms are mainly used by the department's two technicians and sometimes one or two trainees involved with the SOFFŒT diploma program. Because

Fig. 7. Patients' files stored in the corridor. AH, fetopathology department. Author's photograph.

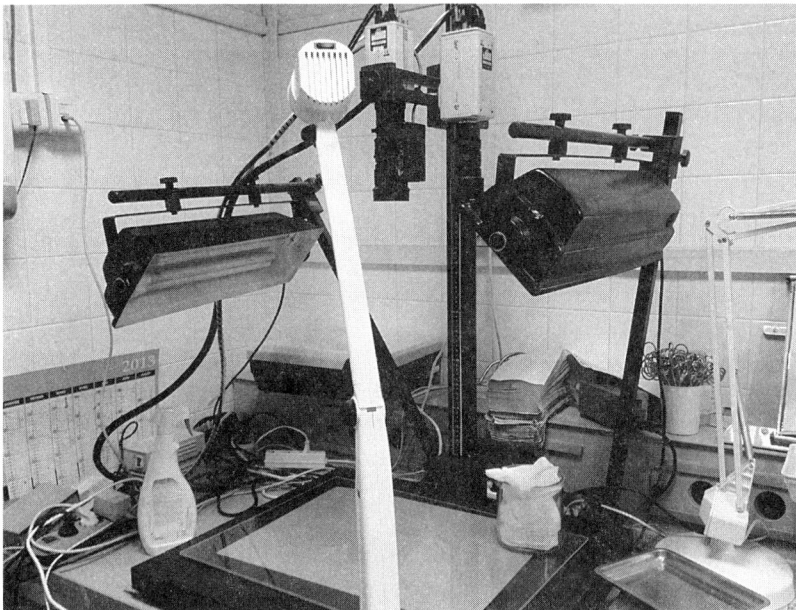

Fig. 8. Photography corner, dissection room. AH, fetopathology department. Author's photograph.

Fig. 9. A workstation filled with a mixture of slides and papers.
AH, fetopathology department. Author's photograph.

of severe space limitations, the rooms may seem chaotic and disorganized, but the fetopathology department's two technicians, the older and more experienced Nora and the younger Leah, seem to know exactly where everything is (fig. 9). Occasionally, when the two fetopathology rooms become too crowded, Joelle shares her office space with other workers. The second pathologist, Genevieve—later replaced by Nadine—did not have an office of her own. She occasionally used Joelle's office, but worked most of the time in one of the common rooms together with the technicians and the trainees.

Senior fetopathologists also work part-time in AH's molecular genetics department. This department, too, is located in a relatively old building, but leading researchers have spacious, well-lit offices, there is a large library-cum-meeting room, and the laboratories are less crowded and better equipped. Although this department also needs to store its instruments, files, and books in corridors, the overall impression is one of more comfortable and less constrained work conditions. One of the senior fetopathologists, Soraya, specializing in malformation of the fetal neural system, has her own comfortable office in this department, while Joelle has the option of working there on a regular basis—a spatial arrangement that mirrors the close relationships between fetal pathology and molecular genetics. In private conversations, AH's fetopathologists—sometimes quite outspoken, as French physicians not infrequently are—occasionally made ironic remarks about colleagues from other specialties. However, at staff meetings and other collaborative ventures, personal likes and dislikes and hierarchical considerations were rarely visible, probably because all the participants share the conviction that their first priority is to get the diagnosis of a fetal condition right.

Weekly fetopathology staff meetings are the main occasion for elaborating on diagnoses. While I was at AH making my observations, all the meeting participants were unfailingly helpful to me. They cheerfully tolerated my ignorance, and patiently answered all my questions, however irrelevant or naive. Besides Joelle, Genevieve (and later Nadine), Soraya, the technicians, and the trainees, the meetings are regularly attended by Myriam, a leading clinical geneticist with multispecialty training in genetics, pediatrics, and fetopathology and an impressive publication record in molecular genetics. She was recently named to a prestigious genetics professorship at the time of my visit, and is especially active in promoting close links between the fetopathology and genetics departments. Her spacious office and laboratory are located in the molecular genetics department.

Myriam is also one of the main genetic counselors at AH, and specializes in complex and difficult cases. Her PhD student, Isabelle, another regular participant at the fetal pathology staff meetings, also works in the molecular genetics department. She is a young pediatrician with parallel training in fetopathology who was working on her thesis in clinical genetics. Myriam, who first specialized in pediatrics before studying genetics and fetopathology, strongly encourages her students' multidisciplinary training.

Chief on the agenda of each weekly staff meeting is a review of all the fetal autopsies held during the past week, with a focus on unusual or difficult cases. The participants collectively produce the exact wording of each diagnosis, which will then be inscribed in the patient's record. Especially challenging cases are discussed over the course of several staff meetings. The most difficult among them are often recalled by the fetopathologists, sometimes years later. I chose to focus on such difficult cases because they display the technical and material aspects of PNDs and highlight the specialists' dilemmas.

Fetopathology staff is one among several regular departmental meetings at AH dedicated to fetal and newborn anomalies. The others are molecular genetics, gynecology and maternity, pediatrics, and a weekly meeting of geneticists and students centered on the investigation of complex genetic conditions. Myriam and Isabelle often participate in all these meetings, with Soraya joining occasionally. In addition, Soraya and Joelle are frequently invited to present cases at the molecular genetics staff meetings and periodically at the maternity and pediatrics staffs. They prepare these presentations very carefully, partly because fetopathology may be seen as a lower-status specialty, and partly because they are keen to receive feedback from other experts.

All the patients' data at the Andral Hospital, including visual ones, are centralized in the hospital's computer system. Their accessibility facilitates contacts among health professionals. Leila, a fetopathologist specializing in the study of fetal brain tissue and usually responsible for retrieving data from cases discussed at the fetopathology staff meetings, might have added, "when the computer system is working as it should," which was far from being always the case. Sometimes, the first five or ten minutes of the fetopathology staff meetings are dedicated to fiddling with the computer to retrieve the appropriate data and project them on a screen.

The weekly fetopathology staff meetings are organized around the projections of these data: data from computerized patients' files, photographs taken during autopsies, X-ray films of fetuses, and images of stained fetal tissues. Fetopathologists at AH also use traditional records: a handwritten casebook and patients' files, which contain copies of letters from the patients' physicians, along with test results and pathology reports. As written reports, the casebook and patients' files are similar to patients' records from the early twentieth century; computer records, with their multiple hypertext links, are of course very different. The simultaneous use of both methods of recording and organizing data may correspond to the "bicultural" aspect of fetal pathology: at the same time a traditional specialty rooted in nineteenth-century autopsy protocols and one that participates in the development of cutting-edge scientific knowledge.[50]

AH's anatomy, cytology, and histology department is situated next door to the hospital's morgue, an austere structure that is painted white and charcoal gray. It has an antechamber, an inner room that can be used as a nondenominational prayer space and a place where parents receive the coffin containing their child's body. There is no decoration, and because the morgue is near the hospital exit, there are no connections to AH's internal green spaces. The morgue is usually empty, and the collection of a fetal body is a relatively rare event. In compliance with the French law of 2008 concerning the disposal of fetal bodies, the decision of what to do with such a body belongs exclusively to the couple. AH's forms for recording such decisions are addressed to "parents" (respectively, "father" and "mother"), and the body is described as that of a "child," independently of the fetus's size. In one of the patients' files I examined, the staff noted that "the parents did not wish to see the child's body"; in that case, the "child" was a fourteen-week-old fetus aborted for a malformation.[51]

The parents are asked to sign a multiple-choice form that specifies whether they want to recognize the child and inscribe him or her in the official family record (*livret de famille*), give this child a name, and receive the body for a burial by the family (otherwise, it is left at the hospital for cremation). In the patients' files I studied, parents who lost a pregnancy in the third trimester (frequently because of a sudden death in the womb) chose to recognize the child, name him or her, and arrange for a burial. A majority of the abortions for a fetal indication were performed during the second trimester of pregnancy, in which case most parents chose hospital cremation of the body, although some recognized the child and had the lost pregnancy entered in the family's official record.

The mother/parents also fill out a second form in which they state whether they agree to an autopsy, a cytological and genetic investigation of the causes of their child's problems, and the use of the child's tissues and cells for research. If the child was born alive, both parents have to sign this form; in cases of miscarriage, induced abortion, death in the womb, and stillbirth, the father's signature is optional. Couples/women who agree to an autopsy usually also accept an extended investigation into the causes of their child's problems, but many refuse to allow the use of their child's tissues for fundamental research on causes of inborn malformations.

Joelle told me that when she started work as a fetopathologist twenty years ago, 90% of the parents agreed to a fetal autopsy. She was convinced that the percentage of permissions received for an autopsy was lower now; recent data indeed support this view. She and Soraya attributed the decrease to a sharp decline in the overall number of autopsies being performed in French hospitals; the growing importance of religion, especially Islam, among people originally from North and sub-Saharan Africa; and the influence of midwives, who increasingly encourage the couple to see the expelled fetus as "their baby." Parents, Joelle added, were always more reluctant to agree to an autopsy of a child born alive. The growing assimilation of fetuses with children may extend such reluctance to nonviable fetuses.[52] Joelle's conviction that persuading parents to agree to fetal autopsies has become increasingly difficult, and that this trend will be amplified in the future, has led her to develop an interest in virtual autopsies.

One of the main preoccupations of the senior members of AH's fetopathology staff is the transmission of expert knowledge. AH is one of the main training centers accredited by SOFFŒT. Joelle and Soraya receive a steady stream of mostly female young physicians who undergo fetopathology training, and dedicate an important portion of their time to such training. Soraya, who at the time of my study was nearing the French obligatory retirement age for public servants (a category that includes physicians in public hospitals), is especially concerned about transmitting her very impressive knowledge of anomalies of the fetal neural system.[53] She organizes teaching sessions for physicians in training at AH's fetopathology department and additional sessions focused on specific diagnostic problems open to outside participants, and often transforms staff sessions into improvised teaching opportunities. Myriam, who is far from retirement age but shares Soraya's concerns for the transmission of expert knowledge in fetopathology, also frequently transforms staff meetings into teaching sessions about fetal anomalies and their origins. Myriam is a very gifted teacher, able to present complex problems in a clear and well-organized manner. She often illustrates her explanations with diagrams and photographs containing images either already present in her computer or rapidly found on the web.

AH's fetopathologists are often very protective and "maternal" toward their trainees, especially those they identify as promising fetopathologists. Their solicitude reflects a worry that young people will abandon a specialty that is difficult to master and has a relatively low standing. Students of fetopathology have to familiarize themselves with a very traditional medical specialty, pathology, and a very new one, clinical and molecular genetics. They need to acquire a "feel" for tissues that they cut and manipulate, and a good visual memory of normal fetal structures and pathological changes in these structures. They have to develop the manual dexterity needed to manipulate tiny fetal bodies, as when learning how to pass a thin probe through fetal heart chambers to check for the presence of blocked passages. Trainees also need to learn how to cope with "difficult" fetal bodies, especially those macerated and decomposed in the womb. Maceration may be the result of late discovery of spontaneous fetal death, or of a long interval between the initiation of induced abortion and the expulsion of the fetus. Macerated fetuses have an unpleasant aspect; they are also more difficult to study, because the pathologist has to distinguish between malformations which existed before fetal demise, and changes which took place after the death, such as the rapid deterioration of brain tissue. To persuade their students to persist in their choice of specialization despite all the obstacles, Joelle and Soraya attempt to convey their enthusiasm about fetopathology. They make special efforts to include students at every interesting professional event in the hospital, take time to explain each unusual case, and strive to find them feasible and stimulating subjects for their dissertation.

One of the tasks of AH's fetopathologists and fetopathology technicians is to prepare a dissected body for burial, and to produce a commemorative photograph. Joelle explained that she started to make such photographs as an informal service for the parents. At first she employed her own very simple camera, but later a bereaved family offered her a better one. Joelle sometimes mentioned parents who wished to dress the fetus/child in fancy or unusual clothes. I saw only one such case: a photograph of a third-trimester fetus dressed in a blue and yellow Superman tee shirt. In another case, Joelle showed me a photograph of a very tiny fifteen-week-old fetus with blackened extremities: the parents had wanted to have prints of the fetus's feet and hands. Sometimes the parents also want to have the hospital identification bracelet placed on the body in the maternity ward. Myriam recalled a case in which the father of a deceased fetus/child who had come for a follow-up visit asked for such a bracelet (the mother, described as highly distressed by the pregnancy loss, had returned to her country of origin). Myriam did not have it anymore, but she fabricated one with the father's tacit consent, so that the mother could have a material trace of her lost child.

Routine Activities of the Fetopathology Department

AH's fetopathologists frequently study fetuses with unusual impairments, but because the hospital has a maternity ward where obstetricians also treat women having low-risk pregnancies, the fetopathologists also dissect fetuses that died unexpectedly in the uterus, stillborn children, newborns who died immediately after birth, and fetuses aborted for a maternal indication. A frequent cause of the latter is preeclampsia—pregnancy-related high blood pressure with a high level of protein in the urine. Preeclampsia can induce thrombopenia (low level of platelets) and liver failure in the pregnant woman, and in severe cases can put her life at risk (HELLP syndrome: Hemolysis, Elevated Liver enzymes, and Low Platelet count). The fetus often suffers, too, but fetal problems are seen as secondary to maternal ones. Other maternal diseases, such as diabetes, may produce severe fetal malformations. In that case, the mother's health is not in danger, but the fetus's survival may be. Joelle remarked to me several times that discussions about causes of fetal malformations tend to focus on abnormal genes and neglect the frequent nongenetic causes of inborn malformations, such as diabetes or placental anomalies.

One of the most frequent causes of an accidental pregnancy loss is failure to thrive, an unexplained arrest of fetal development sometimes attributed to a placental defect. It may lead to the birth of a very small baby, miscarriage, or death in the womb.[54] A severe failure to thrive can justify interruption of a pregnancy. Death in the womb can also be produced by too much or too little amniotic fluid, the leaking of amniotic fluid, and in more dramatic cases the premature rupture of the amniotic sac. Leaking of amniotic fluid increases the danger of abnormal fetal development and of infection. The opposite is also true: an infection can be at the origin of an insufficient production of amniotic fluid, while a combination of infection and a paucity of amniotic fluid can produce fetal malformations. When fetal medicine experts at AH observe a combination of an infection and fetal anomalies, they usually inform the pregnant woman about a "reserved prognosis of the future child"—that is, a high probability that the child will die either in the womb or immediately after birth.

Twin pregnancy is riskier than a single one; hence the relatively high proportion of twins among dissected fetuses. In some cases, the demise of one twin is followed by the death of the other, even when the twins did not share placentas and amniotic sacs. In such cases, autopsy frequently reveals that both twins had multiple malformations. When the twins do share a placenta, they are at increased risk of twin-to-twin transfusion syndrome (TTTS). TTTS is a rare condition (but less rare at a tertiary referral center) in which one twin "appropriates" the circulation of the other. TTTS can occur when the twins have two amniotic sacs (monochorionic diamniotic pregnancy, 70% of identical-twin pregnancies) or share both a placenta and an amniotic sac (monochorionic monoamniotic pregnancy, found in approximately 1% of identical-twin pregnancies).[55] In TTTS, both twins are at high risk of death in the womb and, if they survive, severe inborn defects. In several TTTS cases analyzed by AH fetopathologists, one twin died in the womb, and the second was diagnosed as having severe malformations, usually of the brain, produced by excessive blood flow (twin anemia polycythemia sequence, or TAPS). In these cases, the pregnant woman usually elected to abort the second twin.

When TTTS is discovered late in pregnancy, physicians may induce an immediate delivery. This option has become more viable with the improved survival rate of very premature babies. If the condition develops at a stage at which it is too early to safely induce a delivery, the physicians may attempt fetal surgery in the womb. The most accepted method today is blocking the blood vessels that connect the twins via fetoscopic laser coagulation. According to recent data, in 55–69% of the cases, at least one twin survives after this therapeutic intervention. On the other hand, fetal surgery is linked to a health risk to the mother, and 5–11% of the surviving children develop severe neurological impairments.[56] At the time of my study, laser coagulation was occasionally employed in France to treat TTTS, including at AH, but Myriam, Soraya, and Joelle were skeptical about this approach. They thought that it was not sufficiently reliable and explained its popularity in the United States by that nation's prohibition of an abortion for a fetal indication beyond the viability limit. The impossibility for a woman to terminate a pregnancy after twenty-two to twenty-four weeks led to "desperate" surgeries with a high fetal mortality rate.

Death in the womb, especially late in pregnancy, is frequently an unanticipated event. The demise of the majority of the fetuses dissected by AH's fetopathologists was, however, often the final stage of a complex diagnostic trajectory. Most French women undergo routine screening for Down syndrome at ten to twelve weeks of pregnancy, which includes ultrasound, and a second detailed ultrasound at twenty to twenty-two weeks.[57] Those diagnosed as having a fetal anomaly during the first- or, more often, second-trimester ultrasound scan are usually sent to a secondary center (often a local or regional hospital) for confirmation of the initial diagnosis. If their physicians suspect that the fetus has a rare or complicated condition, they are referred to a tertiary center such as AH, where they are seen by several experts in fetal medicine and clinical genetics.[58] Genetic counseling at AH differs from the lower-status "mopping up" activity described by the sociologist Charles Bosk in his pioneering study of US genetic counseling, *All God's Mistakes*.[59] All of AH's genetic counselors are highly regarded specialists, often, like Myriam, with a solid international reputation. They are integrated into dense networks of interdisciplinary collaborations and rely on the contributions of many colleagues and collective debates about contentious cases.

Genetic counselors are expected to be neutral. In practice, when they diagnose severe fetal anomalies—a frequent occurrence at AH—they indirectly convey their opinion about the desirability of interrupting the pregnancy. When the pregnant woman has not been classified as "culturally different," "recent migrant," or "very religious" (often, but not always, Muslim), and her physicians do not have reason to believe that she is strongly opposed to abortion, AH geneticists employ the formulaic phrase, "the geneticist informed the couple about the severity of the consequences of the detected malformation." The couple is also told that if they decide to interrupt the pregnancy, they will receive information about the results of the postmortem study of the fetus. After the abortion, the couple is invited to a follow-up encounter with the geneticists in order to learn the autopsy results (if applicable) and to receive genetic counseling, that is, advice about their reproductive options and possible risks for other family members.

A majority of the fetuses studied by AH experts are aborted in the second trimester of pregnancy, often between seventeen and twenty-two weeks. However, some women receive a diagnosis of a fetal malformation only in the third trimester, an especially difficult situation. Thus, some brain anomalies are visible only late in pregnancy, when the brain is well developed. Brain anomalies are seen as especially distressing by physicians and parents because of the combination of diagnostic uncertainty and the severity of possible outcomes: neurological troubles and intellectual impairment. French CPDPNs usually agree to a termination of pregnancy when the estimated risk of a severe intellectual disability is at least 10%. Many prenatally diagnosed brain malformations fulfill this criterion. In an especially dramatic case, fetal medicine experts detected a major brain anomaly at forty-one weeks of pregnancy, when the woman was at the point of giving birth. Earlier ultrasounds of the fetus were normal; the severe brain defect was probably caused by a hemorrhage. The woman chose to interrupt the pregnancy at AH; she then declared the fetus as a stillborn child. This case illustrates the difficult negotiation of the boundary between a termination of pregnancy and a postnatal decision about interruption of intensive care. If a brain anomaly is discovered just before birth, French law permits abortion (with feticide) for a 10% risk of severe intellectual disability. If a similar anomaly is discovered after the child is born alive, physicians are expected to do everything to ensure the child's survival, even if the risk of intellectual impairment is well beyond 10%, although unofficially the neonatal resuscitation unit's staff may decide to withhold some such interventions.[60]

AH experts implicitly assume that a couple's "rational" response to an announcement of a "very reserved prognosis" of their future child is to terminate the pregnancy. Deciding to continue the pregnancy despite a poor prognosis generates unease, especially when the decision is not motivated by "custom" or religious faith.[61] On the other end of the spectrum, a woman's decision to have an abortion after a diagnosis of a "borderline" fetal impairment, such as an isolated and not very severe anomaly, also produces discomfort among AH's physicians. Sometimes a diagnosis of "isolated malformation" may be inaccurate. Soraya recalled a case in which a woman asked to terminate the pregnancy after detection of a cleft palate. CPDPNs' members thought that this anomaly did not justify an abortion, but agreed to it for maternal distress. The autopsy of the aborted fetus uncovered many malformations of internal organs.

One way to ascertain that a fetal anomaly seen during an ultrasound is truly isolated is to wait. It is easier to make a precise diagnosis of a fetal anomaly in a more advanced pregnancy. The downside of longer waiting is, however, maternal stress and in some cases a difficult late-term abortion.[62]

Collectively Solving Diagnostic Puzzles

Fetal medicine, like dysmorphology, is a puzzle-solving discipline.[63] In both cases, the term *diagnostic puzzle* may be partly misleading. A wooden or cardboard puzzle has (usually) only one solution, as does a crossword or a mathematical puzzle. Finding the solution is the end of a quest. A diagnostic puzzle also has—ideally, at least—only one "right" solution (the cause of the observed anomaly), but finding this solution may be the beginning, not the end, of a diagnostic trajectory. Many mutations have a variable expression, and many structural anomalies of the fetus have a wide spectrum of consequences for the future child. Finding a mutation that may explain the observed structural anomaly of the fetus, or finding a physiological cause of such anomaly, does not always predict whether the child will be impaired and, if so, to what extent. The solution of a diagnostic puzzle often does not end prognostic uncertainty.

Unlike fetal medicine experts, fetopathologists are not directly concerned with prognosis. Their main goal is to diagnose the cause of an already existing fetal malformation that was found in a dead fetus. In many cases, such a diagnosis is relatively simple. In others, specialists may have initial doubts, but ultimately agree on the most probable reason for the fetal impairment. In some cases, however, they are unable to find a satisfactory reason for the observed anomalies. Unsolved diagnostic puzzles are disturbing. The usual way that AH fetopathologists deal with the accumulation of especially tricky cases is to organize periodic working sessions with physicians, from either AH or other hospitals, specialized in a category of fetal/pediatric anomalies, such as cardiac, skeletal, or digestive system malformations. These meetings are conducted in a friendly and relaxed atmosphere. The invited physicians, who often know the AH pathologists from previous meetings and collaborations, listen to the presentation of the case by AH's physicians, examine images projected on a screen (photographs taken during the autopsy, X-ray images, images of microscope slides), then propose a diagnosis, either with assurance or tentatively and in dialogue with their AH hosts. AH's physicians readily recognize the greater expertise of these guests and are very pleased when they provide a satisfactory solution to a diagnostic puzzle, but at the same time treat them as their equals.

The atmosphere was very different in a meeting with a world-famous expert on fetal and pediatric cardiology who came to AH to participate in a workshop on unusual fetal cardiac malformations. This meeting was held in a tightly packed room in AH's pediatrics department. The participants were leading physicians from the pediatrics and pediatric surgery departments, junior physicians from these departments, geneticists, and fetal pathologists. While many of the attendees—including all the fetopathologists—remained silent during the whole meeting, some of the more prominent, or perhaps more ambitious, AH physicians aspired to impress the visiting specialist and their colleagues with their diagnostic skills. There was a nearly tangible element of competition among these participants. The pediatric cardiology session exemplified in a condensed form the role of diagnostic puzzle solving in producing hierarchies of knowledge and authority.

Specimens observed in the fetal cardiology meeting were hearts and adjoining blood vessels, preserved in formaldehyde. Some of the hearts were from fetuses aborted for severe malformation; others were from children who died shortly after birth. Each heart was displayed by the physician who presented its case, then was manipulated by the invited cardiologist. An image of the manipulated object was projected on a large screen. This was a highly unusual arrangement. In the fetopathology department's staff meetings, all the debates are conducted while viewing enlarged photographs projected on a screen (slides). Bidimensional images viewed in this manner are also the central element of debates about dysmorphological traits in children and adults.[64] In the fetal cardiology session, participants directly observed three-dimensional preparations manipulated in real time, an arrangement close to observations made in a traditional operating theater or in modern operating rooms equipped with closed-circuit television.[65]

The invited specialist used his hands to demonstrate unusual features of each heart, such as the reversed position of its chambers or entanglement of blood vessels. He also demonstrated the relative position of anatomic elements, and at the same time provided tactile descriptions of the tissues—elastic, rigid, stretchable. Such descriptions are often employed by surgeons when dissecting. The main goal of the session was to discover the precise nature of each cardiac malformation and its origin during embryogenesis. In some cases, the answer seemed relatively straightforward. In other cases, session participants had divergent views about the origins of the observed malformations; in still others, they remained perplexed. One heart, from a stillborn child, was described as "primitive" and "reptilian." Another case was described as "chaotic." One of the surgeons, perhaps trained in Israel, employed the Yiddish/Russian term *balagan*. The participants concluded that in that case, they could not determine either the logic of the anatomical arrangement of cardiac elements or its generation.

The session on cardiac malformation also displayed the limits of PND of heart anomalies. In some cases, physicians failed to detect the cardiac defect before birth. In other cases, sonogram and MRI images pointed to the existence of a severe heart malformation, but its unusual character was revealed only during an autopsy. The heart and blood vessels need to be seen, cut, touched, and manipulated. The invited expert was a surgeon who circulated between the dissection room and the operating theater and was credited with saving babies' lives: this may account for his elevated status, different from that of the professionals who deal exclusively with dead bodies.

Fetopathology and New Genomic Approaches

Most of the genetic/hereditary conditions discussed by AH's fetopathologists are rare and unusual, and many are diagnostic puzzles. New genomic approaches are expected to greatly facilitate solving such puzzles. AH was one of the first French hospitals to introduce comparative genomic hybridization (CGH) to the study of fetal cells collected by amniocentesis or chorionic villus sampling. When CGH is employed to diagnose genetic anomalies in children or adults, AH geneticists use a relatively dense "grid" composed from cloned DNA fragments of four hundred kilobases each. Such a grid can detect smaller chromosomal anomalies, but increases the probability of detecting nonpathological variations of DNA sequence and thus of producing false-positive results: fishing with a net with smaller holes will catch more fish, but also more junk, such as discarded plastic bottles. When geneticists study an already existing person, they want to observe as many anomalies as possible, even at the risk of increasing the proportion of false-positive results. These results are, however, much more problematic when a woman has to make a decision about the future of her pregnancy. Therefore, geneticists who adapted CGH to PND proposed a less dense "grid" of DNA fragments of 1.5 megabases. This larger grid, they believe, greatly reduces the proportion of false-positive results, but is still able to detect major chromosomal deletions, duplications, and rearrangements: a net with larger holes will not catch sardines, but most of the items it collects will be fish, not junk.

The introduction of CGH and other new genomic approaches increased the number of dissected fetuses receiving a diagnosis of a genetic anomaly. When she showed me her personal casebook, Soraya proudly pointed to the steady increase in what she called "triple confirmation": a diagnosis first made on the basis of macroscopic observations of the dissected fetus is then confirmed by microscopic and genetic studies. Fifteen years ago, less than a third of the dissected fetuses received a firm diagnosis; in 2015, that amount increased to three-quarters.[66] As far as I know, Soraya's statistics were not official. She decided, according to her own criteria, what should count as a satisfactory solution of a case. On the other hand, the rapid growth of data on genetic anomalies facilitates making more connections between fetal anomalies and the study of changes in the cell's genetic material.[67]

AH fetopathologists are frequently asked to collaborate with geneticists who study rare hereditary conditions, even if such collaborations may be described as long shots. Thus, they were invited to work with a group that studies a recessive hereditary condition, congenital epithelial dysplasia. Children born with this condition cannot absorb food through the intestine, and are nourished intravenously. In some cases, the gut spontaneously matures after a few years; these children can then switch to normal nutrition. When this does not happen, the children often die of liver complications such as hepatic fibrosis and cirrhosis. Congenital epithelial dysplasia is often linked to structural anomalies such as malformations of the nasal cavity and microcephaly (a small brain). Consequently, researchers who study it have wondered whether some of the fetuses aborted for unspecified polymalformative syndrome (the simultaneous presence of multiple and frequently unexplained malformations) suffered from this pathology. Joelle agreed to identify fetuses with multiple malformations that might correspond to a clinical picture of congenital epithelial dysplasia and send tissue samples from these fetuses to researchers studying this disease. She was, however, skeptical about the chances of finding such fetuses: polymalformative syndromes are frequent, while congenital epithelial dysplasia is a very rare condition.[68]

Identifying mutations linked to fetal impairment is important for genetic counseling, especially to reassure parents following a pregnancy loss. This may be particularly important for conditions such as osteogenesis imperfecta (brittle bones disease), which can be produced by either a genetic anomaly transmitted in families or by a new mutation. Since the second case is more frequent than the first, genetic tests can reassure parents who decide to terminate a pregnancy after a PND of such a condition that the fetal impairment was a random occurrence. Myriam recalled a woman who had aborted a fetus for a severe inborn anomaly, CHARGE syndrome. CHARGE syndrome, too, is usually produced by a spontaneous mutation, but it can also be transmitted in families.[69] This woman, described as psychologically fragile, dwelled on the fact that the fetus was grossly misshapen, and was deeply distressed by this knowledge. Later, she was reassured when she was told that genetic studies of her and her partner indicated that the fetal malformation was the result of a new mutation. She then asked whether she could destroy the sonograms of the malformed fetus. This question was interpreted by Myriam as the woman's wish to symbolically eliminate her unhappy past, and also an affirmation that such elimination was legitimate, because past events would not affect her future pregnancies.

Some people readily share information on hereditary conditions in their families, but others are unwilling to provide it. A fetus aborted after a diagnosis of bone anomalies had been diagnosed as having one of the variants of osteogenesis imperfecta. The next logical step was to test the parents, in order to verify whether one of them had a mild form of this condition. The mother tested negative, but the father refused to be tested. Myriam then learned that the couple had split up. She was also told by the woman that her ex-partner already had one child with bone disease. Myriam concluded from the woman's narrative that that child probably did not suffer from osteogenesis imperfecta, so she told her that in her case the fetus's condition was probably a new mutation, and she was at low risk of having another child with this anomaly. Despite the reassurance, Myriam added that this woman was not coping well with the pregnancy loss and was being seen by a psychologist.

Hiding the presence of a hereditary condition is not a rare occurrence. A woman who suffered from a relatively mild form of Steinert disease (an autonomous dominant muscle anomaly with a highly variable expression), and who had aborted three fetuses with a severe form of this pathology, at first categorically refused to disclose her condition to her family. When the geneticists finally persuaded her to do so and she informed her parents and sister about their risk, they rejected the option to undergo genetic testing. Myriam and Soraya described other cases in which people either concealed information about a hereditary condition in the family or denied its presence. Myriam recalled a woman who underwent termination of pregnancy for a severe bone malformation of the fetus; she had told a different story about health problems in her family to each one of her physicians. Later, her physicians learned by chance that her brother's wife had a termination of pregnancy for the same indication a year earlier. Her story illustrates the geneticists' dependency on family history as reported by the patients—and the problems connected with verifying this source of information.[70] It also illustrates the contradictory consequences of receiving an accurate genetic diagnosis—it is very helpful for some women/couples, and distressing for others.

Women who had difficulty conceiving and those who became pregnant thanks to assisted reproduction technology may be especially upset to face a decision about the fate of their pregnancy.[71] In such cases, a genetic diagnosis confirming the existence of a mutation with poor prognosis may facilitate the woman's/couple's decision to have an abortion. In one such case at AH, the pregnant woman was diagnosed as having a large deletion on chromosome 17; in a similar case, the fetus was diagnosed with a chromosome 14 disomy, that is, the presence of two copies of the paternal chromosome and none of the maternal one. The geneticists explained to the pregnant women and their partners that their future child would probably die either in the womb or shortly after birth. Such an explanation allows the woman to perceive abortion as a mere consequence of a tragic event for which she is not responsible.

Genetic analysis can be also helpful when an increased nuchal translucency with a normal number of chromosomes is observed. In one such case, fetal medicine experts detected a heart anomaly of the fetus during a second-trimester ultrasound. Since the woman had already undergone amniocentesis to exclude the presence of an abnormal number of chromosomes (abnormal karyotype) and the fetal cells were still preserved in the hospital, her physicians decided to perform CGH on these cells. They discovered a deletion on chromosome 22 (22q11.2 del, or DiGeorge syndrome), linked to physical malformations and an increased risk of psychiatric pathologies. The woman decided to terminate the pregnancy. In a similar case of nuchal translucency with a normal number of chromosomes, a second-trimester ultrasound revealed many morphological anomalies, tentatively classified as CHARGE syndrome. In that case, the woman elected to have an abortion on the basis of ultrasonographic results alone. Later, a genetic study of tissues from the aborted fetus confirmed the presence of a mutation associated with CHARGE syndrome. AH pathologists argued that such cases indicate that when a fetus is diagnosed as having abnormal nuchal translucency at eleven to twelve weeks, it is desirable to replace karyotyping with CGH. If CGH reveals the presence of a mutation, the woman can decide to have an early abortion and will be spared a long period of uncertainty and a more traumatic late interruption.[72]

Alas, CGH is not a miraculous solver of diagnostic puzzles. In some cases of increase in nuchal translucency, geneticists did not uncover any genetic anomaly. These cases, Joelle explained, especially those in which the nuchal translucency is coupled with a hygroma (a swelling of lymphatic origin that appears on the fetal neck), are very frustrating because of the high degree of prognostic uncertainty. In some cases, the pregnancy ends with either miscarriage or birth of an impaired child; in others, the newborn child looks healthy. The standard advice of AH experts in such cases is to wait and observe fetal development. Some women reject this advice. One of the patients at AH, diagnosed as having significant nuchal translucency and hygroma with a normal number of chromosomes, elected an early abortion by aspiration. This method of pregnancy termination destroys the fetus, complicating a diagnosis of reasons for the fetal anomaly. In the patient's next pregnancy, again the fetus was diagnosed as having significant nuchal translucency and hygroma. This time, the woman elected to wait longer and undergo a medical termination of pregnancy with an expulsion of the intact fetus. She and her physicians hoped that a postmortem would uncover the reason for a recurrent fetal anomaly. This did not happen, however. Macroscopic observations, microscopic investigations of fetal tissues, and CGH all failed to discover the cause of the fetal impairment. As a consequence, AH physicians were unable to tell this woman what her risk would be for a third abnormal pregnancy.

An analysis of fetal cells by CGH is usually accompanied by a parallel analysis of the DNA of maternal and paternal cells. This may lead to uncovering a genetic problem in one or both parents. In one case of late-term abortion for a severe heart malformation in the fetus, geneticists had found out that the father had an equilibrated translocation at chromosome 14.[73] The couple was told that in the next pregnancy, the probability that the fetus will inherit a disequilibrated translocation and display severe anomalies was approximately 30%. The woman could elect to conceive naturally (if she is able to do it; translocation is linked to a high risk of miscarriage), undergo chorionic villus sampling at eleven to thirteen weeks, and if the examination of fetal cells displays a disequilibrated translocation, choose an early abortion. She could also elect a preimplantation genetic diagnosis (PGD), an in vitro fertilization followed by a genetic analysis of the embryos and the selective implantation of an unaffected embryo. The couple was placed on a long waiting list for PGD.[74] In the meantime, the woman tried to conceive naturally. She had two very early miscarriages, a frequent occurrence in fetuses conceived by a carrier of a translocation. The third pregnancy was not immediately miscarried, but chorionic villus sampling at thirteen weeks detected the same genetic anomaly, and she elected an abortion. This woman, Myriam and Soraya explained, was very distressed, but nevertheless decided to undergo voluntary interruption of pregnancy, then attempt an in vitro fertilization with PGD. In the meantime, she requested psychological counseling.

Geneticists know the risk of translocations and selected mutations. However, when performing CGH, they often uncover changes in fetal DNA of unknown/uncertain significance (VUS). When a genetic anomaly is found in a fetus but also in a healthy parent, the geneticists usually conclude that in all probability they are dealing with a VUS—but also that this VUS is benign. In one case at AH, severe fetal anomalies observed during an ultrasound were initially associated with a duplication of part of chromosome 15, found by performing CGH on fetal cells in the amniotic fluid. The woman decided to interrupt her pregnancy for polymalformative syndrome. After the abortion, she and her partner underwent genetic testing, and the geneticists found a similar anomaly in the healthy father. They concluded that the duplication of part of chromosome 15 was a VUS. The severe structural anomalies of the fetus remained unexplained. In a similar case, a fetal structural anomaly, megacolon (an abnormally increased part of the colon), was at first attributed to a deletion of a part of chromosome 10. Later, a similar deletion was found in the healthy father, and was reclassified as a probably benign VUS.

In other cases, the status of a fetal genetic anomaly detected by CGH remains uncertain. AH geneticists observed a duplication of part of chromosome 20 of a fetus aborted for brain anomalies and atypical form of the limbs. This anomaly was not found in the parents. Myriam was not sure what its clinical meaning was, if any. Another fetus, aborted near the end of the pregnancy for a severe brain malformation (the enlargement of cerebral ventricles), was diagnosed as having a deletion on chromosome 17. Usually, such a deletion is linked with Koolen-de Vries syndrome, characterized by a low muscle tonus, problems with feeding, altered facial features, and learning disabilities. But the description of Koolen-de Vries syndrome does not include brain malformations of the kind observed during the autopsy of this fetus. Participants at the fetopathology staff meeting concluded that the deletion on chromosome 17 was probably unrelated to the observed brain malformation, but this was a tentative conclusion only.

When studying rare mutations, only the discovery of similar cases can confirm geneticists' initial hunches. In especially frustrating cases, their initial conclusions are confirmed by future developments. A woman aborted a fetus with severe osteochondroplasia—a bone anomaly produced by specific mutations. She and her partner were told that since they did not carry genes for this condition, the fetal anomaly was in all probability an accident of pregnancy that will not be repeated. However, in her next pregnancy the fetus was again diagnosed as having a severe osteochondroplasia. Again, the genetic tests did not find genes linked to this condition in the parents. In the meantime, the couple wanted to begin a new pregnancy very soon. Geneticists told them that the chances of yet another recurrence of osteochondroplasia were approximately 25%, but privately admitted that this prediction was merely a guess.

At other times, a fetal anomaly observed during ultrasound is strongly suspected to be the result of a mutation, but geneticists are unable to detect the presence of such a mutation. In a complicated case, a sonogram at sixteen weeks indicated that the fetus suffered from arthrogryposis, a rigidity of the joint articulations which is often—but not always—produced by a mutation. CGH performed on fetal cells was normal, and an ultrasound at twenty-two weeks did not disclose additional fetal anomalies. The fetal medicine experts informed the couple that since they did not find a mutation or other possible cause of arthrogryposis such as an infection, the fetal anomaly was probably an isolated malformation. An AH orthopedist told them at the same time that isolated arthrogryposis can be successfully treated after birth with surgery and physiotherapy. The couple was not fully reassured, and asked an outside expert for a second opinion. The second expert gave a more pessimistic evaluation of this child's future, and the couple decided on an abortion. An autopsy of the aborted fetus displayed multiple structural malformations; these had not been detected by prenatal ultrasound. At this point, AH experts strongly suspected the presence of a genetic syndrome, possibly a myopathy (a muscle anomaly), and enrolled the couple in a research program. In another case, a fetus aborted for thanatophoric (lethal) dwarfism diagnosed during ultrasound was found during autopsy to harbor multiple additional malformations not known to be linked with this form of dwarfism. Soraya argued that this was an extreme variant of thanatophoric dwarfism, while Myriam and Joelle thought that this fetus might have been affected by a rare combination of two distinct genetic conditions. Genetic studies failed to confirm either hypothesis.

One of the main aims of the fetopathology staff is a collective collaboration on the precise wording of the final autopsy report. When at the end of a diagnostic trajectory fetopathologists are unable to provide a "complete" solution of a diagnostic puzzle, to use Soraya's vocabulary, they clearly indicate this in their final report. On the other hand, such a report sometimes eliminated traces of hesitations and twists in a diagnostic trajectory, mainly through the suppression of observations that are later perceived as uncertain or irrelevant for the final diagnosis. In a typical case, ultrasound detected a heart anomaly in a fetus. Initially, pathology specialists suspected the presence of DiGeorge syndrome, an anomaly produced by a deletion on chromosome 22. An analysis of fetal cells collected from the amniotic fluid did not confirm the presence of this deletion but displayed a rare, usually lethal mutation, a pericentric inversion on chromosome 2. The couple chose to have an abortion. Autopsy of the fetus confirmed the presence of a severe heart malformation, but CGH performed on the fetus's tissues failed to confirm the presence of a pericentric inversion on chromosome 2, despite the greater sensitivity of the method employed to study fetal cells. Geneticists were unable to explain this discrepancy. The final autopsy report made no mention of the (putative) presence of genetic anomaly in the fetus, and stated only that the autopsy confirmed the original reason for termination of the pregnancy: a severe cardiac anomaly.

In a related case, geneticists who performed CGH on the DNA of a fetus diagnosed as having an increased nuchal translucency found two unrelated chromosomal anomalies: a deletion of the long branch of chromosome 14 and a partial trisomy of chromosome 8. However, the chromosome 14 anomaly was found through a direct observation of fetal cells and in cultured fetal cells, while the three copies of chromosome 8 were found only in cultured cells. One possible explanation is that the fetus was a mosaic (mixture) of cells with two and three copies of chromosome 8; another was that the partial trisomy of chromosome 8 was an artifact of cell culture. A diagnosis of trisomy 8 had no practical consequences, because the confirmed deletion of a part of chromosome 14 is linked to severe inborn impairments, often incompatible with life. The couple, informed about the pessimistic prognosis for their child, elected an abortion. The final fetopathology report did not mention trisomy of chromosome 8: it explained only that the autopsy confirmed the presence of anomalies consistent with a deletion of part of chromosome 14. In fetopathology, as in other areas of scientific and medical activity, the collective production and stabilization of new knowledge and facts amount to a purification process that sometimes erases its own history.[75]

Avoidable Disasters

Fetopathologists deal with dead bodies, but their main goal is to reduce the suffering of the living. They strive to prevent avoidable disasters: a loss of pregnancy or birth of a severely disabled child, a failure to elucidate the causes of pregnancy loss in cases where such elucidation is possible, and professional mistakes. Errors, even in the absence of threatened legal action, are painful to the specialists who made the mistakes, can harm their reputation, and may disrupt the proper functioning of hospital services. A majority of pregnancy losses investigated by AH's fetopathologists are unforeseen tragedies. Women who expect to have an uneventful pregnancy are suddenly confronted with a diagnosis of a severe fetal malformation. Sometimes, however, the pregnancy loss is not an entirely unexpected event. Some women know that they and/or their partners are carriers of a pathogenic mutation. Others had several abnormal pregnancies that, the geneticists strongly suspect, were produced by a hereditary anomaly, despite the fact that they were unable to identify this anomaly. In other cases still, a woman has a history of several pregnancies with impaired fetuses that does not fit the usual pattern of the presence of a hereditary condition, because each time the fetus displayed a different anomaly. In some of these cases, fetal impairments were attributed to the woman's "problematic" behavior; in others, they were attributed to a yet undetected maternal/paternal health problem.

Myriam proposed that when a woman had several abnormal pregnancies, each time with a different fetal problem, both parents may be carriers of genetic anomalies. An alternative explanation may be a maternal pathology that affects fetal development. In discussing the case of a woman who twice had a fetus with spina bifida (an incomplete closure of the fetal neural tube), usually an accident of pregnancy, she mentioned a hypothesis that linked abnormal neurological development of the fetus with maternal blood disorder, the presence of blood clots in the lungs. Soraya disagreed. This hypothesis, she explained, had already been discussed ten years ago, but nobody was able to prove it. Another woman had an especially difficult reproductive history. She was first pregnant with twins produced by in vitro fertilization; one of the twins died in the womb. The next time she became pregnant she had a spontaneous abortion, then an extrauterine pregnancy. Her recent pregnancy seemed uneventful until the discovery of a severe fetal heart defect during a routine ultrasound in the second trimester. AH geneticists told her that in all probability, the most recent problem was an isolated incident, and the chances of another such malformation in the next pregnancy were very small. But the woman failed to be reassured, and was described as suffering from severe mental distress. She was directed to a psychologist and invited to join a support group for pregnancy loss.

Not all fetal anomalies are seen as the result of bad biological luck. In a handful of cases among those I observed at AH, such anomalies are attributed to maternal lifestyle. A woman who suffered from severe alcoholic cirrhosis received a diagnosis of a suspected cardiac anomaly of the fetus. This woman had two healthy children from two different fathers; the new pregnancy was with a new companion. The fetal medicine specialist who observed the heart defect during a second-trimester ultrasound suspected CHARGE syndrome. The specialist also told her that the observed heart defect probably could be treated after the child's birth. The woman decided nevertheless to terminate the pregnancy. Autopsy of the fetus disclosed multiple malformations of internal organs in addition to the heart defect, but fetopathologists concluded that these malformations were probably not produced by CHARGE syndrome. CGH results were normal. Joelle and Soraya were uncertain whether the observed fetal malformations were accidental or represented a severe form of fetal alcohol syndrome; unofficially, they were inclined to accept the second hypothesis.

Pregnancy loss is always an upsetting event, but professionals are especially distressed when an unhappy outcome may be presented as the result of a physician's mistake. A woman whose pregnancy was followed at AH's maternity clinics learned that the fetus had a cardiac anomaly. She was also told that this was an isolated malformation, and that chances were good that this heart defect would be successfully treated by surgery after the child's birth. The couple decided to pursue the pregnancy on the basis of this information. The child was born before term with multiple birth defects in addition to the heart anomaly: abnormal limbs, excessive body hair, and abnormal facial traits. The child was "allowed to die." The parents were described as suffering from a severe shock. They were ready to accept a normal-looking child with an invisible structural anomaly that, if they were unlucky, could lead to the baby's death, but were totally unprepared to see a "monstrous" newborn covered with hair and with a deformed face. The parents were severely distressed, and the father became aggressive with his wife's physicians. The ultrasound expert was very troubled, too; her colleagues described her as being beside herself. Such a mistake, they added, is the fetal medicine expert's worst nightmare.[76]

A fetal medicine expert openly critical of some aspects of the widespread use of PND and selective abortion in France nevertheless fully assumed a decision to perform a very late termination of pregnancy for achondroplasia (dwarfism). Achondroplasia, some French physicians believe, should not be classified as a "severe and incurable" impairment, because it is not incompatible with a fully autonomous life. On the other hand, the pluridisciplinary prenatal diagnostic centers practically always accept a woman's request for abortion for this condition. In one instance, the woman's obstetrician had failed to inform her about the presence of this fetal anomaly until late in pregnancy, despite the fact that this condition is usually diagnosed easily during the second-trimester ultrasound. She was therefore denied an opportunity to elect an earlier, much less traumatic abortion. When the woman finally learned about the fetal impairment, she categorically refused to have a child with this condition. She also rejected the possibility of giving the child up for adoption. Most parents, several French specialists attested, strongly prefer a late-term abortion over the adoption of their impaired child. Both situations are very distressing, but adoption tends to generate much stronger feelings of parental guilt, because the child's existence is a perpetual reminder of their failure to become caring parents of a disabled child.[77] The fetal medicine specialist who performed a late-term abortion for achondroplasia thought that the patient should not be forced to suffer all her life from the consequences of her physician's professional mistake.

French fetal medicine experts may feel uneasy about termination of pregnancy for a nonlethal malformation, but they also believe that such termination, even when performed very late in pregnancy, is sometimes the only acceptable solution for the mother/parents. Their views often put them in a difficult situation. Several French physicians spoke about the contradiction between performing terminations of pregnancy for a given impairment in the morning, and in the afternoon treating children born with the same impairment. Two described this as "a schizophrenic situation"; a third spoke of the "crazy nature of our work." At the same time, these specialists see their contributions to selective reproduction as indispensable, and praise French medical institutions that allow women to decide whether they wish to give birth to a special-needs child, including late in pregnancy.[78] These physicians implicitly defend their right to be incoherent. They accept the lack of internal logic of some of the rules governing legal abortion in France, and recognize the irreducibly complex nature of many reproductive choices. Accordingly, many French physicians perceive their "schizophrenic" situation as a lesser evil among many objectionable options. Collective production of diagnoses of rare or difficult fetal anomalies, communal granting of permissions for pregnancy terminations for a fetal indication, and shared elaboration of final reports of fetal dissections and guidelines for genetic counseling attenuate the risk of diagnostic and prognostic errors, but also the perils of physicians' professional "schizophrenia."

FOUR

Visible Disasters:
Fetopathology in Brazil

Prelude: Floating Fetuses

My observations of French fetopathologists brought me in contact with unfamiliar professional practices. Nevertheless, thanks to my original training as a biomedical researcher, my experience of working in medical research institutions, and my familiarity with the history of and recent developments in French medicine, my observations were made in a familiar territory. Observing a fetopathology department in Brazil was, by contrast, an unsettling experience. I found myself in an environment that was partly familiar, but partly very strange. Trying to make sense of this environment helped me to better grasp the specificity of Brazilian attitudes toward prenatal diagnosis and birth defects, but also to develop a critical distance ("estrangement") toward uses of PND in industrialized countries.[1]

The object that became a symbol, or perhaps a synecdoche, of my observations of fetopathology as practiced in Brazil was a big plastic trash bin.[2] The bin, of the kind often used for collecting garden refuse, was dark green with a well-fitting lid. It was filled with kerosene. From the underside of the bin's lid dangled twenty or more cords of variable length, each with a handmade and handwritten cardboard tag at one end. The other end of each cord was attached to a dead fetus. The bin stood in the corner of the storeroom of the anatomic pathology department at the Maternal Health Center (MHC), a midsized Brazilian public facility that serves lower-class women with problematic pregnancies. Nearly all the fetuses had been miscarried or had died in the womb; with very few exceptions, induced abortion is outlawed in Brazil. Accordingly, fetuses that were illegally aborted in private clinics are not dissected. By contrast, public hospitals can perform autopsies and conduct research on fetuses.

In Brazil, fetuses weighing under five hundred grams are classified as surgical waste and hence are incinerated or discarded. Those weighing over five hundred grams can receive a death certificate and a burial. In those instances, the dissected fetus, made presentable by pathology department technicians, is placed inside a tiny cardboard coffin, then collected by the family or by undertakers from the funeral home. However, many patients treated at the MHC are very poor. Physicians there reported that many women/parents, who feel they cannot afford the undertakers' fees, choose to leave the dead fetus at the hospital; they are unaware of their eligibility for financial aid for burial expenses through the hospital's social services.[3] Unclaimed fetuses are kept at the MHC for a year. The official explanation for this policy is to give women and their families the opportunity to reconsider and reclaim the fetus for a later burial—although I never saw this happen during my observations at the anatomic pathology department. After a year, the fetuses are sent to the pathology department of a large teaching hospital; there they are either discarded and incinerated or used for research or teaching.

This description of a trash bin filled with fetuses may read like the opening sentences in an anti-abortion pamphlet. Alternatively, it may be read as an implicit condemnation of the strange habits of a "developing" (that is, underdeveloped) country that still lacks the moral sensitivity of the more "advanced" industrialized nations. An image of a trash bin filled with fetuses may resonate with attempts to exoticize the tropical other, and add another cliché to the well-worn collection of images of Brazil as a country of violent contradictions: luxurious urban areas and favelas; generous and good-humored people and high levels of brutal urban violence; the fragrance of tropical flowers and overflowing sewage; advanced social experimentation and endemic corruption; idyllic beaches and gardens and huge urban garbage dumps. In fact, the green trash bin is just a convenient, inexpensive, and functional receptacle, and the potentially "scandalous" MHC's collection of fetuses is not very different from similar collections that existed in western European or North American hospitals before the fetus's recent change in status in these countries.[4]

In early twentieth-century embryology laboratories, the anthropologist Lynn Morgan explains, specimens (miscarried fetuses, stillborn children, and sometimes deceased newborns) were stored in large glass jars filled with either formaldehyde or odorless kerosene.[5] Morgan reports that in some laboratories, specimens were kept in kerosene-filled bathtubs that were covered with a lid. Her areas of interest in her work include how fetuses speak (and, one may add, are silenced) in places where they are collected and stored; the layers of meaning of all the forms of prenatal maternity; and complex patterns of conserving and discarding fetal remains. Morgan explains that the old manner of treating embryos and fetuses as exclusively medical waste no longer functions. The proper disposal of embryos has entered the public sphere and is politically negotiated. The containers, Morgan concludes, are leaking.[6] My observations at MHC were shaped by the simultaneous presence of receptacles filled with dead fetuses and the metaphoric "leaking containers" which failed to contain all the aspects of fetal lives within a Brazilian institution.

Technicians at the MHC's anatomic pathology department have a matter-of-fact attitude toward fetal bodies. They treat fetuses with a mixture of indifference, amusement, curiosity, and compassion. The storeroom containing the bin that holds intact fetuses also houses multiple smaller containers of fetal parts along with paraffin blocks that encase tissue samples. It is, fittingly enough, a transitional place that people pass on their way to showers and changing rooms, the dissection room, the department office, the staircase leading to the laboratories, and a meeting room. Despite the potentially distressing nature of the stored fetal material, the room is not a depressing place. Fetal organs and tissues are stacked in white plastic boxes of different sizes, and occasionally in empty boxes of a popular brand of Brazilian ice cream, Kibon (fig. 10). In addition, some fetal parts are stored

Fig. 10. Storage of recent tissue samples. WHC, anatomic pathology department. Author's photograph.

Fig. 11. Storage of older tissue samples. WHC, anatomic
pathology department. Author's photograph.

in an odd assortment of glass jars, some originally intended for laboratory use and others that probably once contained jam and pickles.

The shelves of the room are filled with samples, papers, and cardboard boxes, and the space under the shelves is filled with plastic crates containing older tissue samples (fig. 11).

I observed people walking through the storeroom with a shot of a strong coffee (*cafezinho*) in a tiny paper cup in their hand. In France, fetuses were carefully segregated and kept in a small number of spaces within the hospital: cold room, dissection room, and morgue—all places with rigorously regulated access. The storeroom where the Brazilian fetuses were kept was intermingled with the bustling life of the MHC.[7]

The hospital is situated in an upper-middle-class neighborhood. It is surrounded by elegant high-rise buildings with uniformed doormen and well-tended gardens. Its patients come from a very different background: sometimes the lower-middle class (class C in the Brazilian sociological classification) but mainly the popular classes (class D and E), because as a rule, pregnant middle-class women use private hospitals and clinics. Consequently, the MHC is a hospital for lower-class women in an upper-class neighborhood, a small oasis of popular urban life in a relatively affluent area. This is a curious reversal of the situation I observed in some US cities, where an ultramodern university hospital may be situated in a lower-class neighborhood. When this is the case, the hospital is often carefully isolated from its immediate surroundings. The separation between the MHC and its environment was less obvious, if not less real. Patients of the hospital and the inhabitants of the neighborhood are not formally segregated, but they seldom mix. Affluent Brazilians have no reason to enter the MHC; the only exceptions are a few upper-class women who staff a small thrift shop that raises funds for the hospital.

The MHC's architecture is heteroclite and rambling, with a few main buildings and several smaller ones. Some of the main buildings were constructed in the 1950s and 1960s, and others in the 1980s and 1990s. Smaller buildings house medical and research departments, among them the anatomic pathology department; they are aggregated around the facility's internal courtyard. Occasionally, a black limousine from a burial society enters the hospital and parks in the courtyard. In a typical encounter, the car's driver chats with a pathology technician, both stopping to buy a soft drink from a vending machine. Then the technician heads to the morgue in the anatomic pathology department and returns with a tiny white coffin; he gives the coffin to the driver, who puts it in the back of the car. At the same time, a second man who arrived with the burial society goes to the front window of the anatomic pathology department's office. There a clerk hands him a bundle of official papers. People in the courtyard, some chatting, some eating, many smoking, hardly notice the scene; a few stop for a quick look before moving on. Someone smiles; another person waves to the technician who helped the limousine driver.

Initially, I found the handling of tiny coffins mingled with other mundane activities in the MHC's busy courtyard more reassuring than the impersonal treatment of bodies of fetuses and stillborn children in the white and gray morgue at Andral Hospital, carefully isolated from other hospital activities. The impression of a bittersweet acceptance of the unavoidable and random accidents of pregnancy did not stand up to scrutiny, however. I discovered that for the pathologists who conduct fetal dissections, pregnancy loss is often a preventable tragedy.

Health Services in Brazil: The Public and the Private

The MHC is part of the vast public health system in Brazil. That system is a complex network of interconnected institutions, divided into two large sectors:

public and private. The public system, Sistema Único de Saúde (SUS), provides mainly its own diagnostic and treatment services, but in some cases purchases specific services from the private sector. The private sector manages for-profit hospitals, clinics, and diagnostic centers, but also includes institutions funded by philanthropic organizations. The private-sector health system is funded mainly by health insurance plans. Such plans, used today by approximately 25–30% of the Brazilian population, are highly variable. Those on the lower end of the scale provide only basic health services; more expensive health insurance plans pay for specialists' services and treatment in well-equipped and staffed clinics and hospitals.[8] The most advanced gynecological and obstetrical clinics, for use by very rich Brazilians, are not affiliated with local insurance plans; users pay directly for their services.[9]

SUS was founded in 1988, during a period of national transition from a military dictatorship to a democracy. That year, the Brazilian constitution incorporated the principle that health is a fundamental right, and the state has the obligation to provide health care to its citizens. However, SUS was created during a rapid advancement of neo-liberal ideas in Brazil, complicating the implementation of large-scale public health services. One of the most important successes of SUS has been the diffusion of primary care, especially a highly effective family health program created in 1994. The widespread application of this program led to an impressive reduction of neonatal and child mortality in Brazil. The MHC directors are very proud of their facility's contribution to this endeavor.[10] Besides primary care, SUS has specific domains of excellence, such as select cutting-edge cancer therapies or organ transplantation. The majority of the private health insurance users occasionally use public health services as well, especially for expensive therapies and specialized diagnostic tests. By contrast, the overall coverage of the population's health needs by SUS is uneven and often unsatisfactory. The private-sector health system partly compensates for SUS's insufficiencies.

SUS is tightly regulated by the Brazilian government. The government attempted to regulate the private-sector health system as well, through the creation in 2000 of the National Agency for Supplementary Health, but this agency has limited power to impose its views. Services provided by the private sector remain primarily unregulated.[11] The boundaries between the private- and the public-sector health systems are porous. The wages in the private-sector system are often higher, and work in private hospitals and clinics provides access to resources such as good-quality equipment and advanced laboratory tests, unavailable in the public-sector system. On the other hand, the public sector provides more possibilities for conducting research and advancing one's career. It also offers civic-minded physicians an opportunity to contribute to the treatment of less privileged Brazilians. Many Brazilian physicians, especially high-level experts, combine work in both the public and the private sectors. This arrangement allows them to benefit from the advantages of both. The price may be, however, a permanent confrontation with dramatic examples of social injustice and an active participation in the production of such injustice.

One of the most striking elements of the Brazilian health system is its heterogeneity. This is reflected in its public/private division, but also in the major differences within the public and the private sectors. Some SUS services are at the cutting edge of medicine; others fail to deliver even elementary health care. There are important regional disparities as well. Significant differences exist among health services provided by each of the twenty-six states that constitute the Federated Republic of Brazil (plus the federal district of Brasilia, which has a status akin to that of Washington, DC). Globally, services provided by SUS in the poor Northeast are of poorer quality than those provided in the more affluent South, but differences in health care do not always directly mirror a given state's economy. There are also important disparities within each Brazilian state: between larger and smaller towns, urban and rural areas, and distinct domains of SUS's interventions. Some of these differences may reflect contingent developments rather than planned policies. Parallel differences are found among the nation's private insurance plans. Some plans offer an extensive range of services, while others (usually cheaper) propose limited access to preselected physicians, clinics, and laboratories.

The Brazilian health system is a mixture of cutting-edge and old-fashioned elements. It may be described as a "salad bowl"—a mixture of highly diverse structures in which the intermingling entities are often independent, and their interactions are partly governed by chance. Another image that came to my mind while observing the MHC was fractals—miniature reproductions of structures observed on a much larger scale. Structural heterogeneity observed at a macro level is also reproduced at the intermediary and micro levels, and can be perceived in different institutions and settings. Interventions of MHC's physicians can be described as a complex combination of approaches at the cutting edge of biomedicine and more traditional ones. Some of the heterogeneity of the diagnostic and therapeutic approaches may be attributed to the public institutions' paucity of resources, and some to a hidden division of labor between the private and the public sectors. Others are a consequence of the interdiction of abortion for fetal indications, which affects the practice of obstetricians, clinical geneticists, and experts in fetal medicine.

✶ QUOTE

Abortion for a fetal indication is illegal in Brazil, as is abortion in general. The only exception is an abortion for anencephaly (the absence of
fetal brain), made legal in 2012. The history of the legalization of abortion
for this condition is bumpy. It includes a period in 2004 when the Brazilian supreme court abolished a previously existing tolerance of abortion for
this indication.[12] The adoption of legalized abortion for anencephaly by
the Brazilian parliament was facilitated by the argument that in that case,
the decision about the future child's demise has been already made by God
or Nature. The legalization for anencephaly may thus be interpreted not as
a liberalization of abortion law but as a confirmation of the principle that
women are not allowed to decide the fate of their pregnancy.

The criminalization of abortion does not mean that it is rare; one out
of five Brazilian women is reported to have had at least one abortion.[13]
Women who have illegal abortions often employ the abortive drug cytotex
(misoprostol), purchased on the black market.[14] Others use the services of
clandestine abortion clinics. Those intended for upper-class women provide
good-quality care and, at least until 2016, were rarely bothered by the police. I asked one of the specialists I interviewed, an obstetrician who works
in a luxurious private clinic, whether women diagnosed as having a fetal
malformation in this clinic travel abroad for an abortion. Terminating the
pregnancy in a country where abortion is legal, I thought, could be a good
solution for Brazilian women who can afford it, as it is for women from European countries that criminalize abortion, such as Ireland or Poland. The
obstetrician was, however, surprised by my question: "Why should they?
There are very good clinics in Brazil."[15]

Lower-end abortion clinics provide less safe services, often in stressful
conditions. These clinics are usually tolerated, but such tolerance is fragile.
In the fall of 2014, Brazilians were shocked by the story of two women who
died following botched abortions. The charred body of one of the women
was found a few days later in a municipal trash dump. Both women were
probably in relatively advanced stages of pregnancy; they were also able to
pay the clinics' fee (the media mentioned 3,000 reais, at that time approximately US$1,300; in 2014, the minimum monthly salary in Brazil was about
700 reais, at that time about US$280). The police then raided an especially
infamous abortion clinic in a suburb of Rio de Janeiro. A few weeks later,
a similar clinic was dismantled in the state of Rio Grande do Sul. The Rio
de Janeiro raid was named Operation Herod, indicating that the main issue was not physicians' criminal incompetence but the "massacre of the
innocents," that is, abortion.[16] In 2014, the media focused on corrupt and
criminal abortionists.[17] Then in April 2016, Rio de Janeiro's police arrested

a well-known gynecologist and a woman who had just undergone an abortion. This story, which had lower visibility in the media than Operation Herod, was interpreted by medical professionals as an attempt to end the tacit tolerance of activities conducted in high-end gynecological clinics in Rio de Janeiro.[18]

Abortion for a refusal to have a child is widespread among all social classes in Brazil. The main difference between affluent and less affluent women is the conditions in which they terminate unwanted pregnancies.[19] Abortion for a fetal indication is different. Fetal anomalies are frequently uncovered relatively late in pregnancy, especially in the absence of a systematic screening for such anomalies. Upper-class women who elect to terminate a pregnancy with a malformed fetus usually are able to do so, but such an act is difficult and risky if the woman does not have considerable financial resources.

Medical Genetics in Brazil

The development of medical genetics in Brazil mirrors the many contradictions of this country. Brazil has achieved a high level of research in many areas of biomedicine and biotechnology, including genetics. Top-level geneticists are often trained in cutting-edge US and European laboratories and publish in well-known international scientific journals. Brazilian scientists have had a long-standing interest in the role of heredity, understood mainly as racial and ethnic origins, in shaping individual and collective health. In the first half of the twentieth century, they developed a specific variant of "Latin eugenics," an approach that downplayed the role of invariable hereditary traits and selective breeding and foregrounded the importance of environmental factors in bettering future generations. At that time, Brazilian experts believed that their goal, the rise of a vigorous Brazilian nation, would be achieved through the elimination of transmissible diseases and the improvement of mothers' and children's health.[20]

The Brazilian Society of Genetics was founded in 1955 and the Brazilian Society of Medical Genetics in 1986. The latter society issues board certifications in medical genetics. Genetic services, both public and private, are unevenly distributed, however; professionals agree that they are woefully insufficient to meet the population's needs. There are no official data on the number of genetic tests offered in Brazil and their geographic distribution. Advanced genetic tests are often performed in the private laboratories; some tests are shipped by these laboratories to the United States. At the same time, Brazilian geneticists provide services for physicians in other

Latin American countries.[21] Bianca, one of the pathologists at the MHC's anatomic pathology department, previously worked in one of the leading private hospitals in the country, and could compare anatomic pathology and genetic services between an affluent and a resource-deprived institution. She often repeated, "You cannot imagine how I suffer seeing the difference between what is done here, and what can be done with the right funding and equipment." At the same time, she, like many other experts at the MHC, was very proud of what the staff was able to achieve despite the facility's limited means. Bianca was acutely aware of the importance of maternal and pediatric services provided by SUS and strove to improve them, but, like other Brazilian health professionals, would have never considered using these services herself, or recommending them to her family and friends.[22]

Brazil has a National Newborn Screening Database (Programa Nacional de Triagem Neonatal) that collects epidemiological data on phenylketonuria, congenital hypothyroidism, sickle-cell anemia and other hemoglobin anomalies, cystic fibrosis, and congenital deafness; the implementation of such screening is, however, very uneven.[23] Brazil also has a centralized birth defect notification system. Information gathered from birth certificates (paragraph 34) is collected in newborn morbidity and mortality databases, but birth malformations may be underdeclared.[24] A nongovernmental Latin American network, ECLAMC (Estudio Colaborativo Latino Americano de Malformaciones Congénitas), also collects data on inborn impairments via blood samples of newborns. Despite its unofficial status, ECLAMC is often regarded as a more reliable source of information on birth defects in Brazil than the health ministry data.[25]

One of the major obstacles to the development of medical genetics in Brazil is the nation's criminalization of abortion. The inability to terminate a pregnancy for a fetal indication is the "elephant in the room" in debates about genetic diseases in Brazil. Pregnant women treated in private clinics— virtually all middle-class and affluent women—have access to screening for fetal anomalies such as Down syndrome, and many among those who receive a positive PND elect an abortion. But since abortion is illegal, obtaining reliable data on the frequency of such abortions is impossible.[26] On the other hand, private maternity clinics report very low numbers of birth defects, an indirect indication that many women who use their services terminate pregnancies with impaired fetuses.[27] The interdiction of an abortion for a fetal indication and its consequence, a lack of systematic dissections of fetuses aborted for this reason, affect the quality of genetic counseling, including for middle- and upper-class women, because physicians cannot improve their diagnostic skills through the comparison of genetic, ultrasound, and dissection data.

Clinical genetics, like other areas of medical activity in Brazil, is shaped by differences in the provision of health care between the poor and the affluent. A patient-oriented approach in the private-sector health system contrasts with "mass-produced" medicine available for the lower classes. The structure of genetic services in the public-sector health system also mirrors the complicated negotiations involved for members of the rapidly growing lower-middle class to access medical services. Members of this class usually cannot afford expensive private health insurance. They use "health plans," which provide a more limited access to specialized services, and may need to supplement the services provided by these plans with those provided by SUS. Moreover, when people with genetic diseases need an expensive lifelong treatment, they, or their parents if the patient is a child, regularly turn to the courts to obtain access to costly medication in the name of the "right to health" inscribed in the Brazilian constitution. Their efforts sometimes succeed and other times fail, or they may be granted the right to treatment, then lose it because another judge overturns a previous court decision.[28] The Brazilian way of "improvising medicine" is very different from the one in Botswana described by the US anthropologist Julie Livingston.[29] However, Brazilian physicians, like their colleagues in Botswana, creatively tinker with limited medical resources. Facing local shortages and inequalities of access, health professionals often find solutions, abandon some directions, and favor others. Their situated improvisations increase the overall heterogeneity of genetic services provided by SUS.

One factor that promotes the extension of SUS services is patients' activism. Many disease-centered associations use this principle of the right to health to fight for better access to specialized services and treatments. Rare diseases, mainly genetic ones, are a significant part of patients' struggles. In January 2014, the Brazilian parliament passed a new law, National Policy on Comprehensive Care for Persons with Rare Diseases (Política Nacional de Atenção Integral às Pessoas com Doenças Raras). This law was elaborated with the participation of geneticists, but also with major input from patients' associations.[30] A previous law, the National Policy for Integral Care in Clinical Genetics in SUS (Política Nacional de Atenção Integral em Genética Clinica no SUS) of 2009, existed mainly on paper, because it was not accompanied by an ordinance regulating the implementation process for the SUS-provided genetic services.[31] Brazilian geneticists hope that the new law, accompanied by an ordinance detailing its implementation and funding, will improve the detection and management of rare genetic conditions, expand patients' access to clinical genetic services, and support the growth of such services.[32]

The law on comprehensive care in clinical genetics discusses provisions for patients and screening of newborns, but it does not mention PND of hereditary conditions. Such an omission was, in all probability, deliberate. All the stakeholders—government officials, geneticists, and activists—did not want to tackle an issue that might have hampered the law's adoption. A middle-class Brazilian woman who knows she is a carrier of genetic disease can undergo genetic testing, and if the fetus has been affected by the disease, can elect to terminate the pregnancy. She can also, especially if she is affluent, opt for in vitro fertilization with preimplantation genetic diagnosis: this approach is proposed in several Brazilian clinics. Poor patients who use SUS's services do not have access to these options.[33] Geneticists who provide counseling in public hospitals for families with hereditary conditions argue that couples benefit from understanding their reproductive risks. At the same time, they often also provide genetic counseling in the private sector. They know therefore that poor women can only choose not to have children or risk the birth of an affected child; affluent women have other choices besides these. Geneticists also know that occasionally a woman who received a diagnosis of a fetal malformation in a public hospital does not visit the hospital anymore, possibly because she has elected an illegal abortion.[34]

When the options for an effective prenatal intervention are limited, women may have difficulty assimilating information about their risks. After a pregnancy loss, women being treated in a public hospital are informed about the chances that a similar fetal malformation will recur. Many of these women can correctly recall the "risk number" they received during a genetic counseling session, but rarely act on this information.[35] They either believe that the "right" behavior during the next pregnancy, such as better nutrition, will eliminate the danger for their future child, or reject altogether the notion of quantifiable risk. A few women express the hope that in the next pregnancy they will receive a "genetic treatment" that will help them to have a healthy child. Other women reaffirm their religious faith. God, not the physicians, decides what the child's fate will be. Their reproductive decisions are shaped above all by the strength of their desire to have children, not by risk numbers they receive from geneticists. Genetic counseling provides them with information, not with knowledge. The women's main complaint is that when they learned that the fetus had died in the womb or would die immediately after birth, they were still obliged to give birth vaginally and to suffer (women in the public-sector health system often do not receive pain relief during childbirth), and were not entitled to a C-section. This complaint refers in all probability to the fact that the great majority of Brazilian middle-class women treated in the private clinics elects surgical births, and that many poor women see an access to a painless and rapid surgical birth as a class privilege.[36]

A Brazilian PhD dissertation on data collected by fetopathologists in a public hospital is dedicated to "parents of my tiny patients, who went through so much grief and suffering. To you, I offer the hope of a new genetic counseling."[37] The statement that genetic counseling will end parental suffering may seem to be a surprising conclusion of a dissertation reporting that only in 5% of the studied cases was the pregnancy loss attributed to a hereditary condition transmitted in the patient's family. All the other cases were random accidents of pregnancy that could not have been prevented by conventional genetic counseling. The key term is, however, the adjective *new*. The author probably hints that in order to reduce grief and suffering, women/couples need a new kind of genetic counseling, one that is not available for women who use SUS. "New genetic counseling" may be shorthand for an approach that will give more choices to women.[38] Physicians at the MHC, all in favor of liberalizing the Brazilian abortion law, were not very optimistic about its happening in the near future.

Fetal Medicine in Brazil

Prenatal care in Brazilian public health clinics and hospitals does not include routine ultrasounds or serum tests that detect an increased risk of fetal anomalies. Official directives explain that ultrasounds during pregnancy are unnecessary, because they do not improve pregnancy outcomes, defined as the "reduction of perinatal or maternal mortality."[39] The text does not mention serological tests for an increased risk of fetal anomalies.[40] Many of the women followed within SUS do not have any prenatal tests until late in pregnancy. In Rio de Janeiro, only 1% of the fetal problems among SUS users are detected in the first trimester of pregnancy, 70% in the third trimester, and some only near the woman's delivery date.[41]

The absence of an official endorsement of obstetrical ultrasound does not mean that Brazilian women do not use this technology. Just the opposite is true. Brazil has exceptionally high rates of ultrasounds during pregnancy. It is seen as one of the countries with the highest density of ultrasound specialists and ultrasound facilities worldwide. But the distribution of ultrasound clinics is uneven: there are many in the cities and few in the countryside.[42] The goal of ultrasound, often offered by an inexpensive street-corner clinic, is to confirm the pregnancy, identify the fetus's sex, and produce "baby's first photograph." Such facilities are often weakly regulated, and the skills of their operators may vary greatly. Women who frequent the lower-end private clinics sometimes visit such a facility before their first (and, not infrequently, belated) prenatal consultation. When an ultrasonographer in a private facility detects a fetal anomaly, s/he may advise the woman to undergo a more advanced examination in an accredited SUS center.[43] The MHC specializes in the follow-up of difficult cases, and women treated there receive individualized attention. In other public hospitals, ultrasound often takes place in crowded and impersonal facilities, and women who are told about an abnormal finding may fail to receive medical and psychological support.[44]

Affluent women who use expensive private maternity clinics have access to well-trained ultrasound experts who use advanced equipment.[45] These women, like pregnant women in France, usually have their first routine diagnostic ultrasound at eleven to thirteen weeks of pregnancy, and the second in the mid-second trimester; many also have serum tests to screen for fetal anomalies. The use of serum tests is not codified; the type of test and its timing vary among clinics and physicians. The main goal of serum tests and an early diagnostic ultrasound is the detection of Down syndrome. Women are rarely prepared for the possibility of a diagnosis of another fetal problem, such as an abnormal number of sex chromosomes. The decision to interrupt the pregnancy for a fetal anomaly may be made in haste, without adequate genetic counseling. Illegal activities are not conducive to lengthy deliberations about whether and how to perform them.[46]

Women who undergo ultrasound at an inexpensive facility (some of which charge less than US$10 for a test) are especially interested in learning the fetus's sex. Knowing the "baby's sex" (the term *fetus* is never employed when Brazilian professionals speak with pregnant women) is a key step in recognizing the future child as part of the family. The child receives a name, and Carlinho or Clarita is included in the family's life, in middle-class families mainly though targeted consumption: buying sex-appropriate clothes or painting the future child's room.[47] Pregnant women are unprepared to receive bad news, and many minimize the gravity of fetal malformation by assimilating only a part of the ultrasound expert's message.[48] Those referred to a specialized center such as the MHC for a further assessment frequently believe that the referral's aim is to fix the observed problem. This belief may be intermingled with religious ones. Pregnant women hope that a combination of medical expertise and prayers will cure their future child; at the same time, many also express apprehension about their child's future.[49]

Women's reactions may be reinforced by the physicians' tendency to accentuate diagnostic uncertainty. Physicians who work in the public sector often explain that they will know for sure what the child's problems are only after birth.[50] European or North American physicians sometimes propose to the woman and her partner who just received a diagnosis of a "manageable" fetal pathology to meet affected children and their families. The underlying assumption is that such a meeting can help the prospective parents to better understand this pathology, attenuate their overly pessimistic view of life with a given impairment, and help them to make a truly informed decision about the pregnancy's fate. In Brazil, women/parents have no (legal) option to terminate the pregnancy, and physicians may fear that showing prospective parents a child with the same condition as the one diagnosed in the fetus may unnecessarily scare them and deprive them of hope. Veronique Mirlesse reports a conversation between a consultant and a couple who just received a diagnosis of osteogenesis imperfecta (brittle bones disease):

MAN: "Is there a baby with the same problem that we could [see]?"
CONSULTANT: "Ah, not today there isn't. But even if there were, I would not show you. You want to know why? If you want to see a baby with dwarfism, you've already seen them on the internet. . . . They are other people's babies, not yours."[51]

The focus on diagnostic uncertainty, even in many situations in which ultrasound clearly shows a serious malformation with potentially severe consequences, is probably a coping strategy produced by the absence of choice to terminate the pregnancy. As a fetal medicine specialist explained to a woman who just received a diagnosis of a fetal anomaly,

We do not really know what the situation is . . . ultrasound is a picture, like a photograph. If you bring today my photograph to show it to your family, you are going to say: "Here is the pediatrician that we have met." From the photograph they can conclude that I have vision problem, because I wear glasses, but can they know whether I speak French? Ultrasound shows a structure, not a function. The function is visible only when you directly face a given situation.[52]

Several Brazilian fetal medicine experts affirmed to me that the interdiction of abortion for fetal indications is regrettable, but at the same time explained that in all probability, the decriminalization of abortion will not lead to a major increase in terminations for fetal indications. Legalization of abortion will help middle-class women decide about the fate of their pregnancy in less stressful conditions and provide a solution in a small number of truly dramatic cases, but most of the poor women treated in SUS facilities will reject a termination of pregnancy for a fetal anomaly, especially a nonlethal one.[53] This argument mirrors Brazilian physicians' tendency to make a radical distinction between educated middle-class pregnant women who use private clinics and uneducated poor and deeply religious women who use SUS's facilities.

Historical and sociological studies may lead to a very different conclusion. In the 1960s and 1970s, when international organizations first attempted to introduce modern contraceptive methods in Brazil, their activists reported that this would be a very difficult task. The Roman Catholic Church and other Christian denominations opposed "modern" contraceptive methods such as hormonal contraceptives, intrauterine devices, and sterilization; Brazilian physicians were reluctant to play an active role in the circulation of these methods; and the women interviewed, especially those from lower socioeconomic strata, explained that they were not interested in birth control because the church forbade it, and because they believed that all children are sent by God.[54] Religiosity has not diminished in Brazil; it might even have become intensified through the growing popularity of evangelical Protestant groups. In addition, some physicians continue to have a reserved opinion about contraception, and until the late 1980s, the Brazilian government did not support family planning. Nevertheless, in the last third of the twentieth century, Brazil underwent one of the most rapid demographic transitions in the world. In the early 1960s, the Brazilian fertility rate was 6.3 children per woman; it dropped to 2.6 children per woman in 1990 and 1.8 in 2006. This development is attributed mainly to a widespread use of hormonal contraceptives and female sterilization.[55]

When interviewed about their attitude toward an abortion for fetal indications, Brazilian women treated in SUS clinics frequently answered that they would never consider it, because abortion is a crime.[56] On the other hand, many studies attest that when faced with an unwanted pregnancy, Brazilian women from all social strata and religious persuasions often elect to terminate it.[57] Abortion is perhaps a crime, but many Brazilian women are willing to commit it. It is reasonable to assume that in the hypothetical case of decriminalization of abortion in Brazil, the choice to terminate a pregnancy for a fetal indication would be affected by sociocultural variables, including religious faith. It is less certain that were this the case, the great majority of Brazilian women will choose to continue a pregnancy with a severely impaired fetus. The undoubtedly sincere conviction of Brazilian fetal medicine specialists that the decriminalization of abortion will not have practical consequences for patients treated in the public sector may help them to reduce their unease rooted in their awareness of the gap between choices open to poor and middle-class women.

Brazilian women diagnosed as having a very severe fetal anomaly can ask a judge for exceptional permission to terminate the pregnancy. As a rule, women who ask for such permission are encouraged to do so by their physicians. Some judges are known to be more willing to accept such a request than others. However, a pregnant woman cannot know in advance who will be judging her case, and thus her chances of receiving an affirmative answer to her petition to the court. Conjoined twins is a severe and easily observable malformation that in the majority of cases is incompatible with life.[58] This anomaly was chosen by a group of Brazilian physicians to test judges' willingness to agree to an interruption of pregnancy for a fetal impairment. Obstetricians from São Paulo University Medical School proposed to all women diagnosed as having a poor-prognosis conjoined twins pregnancy in a public facility to ask a judge's permission to terminate the pregnancy. All the women invited to make such a request were less than twenty-five weeks' pregnant, and all underwent a psychological assessment to verify that they could make an informed decision.[59] The São Paulo team identified 30 cases that corresponded to these criteria. Among these women, 19 decided to ask permission to terminate the pregnancy. Twelve women received permission, and 5 failed to receive it; 2 women were lost from sight. In a follow-up text published by the same group, the 12 permissions to terminate the pregnancy were presented as an important success for the medical team and proof of the feasibility of this approach—not as an illustration of the arbitrary nature of the juridical process, which obliged one-third of the women pregnant with nonviable children to continue the pregnancy against their will and undergo a difficult and traumatic birth.[60]

Fetal medicine experts who treat lower-class women at the MHC may have complicated relationships with their patients: they may have a disparaging attitude toward their "ignorance," and at the same time admire them for their stoicism, religious faith, and willingness to love an impaired child. They may also have complicated relationships with the affluent women they treat in private clinics, respecting them for their social status but perceiving them as "spoiled," interested mainly in consumption and unable to handle difficult situations. As a consequence, physicians may have a paternalistic (but not identical) attitude toward lower-class and upper-middle-class women. In private discussions, fetal medicine specialists, clinical geneticists, and fetal pathologists complain about the negative effects of the

this is where the law bites.

interdiction of abortion for fetal indications on exercising their profession, including in the private sector. Criminalization of abortion makes impossible any open discussions with pregnant women about the consequences of a positive PND. Some physicians are nevertheless willing to help their private patients end a pregnancy with a severely impaired fetus, occasionally operating on the margins—or beyond the margins—of strict legality.[61] One imaginative alternative, especially when a fetal anomaly is discovered relatively late in pregnancy, is the use of a diagnostic procedure to terminate the pregnancy. A physician can propose to the woman that she undergo an invasive diagnostic procedure known to have a high risk of complications which lead to death of the fetus, then "allow" such a complication to happen.[62] Physicians who offer such an approach—reserved for patients they fully trust—are, as a rule, excellent specialists, who in a different setting are proud of their skill in performing risky procedures with minimal fetal loss. In the topsy-turvy world of Brazilian fetal medicine, they apply their professional skills to successfully "mismanage" a diagnostic test, an ironic variant on uses of embodied expertise.

Female Pathologists, Social Justice, and Care

Physicians at the Maternal Health Center whose work I observed—and, it is reasonable to assume, their colleagues in other MHC departments as well—are sincerely devoted to their patients' well-being. On the other hand, they belong to a very different social class from the women they treat. Moreover, all of them are white, while many of their patients are either black or, more frequently, brown (*pardo*).[63] A difference between the social class of physicians who work in a public facility and that of the majority of their patients is far from being unique to Brazil. However, the contrast between the aspiration of producing a universal health system that will promote greater social justice and the reality of the key role played by unequal access to health services in reinforcing class stratifications may be especially great—and especially visible—in that country. At the same time, the growth of the nation's middle class along with government initiatives favoring students of color and of modest origins, which culminated in a promulgation of the affirmative action law of 2012, are slowly changing the profile of university graduates and the professional classes.[64] The affirmative action law formalized practices of positive discrimination voluntarily adopted earlier by many public universities. The diversification of students' social origins and skin color, Brazilian colleagues reported, is especially visible in technical professions such as computer sciences and engineering. Since affirmative action, university departments that train students in these areas have received more "black" and "brown" students, often from lower-middle-class families. By contrast, medicine seems—as for now—to resist such diversification. Physicians, male and female, continue (in 2018) to originate nearly exclusively from white middle-class families.

Foreign social scientists who observe Brazilian medicine have noted that their Brazilian colleagues, including those sincerely committed to the improvement of the national public health service, do not use this service themselves. Scholars from western Europe or Canada, where public hospitals and clinics are often synonymous with quality care, and those from the United States, where the segregation of users of private health insurance from users of public/semipublic plans is less drastic than in Brazil, may view as problematic the avoidance of public services by those promoting them. Yet the MHC's physicians were surprised when I asked them how they reconciled their sincere commitment to distributive justice in the Brazilian health care system with the fact that they and their families and friends do not use SUS facilities. Some were uneasy with what they might have perceived as a hint that, because of the entrenched class differences between physicians and patients at the MHC, they might not care enough about the women they treat. For many Brazilian physicians, including the progressive ones, major class differences in the quality of health care delivery are a "normal" situation, and their acceptance of this situation has nothing to do with their sincere dedication to their patients' well-being. Improving basic medical care for the underprivileged is perceived by progressive physicians as much more important than equal access to health services. Such an attitude promotes the reproduction of affluent patients' privileges and the invisibility of the lower class. The paucity of professionals' efforts to promote equality in the Brazilian health care system and their focus on dramatic situations and stopgap measures also reinforce their perception of poor patients as being resigned to their fate.

In her study of the use of sex hormones in the Brazilian state of Bahia, the anthropologist Emilia Sanabria describes the production of two distinct forms of citizenship by the national health system. One, available for users of the private-sector health system, focuses on personal autonomy and individualized choices, while the second, available for users of SUS, promotes inclusion through access to standardized medical services.[65] Middle-class women who had private health insurance received (or, to be more accurate, were told that they would receive) services and medications tailored to their unique needs. Poor women who went to SUS clinics received a standardized treatment.[66] For the poor, access to health care at SUS clinics was a means of attaining a socially recognized status as a person.[67] The achievement of such a status was, however, not automatic. It was conditioned by the patient's responsible behavior. Physicians mobilized notions of citizenship and rights to promote patients' compliance with medical regimes; at the same time, they mobilized the notion of free choice to attract affluent clientele.[68]

Rede Cegogna (the Stork Network), introduced in 2011 by Brazil's federal government, aims to improve pregnancy outcomes among poor women through the widespread provision of a basic prenatal care: monitoring of the pregnant woman's weight, blood pressure, and blood sugar, and identification of women at high risk of pregnancy complications.[69] Ultrasounds were not included in interventions promoted by Rede Cegogna. One of the main goals of the new program has been to reduce maternal mortality in Brazil, estimated in 2006 as 54 per 100,000 live births, with higher rates among very poor women, and to bring it to a level comparable with maternal mortality in industrialized countries (in 2008, maternal mortality in France was 10 per 100,000 live births).[70] Another important goal has been to limit preventable causes of newborn mortality and morbidity: infections and complications from maternal pathologies, mainly hypertension and diabetes.

The implementation of Rede Cegogna, the MHC's physicians often complained, was uneven. Women who arrived at the center with severe pregnancy complications were sometimes criticized for their presumed ignorance and neglectful attitude, but their predicament was more often seen as reflecting the chaotic and ineffective introduction of government and state directives, especially in the favelas (shantytowns) and poor rural areas. In the United States, those reviewing cases of fetal and newborn deaths have often held "neglectful mothers" responsible for poor pregnancy outcomes, disregarding their life circumstances or difficulties of access to health care.[71] The MHC's physicians frequently blamed the health care system, not the woman, for her pregnancy-related problems and their severity.[72] Too often, women who needed to be treated early in pregnancy, such as those at risk of hemolytic disease of the newborn, were diagnosed too late, when treatment of this condition was complicated and its success uncertain.[73]

The MHC's fetopathologists are directly confronted with the dramatic consequences of social injustice. In Brazil as in France, fetopathology is a strongly feminized subspecialty. Women tend more than men to integrate care as an important part of their professional work. How exactly they do it, however, depends on the context in which they are using their expertise. In describing the work of Colombian female forensic geneticists who employ their skills to identify victims of violence and return their remains to their families, Tania Pérez Bustos and her colleagues show how these women struggle with their emotional involvement in their work, and at the

same time regard it as a mission to heal the nation's wounds. In a context
marked by inequalities and an ongoing armed conflict, Colombian forensic
geneticists employed by a government-sponsored criminal investigation
laboratory developed an ethics of reparation for wrongful acts through the
identification of victims and the gathering of evidence against perpetrators.
They work with human remains, and are in close contact with tragedies
they cannot prevent, but they hope to diminish the suffering produced by
these tragedies or at least help the survivors achieve some sense of closure.[74]
Colombian forensic geneticists explained that they are especially proud of
their ability to help the underprivileged.

Brazilian pathologists who dissect fetuses and stillborn babies are less
fortunate. They, too, work with human remains, and their work brings them
directly in contact with acute social problems. Unlike their Colombian colleagues, however, they can rarely attenuate the consequences of these problems. In some cases they study, fetal demise is a truly random event that
would have occurred even if the woman had access to excellent health care.
In many other cases, a timely intervention might have prevented the pregnancy loss. In other cases still, a fetal anomaly could not have been prevented, but a more affluent woman might have had the opportunity to elect
an early termination of pregnancy. Fetopathologists also know that middle-
class women who learn that they carry mutations that can be transmitted
to offspring have access to solutions reducing their risk of giving birth to an
affected child. The caring attitude of the female Brazilian fetopathologists
and their striving to help their patients are expressed through their dedication to their work, but also through their dissatisfaction, impatience, and
frustration. These qualities are also articulated through their hope for a new
institutional and legal framework in which the tragic consequences of "nature's mistakes" will be not amplified by harsh laws and an unjust society.

Genetics and Fetopathology at the MHC

The genetics department at the MHC was founded in the 1960s, when the
center was one of several municipal maternity clinics. At that time, there was
only a limited interest in genetic diseases in newborns, so the small genetics department at the MHC was mainly dedicated to fundamental research.
However, this interest increased greatly in the 1980s, when inborn defects
replaced infectious diseases as the main cause of newborn mortality in Brazil. Yet the new focus on inborn defects did not entirely replace the interest in infectious diseases. Such diseases continue to be a significant cause
of pregnancy loss and newborn mortality and morbidity. An epidemic of

microcephaly (abnormal smallness of a newborn's head, frequently linked with severe neurological and cognitive impairments) first observed in the North of Brazil in the fall of 2015 was attributed to infection by the Zika virus, transmitted by the mosquito *Aedes aegypti*.[75] By contrast, at least before the Zika epidemics, the proportion of birth defects attributed to infectious causes was decreasing steadily, and the proportion of those associated with genetic anomalies was increasing. In the 1990s, when the MHC became a national referral center for pathologies of pregnancy and childbirth, its genetics department semiofficially began to offer services to pregnant women and to families of affected children while maintaining its research activity.

Unlike the prestigious molecular genetics department at the Andral Hospital in France, the genetics department at the MHC is small and modest. Nevertheless, the geneticists who work there, especially the department's senior scientists, Xavier and Louisa, are internationally recognized specialists in their domain; they strive to promote clinical genetic services within SUS. They participate in governmental and regional consultations on rare hereditary conditions, and investigate links between genetics and public health in Brazil, with a focus on problems caused by insufficient genetic services and unequal access to these services. Yet many of the MHC's geneticists are trainees—young physicians who want to specialize in medical genetics—and have a relatively light workload.

The department usually holds two half-day sessions of genetic counseling per week. During these sessions, they receive women who have suffered a pregnancy loss; they also provide counseling for families with a hereditary condition. Advice to such families often focuses on the evaluation of their probability of having an affected child. Such an evaluation is especially tricky when a given condition has a variable expression. Louisa mentioned a woman with a hereditary holoprosencephaly (a brain anomaly). In her case, the anomaly was an inherited dominant trait, and she had a 50% chance of transmitting it to her children. The woman had a very mild variant of this malformation; her daughter was severely affected. When she was pregnant again, an ultrasound displayed anomalies of the fetal brain, but sonograms cannot predict the severity of the future child's functional deficit.

The MHC's geneticists openly support women's right to interrupt a pregnancy for a fetal indication. They encourage women distressed by having been diagnosed as having a serious fetal anomaly to seek permission for a legal interruption of pregnancy. For women deciding to do so, MHC geneticists prepare the file destined for the judge. Before the legalization of abortion for anencephaly, this malformation and a few others, such as cyclopism (a severe malformation of the skull, with a single eye), were among

Criminalization prevents/blocks a public health perspective.

the rare cases in which the judge was inclined to allow an abortion: sono-grams of truly "monstrous" fetuses were a persuasive argument. In contrast, women who receive a diagnosis of a lethal condition that does not cause gross malformations, such as a severe kidney or heart defect, have a lesser chance of receiving permission for a legal abortion. After the legalization of abortion for anencephaly, the MHC's physicians occasionally tried to extend the definition of this anomaly to other severe malformations of the neural system, such as exencephaly and encephalocele (conditions in which the brain is completely or partially localized outside the skull). Such requests had reasonable chances of success, especially if the fetal head looked grossly misshapen. The success of requests for abortion for other fetal indications continued to be variable. In 2016, Xavier's impression was that the 2012 law legitimating abortion for anencephaly did not change the judges' attitude toward other fetal anomalies incompatible with life.

Juridical decisions, Louisa explained, are impossible to predict: "Judges do what they want." Before the legalization of pregnancy termination for an anencephalic fetus, requests for abortions for this indication were reviewed by one of four judges of the state court specializing in these cases. Three among them frequently granted permission to terminate the pregnancy; the fourth judge almost never did. If the anomaly was not lethal, the chances of obtain-ing permission for an abortion were very low and did not depend on the predicted severity of the future child's impairment. A woman who applied for permission to interrupt a pregnancy had no way of knowing who the judge assigned to her case would be or how long it would take to obtain a decision. I was told about the case of a fetus without kidneys—an invariably lethal condition—in which the juridical process dragged on so long that the woman received permission only after she was thirty-three weeks' pregnant. She chose to give birth naturally; the child lived for one hour. Louisa added that a few women diagnosed as having a severe fetal malformation in advanced pregnancy never returned to the MHC, and some might have attempted to interrupt the pregnancy in an illegal—and often unsafe—abortion clinic.

Louisa and Xavier affirmed that the criminalization of abortion in Brazil prevents the development of a public health perspective on birth defects. It also diminishes incentives for interdisciplinary collaborations in fetal medi-cine, because an accurate diagnosis is not very important.[76] When physicians who work in a public facility detect a fetal anomaly, they usually propose that the woman wait until the child is born in order to learn the severity of the suspected problem. When a pregnant woman diagnosed as having a fe-tal anomaly is treated in a private clinic where abortion is an (unofficial) op-tion, her physicians may be more interested in evaluating the severity of the fetal impairment, but the decision of how to reach an accurate diagnosis is left to each physician. Francisco, a fetal medicine specialist who works in a high-end private obstetrical clinic, told me that when confronted with a difficult case, he sometimes asks the advice of a clinical geneticist, but this is not a systematic practice among other experts in this clinic. He and his colleagues more often consult pediatric neurologists, because a diagnosis of fetal brain malformations can be very tricky; but here, too, the decision whether to consult another specialist belongs only to the pregnant woman's physician.[77]

Geneticists at the MHC, like the geneticists at AH, are expected to have a weekly meeting with the center's fetopathologists to review difficult cases. In practice, during the period I observed the MHC's fetopathology department, meetings with geneticists and genetics trainees usually took place only every three or four weeks. Scheduled meetings were frequently canceled, either because there were few "interesting" cases, or because several participants were busy elsewhere. The MHC staff meetings, like those at AH, are organized around photographs of dissected fetuses projected on a screen (slides), but the number of examined photographs is much smaller than during an AH staff meeting, and they are made with non-professional photographic equipment. Participants in the MHC's staff meetings systematically attempt to compare their photographs of affected fetuses and stillborn infants with those printed in textbooks, because they did not have access to Internet databases on human malformations. Such databases are available only to institutions that can afford the relatively high fees for their use. Moreover, since Internet access at the MHC was—to put it mildly—not fully reliable at the time of my observations, participants in staff meetings seldom used their laptops to look for additional information.

Occasionally and irregularly, fetopathologists and geneticists meet with pediatric surgeons and their students. I assisted at one meeting like this which was mainly a teaching session. In it, the pediatric surgeon, Marcos, one of the best-known experts in his field and a previous head of surgery at the MHC, presented each case as a puzzle, and invited his trainees to solve it. The first presented case featured photographs of a living newborn with a highly distended intestine. None of the young physicians were able to provide the correct answer for the newborn's condition: cystic fibrosis, an autosomal recessive hereditary condition. In that case, the diagnosis was especially difficult, because it was probably a new mutation: there was no family history of cystic fibrosis. The intestinal anomaly was not detected during prenatal ultrasound and was observed only at birth. The diagnosis of cystic fibrosis was confirmed with a sweat test and biochemical tests for pancreatic enzymes. The child did not undergo genetic testing. Marcos explained that a severe intestinal blockage was the worst-case scenario for this pathology. Sometimes the intestinal blockage can be eliminated by surgery, but in light of the severity of the anomaly, it is highly probable that the child would be sick all his life and not survive long.

The second case, presented by a trainee, was a newborn with a diaphragmatic hernia, a condition requiring surgery immediately after the birth to save the child's life. AH's surgeons affirmed (in 2015) that their success rate of operations for this indication was close to 70%. The success rate at the MHC was lower, but nobody was able to provide precise data. The child, a girl, was diagnosed as having a diaphragmatic hernia during a prenatal ultrasound. She was scheduled for an operation immediately after birth, but surgery was delayed because her physicians decided she was strong enough to undergo an extended surgery. She died fourteen days later. An autopsy indicated that the initial diagnosis had been inaccurate. The girl had suffered not from a diaphragmatic hernia but from a related condition, the absence of muscles of the diaphragm.

Marcos, Louisa, and a pathologist, Adriana, discussed whether this diagnostic error affected the child's fate. They concluded that with an accurate diagnosis, surgeons probably would have decided to operate immediately after birth. Louisa was not sure, however, if this would have improved the child's relatively low chances of survival. Marcos thought that the surgery might have been successful, but only if it had been performed by a very good pediatric surgeon. Adriana added that it's a pity that the pathology department did not receive the child's placenta for analysis. Diaphragmatic hernia produces a "typical" image of placental anomaly. The absence of such a typical image might have attracted the pediatrician's attention to the possibility that the original diagnosis was inaccurate. It is not clear why the pathologists did not receive the placenta: perhaps this was another example of the difficulties of collaboration between the MHC's departments.

Anatomic Pathology at the MHC: A Cheerful, Ghastly Kitchen

I visited the anatomic pathology department for the first time in order to speak with a senior pathologist, Adriana. Adriana, I had been told, was interested in the social sciences approach, and thus may be willing to help with my research on PND in Brazil. AT the MHC, as in many other Brazilian public institutions, more formal ways of contacting a researcher—sending a letter or an e-mail requesting a meeting or addressing the department's administrator—may not work. It is often simpler to try to knock on the person's door. When I asked people who worked in the anatomic pathology department's office where I could find Adriana, adding that I was a foreign researcher visiting the MHC, one of the secretaries waved in the direction of the door of the dissection room. In contrast to the rigid regulation of access to the dissection room at AH, the MHC's dissection room is freely accessible, though very few people venture there. I found Adriana busily dissecting a stillborn child, who looked black and grossly abnormal, a rather unsettling sight. Adriana listened to my short introduction about the goals of my study and immediately offered her help without interrupting her work. She then ended the dissection, leaving the task of closing the cadaver and making it as presentable as possible to the dissection room's technician, Maria. After removing the rubber apron that protected her clothes and washing her hands, Adriana invited me to her office to discuss my request to observe the department's work. Later, Adriana and her colleagues, Simone, Bianca, and Raquel, provided unfailing support for my study.

The MHC's anatomic pathology department is located in a small building at the end of an internal courtyard. The dissection room is on the ground floor, next to the department's office. To enter it, one needs to pass the two storerooms containing fetal remains; near the entrance is also a small room with a coffee machine, where the dissection room technicians, Maria and Ricardo, prepared endless tiny cups of strong *cafezinho*. The first floor houses the cytology laboratory, where tissue specimens are preserved, cut, and stained, along with additional storerooms for collections of slides; cytology technicians have workspaces on that floor. Pathologists' offices are on the second floor. These offices, like their counterparts at AH, are small and crowded (fig. 12). They are filled with microscopes and slides, computers, papers, files, and books.

One of the pathologists has a clay figurine caricaturing a pathologist, made in typical Brazilian folkloric style. It sits on her shelf, a specifically Brazilian representation of a universal medical specialty (fig. 13).

Some of the books are very old: on one of the shelves, I found a collection of partly crumbling French histology atlases dating from the late nineteenth century (fig. 14).

The dissection room is spacious and well illuminated. It is organized around a huge dissection table covered by a metal sheet, with a metal sink and faucet at its head. The department's pathologists sometimes wash fetuses under the faucet. Near one wall is a laminar hood for the manipulation of chemical substances that generate toxic fumes; also there is a small elevator, used for transporting tissue samples to the first floor, where they are preserved and stained. The hood and the elevator are hardly ever used. A

Fig. 12. Workstation. WHC, anatomic pathology department. Pathologists'
workplaces in France and Brazil are very similar. Author's photograph.

Fig. 13. Figurine caricaturing a pathologist in the office of a WHC's anatomic
pathologist. The "pathologist" is crafted in the style of folk figurines
from the Brazilian Northeast. Author's photograph.

Fig. 14. Old textbooks. WHC, anatomic pathology department. Some of the
textbooks are from the late nineteenth century. Author's photograph.

dissection room technician, usually Ricardo, uses the stairs to take preserved
tissue samples to the micropathology laboratory. On the opposite wall, two
scales stand on a long, white-tiled bench: a small precision scale, used for
weighing small fetuses and surgical samples, and a large mechanical scale,
of the kind used in shops before the advent of electronic scales, for weighing
larger fetuses, dead newborns, and large dissected organs such as a uterus.
At the end of this bench is a stand holding surgical tools and a set of large
dissecting knives. A section of the bench is used for photographing fetuses
before a dissection: this part is covered with blue paper to provide a better
contrast (fig. 15). The photographs were usually taken by Maria with an ordinary compact camera.

A large window in the dissection room contains opaque glass that blocks
outsiders from viewing the room's interior. The room also has, at its far end,
a much smaller window that can be partly opened. This window, which
even when opened does not permit a view inside the dissection room, is
the department "communication center." Maria and Ricardo use it to call
to a technician or a nurse spotted in the courtyard; to receive samples from
the maternity ward or the operating room and documents brought from

Fig. 15. Photography corner. WHC, anatomic pathology department. Author's photograph.

the MHC's patients' files office; and sometimes to chat with a friend. The dissection room is illuminated by strong lamps which make it very warm on hot days; it also receives natural light from the large window. Next to the dissection room is a much smaller room for the storage of bodies and tissue samples awaiting analysis. It contains a domestic-type refrigerator for storing tissue samples stacked in plastic boxes of different sizes, shapes, and origins, and a much larger refrigerator for the storage of fetal and newborn bodies. Fetuses are swaddled in blue hospital cloth. Dead newborns are frequently dressed in clothes brought by the family, typically a blue or pink knitted outfit with matching tiny socks. The door between this room and the dissection room is usually left open. The tissue and bodies storeroom also has a door leading to a very small mortuary room, where families receive the bodies of their dead children.

The MHC's dissection room, even more than the one at AH, brought to mind Claude Bernard's description of the laboratory as a "long and ghastly kitchen."[78] At the MHC, this impression was reinforced by the presence of equipment and utensils found in an ordinary kitchen: a refrigerator, a collection of plastic boxes of various shapes, dissection knives that looked very much like normal kitchen knives, an old-fashioned scale, and plastic trash bins. Fetuses and organs were often placed directly on the scale or the dissection table. The "ghastly kitchen" impression of the dissection room was especially strong at the end of a busy day, when the dissection table and tools were covered with blood and bins overflowed with trash.

Early operating theaters probably resembled a busy kitchen themselves, or sometimes even a butcher shop.[79] Then the introduction of aseptic surgery in the late nineteenth century imposed very different rules for the management of bodies, body parts, and bodily fluids. In a contemporary operating room, organized around the rigorous maintenance of sterility, flesh and blood are carefully contained in well-defined spaces. Pathologists who work with dead bodies do not need to adhere to sterility rules, however. The MHC's anatomic pathologists did not seem to be preoccupied by the external aspect of their workplace while they were working. Only at the end of the day did technicians clean and tidy the dissection room, preparing it for the next day.

Sociologists have linked "dirty work" with a lower occupational status. Blue collar workers have dirty hands, but not white-collar ones; cleaning jobs are less prestigious than managerial ones. In the MHC's anatomic pathology department, this hierarchy is reversed. Pathologists frequently have bloody hands and wear thick rubber aprons over their laboratory coats, while technicians often have clean hands and wear immaculately white coats. As a rule, the technicians do not deal with fetal bodies. The only moment they handle bodies is at the end of a dissection, when a technician, usually Maria, closes the dissected body and, if applicable, dresses it up and prepares it for burial. Only at that moment does a rubber apron get put on. The technicians' main job is to prepare dissection tools, organize and mark tissue samples, photograph dissected fetuses, and write down the pathologists' observations.

One of the main preoccupations of pathologists and technicians is the avoidance of a mix-up with the studied samples. Each dissected fetus receives an anatomic pathology department number, which is different from the pregnant woman's patient number. This number gets written on a single piece of white adhesive tape, which is then stuck on each sample of tissue from that fetus, in sequence: before staining, after preserving and staining, and then on microscope slides prepared from the fetal tissues. In 2012, 2013, and 2014, when I made my observations in the MHC's anatomic pathology department, all the identifying numbers—on containers of fresh tissue samples, containers of preserved tissue samples, microscope slides, and cardboard tags attached to fetuses preserved in the storeroom—were handwritten, as were their entries in the department's casebook and in the autopsy report in each patient's file. Patients' records were not computerized, nor were the department's statistics. Computers were used only by the departments' secretaries for their official correspondence.[80]

One of the technicians' main tasks is to take notes during a dissection. A typical scene from the dissection room is the pathologist, in an apron and rubber gloves, cutting through a fetus while dictating her observations to a technician standing on the other side of the room. The pathologists carefully consider each term they dictate, because the dissection notes are nearly always directly incorporated into the patient's record.[81] When more than one pathologist is present at a dissection, they sometimes discuss which word should be used to describe the fetal structures they observe, and how exactly to define the deviations from the norm. I was impressed by the acuity and precision of the MHC's pathologists' observations, their great attention to anatomical details, and their striving to get descriptions of all the details right. On the other hand, the MHC's pathologists rely above all on their individual experience to produce accurate descriptions and tentative diagnoses. Pathologists' knowledge always strongly relies on the embodied knowledge of the expert, but at AH such incommunicable knowledge becomes partly communicable through intensive exchanges with colleagues and a collective process of producing the definitive report of a postmortem. Such a shared elaboration of the results of a pathological examination was rare at the MHC.[82]

The MHC's pathologists usually dissect three to ten fetuses or newborns every week, and about twice as many placentas. The fetopathology department at AH has a similar volume of activity, but most of the fetuses dissected there were aborted for a fetal indication, while the fetuses dissected at the MHC either were miscarried or died in the womb. In some of the latter cases, the postmortem diagnosis is made more difficult by the partial decomposition of the fetus. The sole exceptions are fetuses aborted for anencephaly. Yet the anatomic pathology department at the MHC, unlike the fetopathology department at AH, does not specialize in fetuses and placentas, but also studies tissue samples from surgeries performed in the pediatric and gynecological wards. The gynecological samples are often segments of tumors.

Pathologists analyze preserved tumor tissues, but also frozen sections: small pieces of tissue excised during a surgery that have to be analyzed very rapidly, because the patient is on the operating table and the surgeons are waiting for the pathologist's verdict in order to know how best to continue the operation. Frozen sections are always emergencies. The pathologist abandons whatever she is doing in a given moment to study the sample sent from the operating room. The MHC's pathologists perform macroanatomical work (that is, dissections and preparation of tissue samples) in the morning, and microanatomical work (the study of microscope slides) in the afternoon. Typically, the pathologist (nearly always only one) starts her morning with a study of surgical samples, then she examines placentas, and ends with dissecting fetuses—often the most complex and demanding part of her day, and not infrequently the most distressing one as well.

Unavoidable Disasters, Avoidable Tragedies

The rules that govern a fetal postmortem in France are very different from those rules in Brazil. At AH, mothers/parents decide whether the fetus will be "recognized" or "unrecognized," and whether it will be treated as the woman's/couple's child. At the MHC, the only criterion for treating a fetus as the couple's dead child instead of surgical waste is its weight. When the fetus's weight is less than five hundred grams, the mother/parents have no rights whatsoever to this fetus. Beyond this weight, they can decide whether the anatomy department will be allowed to dissect the body. They also have the right to claim the fetus (which at this point is always called *neomorto*—a stillborn child) for burial. By contrast, there is no need for the MHC to ask permission to study the placenta, including those from live births. As Adriana often repeated, "The placentas are all ours." The dissection of a placenta, especially when combined with a detailed external examination of the fetus, can yield important information about the reasons for fetal demise. In addition, each placenta has a maternal and a fetal layer (the free fetal DNA in maternal circulation, used in noninvasive prenatal testing, is placental DNA). Since the anatomic pathology department "owns" the placentas, its experts do not need to ask the woman's permission to investigate them. In principle, they can have access to "interesting" fetal DNA even when they did not receive an authorization to conduct an autopsy. This is, however, a purely theoretical possibility: genetic studies of fetal material are rare.[83]

The study of placentas by fetopathologists may have legal implications. The family of a woman whose pregnancy was being followed at the MHC claimed that she arrived at the MHC's maternity ward on Tuesday in the early stages of labor and was sent home, despite the fact that she did not feel well. She was readmitted to the hospital on Monday, complaining about severe pain, but a nurse told her to be quiet, because everything was normal. On Wednesday morning she gave birth to a stillborn child. The MHC's obstetricians affirmed that they had examined the woman several times and did not find anything wrong with either the baby or the progress of labor. The family decided to sue the hospital. They authorized an autopsy of the fetus, which was to be conducted at the medicolegal institute and not the MHC to avoid a potential conflict of interest. The placenta, however, was given to Adriana, who exclaimed, "Nossa!" (short for "Nossa senhora," Our Lady) upon seeing it. The placenta was not red and filled with blood as usual, but grayish blue because of an accumulation of meconium (the content of the fetal intestine), a well-known sign of fetal distress during labor. The case was reported in detail in a national newspaper under the headline "Parents of a Woman Who Lost a Baby Are Suing MHC for Negligence." The police asked the hospital to give them documents relative to this case, and transported the baby's body to the institute of legal medicine. MHC pathologists were reluctant to talk about this case; I never learned how this story ended.

MHC fetopathologists, like their French colleagues, complained about a growing number of rejected requests for fetal postmortems, but had varying ideas about the extent of this phenomenon. Raquel was convinced that there was a recent sharp decline in the number of permissions, while the head of the anatomic pathology department, Doctora Teresa, believed that such a decline was not very significant, and might reflect fluctuations in the MHC's activity.[84] In France, permission for the autopsy of a fetus or stillborn child requires the mother's signature. If the child was not born alive, the father's permission is not obligatory. The local rule at the MHC is that for forty days after birth, the woman is not allowed to sign an autopsy permission for a fetus, a stillborn child, or a child who died after birth. This rule is justified by the principle that childbirth diminishes a woman's legal responsibility. This principle led to the supposition that the tragedy of pregnancy loss may further diminish the woman's capacity for rational judgment. It is possible, Adriana explained, that a distressed woman will first give permission for an autopsy, then change her mind. In such a case she may sue the MHC, arguing that its researchers exploited her vulnerability in order to conduct research. Maria added that some women become "crazy" (*louca*) after the loss of a child.

I was surprised to hear that a woman is not allowed to decide the fate of her child's body, and asked to see the legal text that supports this decision. Adriana showed me a resolution of the Conselho Federal de Medicina of 3 December 1982, number 1.081/82, relative to obtaining permission for medical tests and necropsy. We read it together, and found out that the resolution does not deny the woman who gave birth the right to make legal decisions about the fetus's fate. Just the opposite is true: it states that the mother is expected to sign the permission form for medical tests made on the body of the dead newborn or stillborn child. The text adds, however, that when the woman is not in possession of all her faculties, such permission should be signed by a family member or, for a legally incompetent woman, her legal guardian. In the file dedicated to this question in the anatomic pathology department, this statement is accompanied by a copy of a Brazilian court decision grounded in the *DSM-IV* definition of *postpartum condition* (*estado puerperal*), an acute transitory stress that may arise following childbirth.

The MHC's internal rule to always ask a family member to sign the autopsy permit was, I assume, a local extension, to *all* the women who gave birth, of the principle that *some* women suffer from a postnatal depression that may diminish their mental competence.[85] I was also unable to uncover the exact origin of the conviction that women are not in possession of their full mental powers for forty days after birth. This lapse of time resonates nevertheless with the Catholic Church's custom (now abandoned) according to which a woman who gave birth was not allowed to attend Mass for forty days. At the end of that period, the woman underwent a "churching ceremony": she was blessed and readmitted into the church. This exclusion of women from participation in church ceremonies is related to Old Testament purity rules that see women who gave birth—and menstruating women—as impure (Leviticus 12:1–8).[86] MHC's regulations and Maria's statement that some women are "crazy" during that period may indirectly mirror similar beliefs.

In some cases, Adriana explained, inexperienced physicians asked a woman who had lost a child to sign an autopsy permission form, an intervention perceived at the MHC as a professional mistake. I did not find such mistakes in the autopsy permits in the patients' files I examined—all had been signed by other persons, not the woman who gave birth. No file indicated who this person was. In about half the files, the permission had been granted by a man who might or might not be the child's father, and in the other half by a woman. When the patient was very young (a significant portion of women who gave birth at the MHC were under the age of twenty, and some were under sixteen), permission to conduct an autopsy was often signed by a woman with the same family name as hers, possibly her mother. When a woman was over the age of twenty, the autopsy permission was more often signed by a man. This man frequently had a family name different from that of the patient, but this did not necessarily mean that they were not spouses, since many Brazilian married women keep their maiden name. Alternatively, he could have been her partner or a relative.

In one case, a very young woman miscarried a fetus with severe malformations of the central neural system. The fetus remained for several days inside the anatomic pathology department's refrigerator while the department waited for an autopsy permission. The woman's physicians believed that an autopsy was important for understanding the causes of the fetal demise. The woman agreed to an autopsy, but refused to give the name of a relative who could sign an autopsy permission. Maria speculated that perhaps she had not informed her family about her pregnancy; alternatively, she might have wanted to conceal the pregnancy's outcome, and decided to tell her family that she had miscarried without mentioning the fetal malformation. For several days, a period that might have corresponded to the length of the woman's stay at the MHC's maternity ward, her physicians attempted to persuade her to find a family member to sign the autopsy permission. The woman, dubbed "the lonely one" (*soazinha*) by the maternity ward staff, obstinately refused. The fetus did not undergo an autopsy. The mother later claimed the body for burial.

Many autopsies at the MHC are conducted on third-trimester fetuses that either were miscarried or died in the womb. The average age of fetuses dissected at AH's fetopathology department is decidedly lower, because many were aborted for a fetal indication in early or mid-second trimester. In addition, maternal causes of late fetal demise (infections, diabetes, heart disease) are relatively rare at AH. Another important difference between France and Brazil is the great number of "monstrous" fetal bodies seen at the MHC. A majority of French women have a first diagnostic ultrasound at eleven to thirteen weeks of pregnancy. At that stage, it is often possible to either detect a major fetal malformation or suspect that something is seriously amiss, and advise the woman to have an additional ultrasound a few weeks later. Early abortion of fetuses with severe structural anomalies may explain the relative paucity of dissections of "monstrous" fetuses at AH's fetopathology department. During my observations at that department, I saw several unusual fetuses (usually aborted in the early second trimester), including one small "acardiac monster" (a parasitic twin), but no cases of major trunk malformations, anencephaly, or conjoined twins.

Despite the efforts of the organizers of Rede Cegogna to reach all pregnant women of modest background early in pregnancy, some women treated at the MHC are seen by a health professional only after they miscarried an abnormal fetus or when the child was stillborn. The MHC's physicians have divergent evaluations of the efficacy of Rede Cegogna. Its advocates, often the center's public health specialists, stressed its achievements, above all

swatched

the enrollment of a greater number of pregnant women in primary care programs. Its critics, who include many of the MHC's pathologists, point to its shortcomings: pregnant women's uneven access to health services, insufficient follow-up of problematic cases, and paternalistic attitudes of some of the network's organizers. While dissecting a fetus/newborn that, they estimated, could have survived with better prenatal care, Adriana and Simone occasionally murmured that their job is "mopping up" after Rede Cegogna failures.

Sometimes, the pathologists' initial supposition that fetal loss was induced by a preventable cause is disproved by further investigation. A teenage woman gave birth to a stillborn child with obvious signs of fetal infection. The placenta, too, displayed typical signs of infection. Adriana and Simone initially thought that the woman failed to be tested for one of the common infections that produce fetal demise. This opinion was reinforced by data in the patient's file that indicated that she did not receive adequate prenatal care. All the blood tests for congenital infection of the fetus—syphilis, toxoplasmosis, cytomegalovirus, and rubella—were, however, negative. Through an exclusion process, Adriana concluded that the fetus was probably infected with parvovirus B19. This virus usually induces a very mild disease, but in pregnant women it can lead to death of the fetus. Infection with parvovirus is rarely detected in routine prenatal visits. The young woman had a patchy prenatal monitoring, but the pregnancy loss was probably an accident that could not have been prevented.

Fetal deaths attributed to infections, especially "preventable" ones, are seen as especially tragic, but they are relatively rare. The great majority of the fetuses dissected by the MHC's pathologists displays multiple structural malformations; the pathologists also dissect newborns with such malformations. One of the reasons that so many severely impaired children are born at the MHC, Adriana proposed, is the reluctance of many of the women treated at this center to consider an abortion for a fetal malformation. Xavier, who promoted many requests for juridical permission to terminate pregnancy, disagreed. He argued that some women do categorically refuse to consider termination of pregnancy, even when they learn that the fetus is not viable, but many others are willing to consider an abortion. One of the main obstacles to receiving permission for an abortion for a fetal indication, besides the judges' obstruction, is uneven institutional support for such requests. The MHC's genetics department systematically informs women of their rights and helps those willing to ask for permission to terminate the pregnancy, but not all the center's departments and not all the physicians have a similar policy. Many physicians—aware of the impossibility of predicting the judge's decision, the elevated probability of the petition's failure, and the potentially negative effects of such failure on the pregnant woman—are unwilling to undertake this time-consuming task.

Each year, the MHC's anatomic pathology department's files record several cases of especially severe malformations, such as sirenomelia (fusion of the lower limbs), cyclopism (severe anomaly of the skull, with a single eye), anencephaly, and conjoined twins. They also record other severe malformations diagnosed in third-trimester fetuses and stillborn children, such as extended gastroschisis (defect of the abdominal wall that leaves part of the intestine outside the body), exencephaly (exteriorization of parts of the brain), severe cranial and facial malformations, hydrocephalus (fluid in the skull), and microcephaly (abnormally small head). In the majority of these cases, the fetal defect is attributed to an accident of pregnancy: abnormal number of chromosomes, abnormal embryonic development, amniotic bands (protein "cords" in the placenta that restrain fetal growth), and placental anomalies of unknown origin.[87] Many of the MHC's autopsy reports end with a quasi-tautological conclusion: the cause of fetal/newborn death was the presence of multiple inborn malformations. In some cases, the pathologists propose a specific diagnosis, such as a known genetic anomaly. Only in approximately 10% of the investigated cases is a suspected genetic anomaly confirmed by further genetic investigations, limited nearly always to the study of fetal chromosomes (karyotype).

In two recent cases, one from 2013 of a fetus diagnosed as having holoproencephaly (anomaly of the brain), another from 2012 of a fetus diagnosed tentatively as having Meckel-Gruber syndrome (an autonomous recessive hereditary malformation producing a characteristic lethal combination of structural anomalies that include neural tube defects, kidney and lung anomalies, and abnormal number of fingers), the MHC's geneticists looked for disease-specific mutations.[88] Finding such a mutation may be important for genetic counseling. Holoprosencephaly is often produced by a new mutation, but there are also cases of transmission of this condition in families. In the latter case, the MHC's geneticists can inform the family about an existing risk, but often cannot do much more to help them. Meckel-Gruber syndrome is always hereditary. The initial diagnosis is made based on pathological findings, but because some of the anomalies linked with this syndrome are found in other inborn conditions, too, it is important to confirm the diagnosis by genetic studies. If Meckel-Gruber syndrome is confirmed, geneticists can inform the parents about the 25% chance of recurrence of this condition in the next pregnancy—or, in a more optimistic version of this statement, that they have three chances out of four of giving birth to a healthy child.[89]

Tolerated Risks

Brazilian physicians have been described as less preoccupied—and, for some, less obsessed—than their French colleagues with a desire to eliminate all the possible risks for the future child. They are more willing to recognize the limits of their power, and more inclined to advise pregnant women to stop worrying and enjoy their pregnancy.[90] Before studying PND, I had investigated the problematic aspects of screening for precancerous conditions and cancer risk: overdiagnosis, iatrogenic effects of preventive interventions, and stress.[91] I was also aware of problems produced by the rapid expansion of prenatal testing and screening in western Europe and North America. I was therefore initially sympathetic toward Brazilian physicians' more relaxed attitude concerning the risk of fetal malformations, and their claim that the existing diagnostic approaches have a limited scope, cannot ascertain that the child will be perfectly healthy, and can produce insoluble dilemmas which increase pregnant women's anxiety.

In comparing the practices of Brazilian and French physicians, I gradually realized that the argument in favor of greater tolerance of uncertainty and risk in Brazil can be turned on its head. Brazilian physicians' lesser preoccupation with dangers for the fetus may mirror a lower concern about their own professional risks. French physicians are responsible for all aspects of maternal and child health, including an accurate detection of fetal anomalies. Close interactions between French specialists from different disciplines are supported by the need to provide such an accurate diagnosis. Cooperation among professionals aims to reduce preventable diagnostic errors. Providing the correct diagnosis of fetal impairments is seen as a heavy responsibility, because it shapes the woman's decision about the future of her pregnancy. A diagnostic mistake can have disastrous consequences for the patient; it can also affect the physicians' reputation and their self-perception as a competent professional and healer.

Leading Brazilian gynecologists and fetal medicine experts are no less competent than their French colleagues, but their attention is focused especially on the pregnant woman. Physicians' duties include the detection and treatment of pregnancy-related pathologies that affect the woman but can also harm the future child, such as preeclampsia, high blood pressure, and sickle-cell anemia.[92] Physicians are also responsible for the management of childbirth and the prevention of maternal and newborn mortality and morbidity. Complications of childbirth that lead to an irreversible disability in the child are seen as the obstetricians' fault. Some obstetricians legitimate the exceptionally high rates of C-section in Brazil by their concern for the safety of newborn children. Physicians' responsibility does not extend, however, to the birth of a child with impairments produced by genetic anomalies or unpredictable accidents of fetal development. Brazilian physicians can emphasize the difficulty in predicting the child's future, because they face fewer risks if they fail to correctly diagnose a fetal problem. A disastrous outcome of a pregnancy can be blamed on Nature, God, or, for some, the conservatism of Brazilian legislators, but usually no blame can be attached to the physicians who took care of the pregnant woman.

Brazilian physicians' failure to detect a fetal problem will rarely have major consequences in a public health care facility because, no matter the fetus's prognosis, the woman is expected to continue the pregnancy.[93] Physicians' accountability for diagnostic mistakes is limited in the private-sector health system, too. If the pregnant woman elects an illegal abortion, the accuracy of the original diagnosis is seldom verified after this intervention. If she decides to continue the pregnancy and finds out that the child's problems are more severe than she was told, she cannot sue her physician for a lost opportunity to commit an illegal act.[94] Moreover, when the child's malformation does not correspond to the anomalies detected by the fetal medicine expert, such a discrepancy can be employed, especially in public hospitals, to consolidate the discourse on the uncertainty of prenatal predictions. Such a discourse helps fetal medicine experts to cope with their inability to propose a solution to a woman who learns about a severe fetal impairment. Only in a handful of treatable cases in which it is important to ascertain that the birth will take place in a hospital that will provide adequate care to an impaired newborn (for example, hemolytic disease of the newborn; operable gastroschisis) can a physician be held responsible for failing to detect a fetal anomaly during the pregnancy. In all other cases, the role of the Brazilian fetal medicine expert may be seen as akin to that of an oracle. The scrutiny of the unborn child, like the oracle's verdict in the ancient world, frequently reassures, sometimes predicts an unavoidable disaster, and in many other cases produces a murky statement, deliberately left murky.

FIVE

Balancing Risks:
PND and the "Prevention of Disability"

"Flawed" Fetuses and Disabled People

The aim of seeing "what is about to be born" is preparing for the future. Prenatal diagnosis evaluates the probability of giving birth to a disabled child, and describes the kind of disability—or frequently, the spectrum of disabilities—the child may have.[1] The official discourse of PND experts affirms that women's/parents' awareness of their future child's potential problems helps them to be ready for the care of a special-needs child. In practice, when abortion for a fetal indication is legal—or available for those who can afford it—such preparation is not infrequently a preparation for the child's nonbirth. Selective abortion of impaired fetuses has become the main focus of public debates about PND. Termination of pregnancy for a disability risk, critics of this approach argue, gravely harms people living with disabilities by sending a strong signal that their lives have less value than the lives of able-bodied persons.

One of the consequences of the new focus on links between PND and disability rights is a growing difficulty in speaking openly about "selective reproduction." In the twenty-first century, the professionals' official discourse, especially but not exclusively in the United States, no longer mentions the "prevention of disability" as a desirable individual or social goal. Even when all the stakeholders know that an early diagnosis of a given fetal malformation will nearly always lead to the termination of pregnancy, it is not acceptable to say this explicitly. In discussing the inclusion of chromosomal microdeletions in non-invasive prenatal testing, the CEO of Natera, the company that produces this test, explained that "nobody in the twenty-first century should have a pregnancy without being screened for these microdeletions. Routine checking for microdeletions could enable more families to prepare for children with special needs."[2] It is, however, highly unlikely that Natera's CEO did not know that several among the microdeletions included in Natera's new test (cri du chat [cat's cry] syndrome, Angelman syndrome, 1p36 deletion) cause very severe inborn impairments, and that nearly all the women who receive a diagnosis of one of these conditions would elect an abortion.

Pioneers of PND had a dramatically different point of view. They developed the new diagnostic technology with the explicit aim of helping women to have healthy children by giving them the option to abort impaired fetuses.[3] Contributors to early discussion about PND viewed the termination of pregnancy as the normal—and, for many, desirable—outcome of a diagnosis of genetic anomaly of the fetus. As one of the participants of a 1970 conference on scientific and ethical aspects of PND, organized by the US National Institutes of Health, put it, "It seems to me that all the gentlemen agree, some more explicitly than others, that to abort is a good thing and should be encouraged."[4] In 1975, in discussing the results of a report on the safety of amniocentesis, Theodore Cooper from the US Department of Health, Education, and Welfare strongly defended extending the use of this technique: "Few advances compare with amniocentesis in their capacity of prevention of disability . . . with this technique we can assure the older woman who is pregnant that she need not fear the birth of a child with Down's syndrome and her consequent lifetime devoted to the care of a handicapped child."[5] When the social epidemiologists Mervyn Susser and Zena Stein proposed in 1973 to generalize the use of amniocentesis for the detection of Down syndrome, they argued that "the lifelong care of a severely retarded person is so burdensome in almost every human dimension that no preventive program is likely to overweigh the burden."[6] In the early days of PND, this technology was presented by its promoters as indisputable progress. This view was, however, rapidly contested.

In the 1970s and 1980s, the opposition to PND and selective abortion of "impaired" fetuses came mainly from two groups: pro-life activists, opposed to abortion in general but especially the selection of "fit" fetuses and the disqualification of "unfit" ones, and a fraction of feminist activists opposed to the medicalization and instrumentalization of pregnancy. In the 1970s, opponents of legalized abortion employed "extermination" language and an analogy with the Nazi regime to describe pregnancy termination for a fetal malformation. Thus, in 1974 French parliamentarians opposed to the decriminalization of abortion invoked Nazi physicians, genocide, racial eugenics, and concentration camps. Legalized abortion, they argued, would be the first step in a "monstrous regression" that would lead to the euthanasia of the disabled and the murder of those defined as "useless people."[7] At that time, Christian theologians in Europe—Catholic, Protestant, and British Anglican—did not have a uniform position on abortion. Some justified termination of pregnancy when its continuation threatened the well-being of the mother or her family. Others explained that the element which makes life as "human life" possible is not the physiological conception but human acceptance.[8] Moreover, in the 1960s and early 1970s, progressive Catholics in the United States did not unconditionally reject termination of a pregnancy as murder. Their opposition to abortion was rooted in their equating the protection of the fetus with the defense of the weak and the vulnerable.[9] The sociologist Kristin Luker, who studied US pro-life and pro-choice groups in the early 1980s, discovered that while public opinion polls indicated that an abortion to prevent the birth of a severely impaired child was acceptable to more than four-fifths of Americans interviewed and had much higher approval ratings than an abortion for "social" reasons, this practice was seen as the least acceptable by pro-life activists. For them, an abortion for a fetal anomaly meant that a human being could be ranked along a scale of perfection, and that people who fall below a certain arbitrary standard of "fitness" could be excluded. Amniocentesis, they claimed, was a "selective genocide against the disabled."[10]

Feminists opposed PND from a different point of view. They supported women's aspiration to control their fertility, including through abortion, but at the same time viewed the selective termination of pregnancy as tantamount to transforming women into "producers" of healthy children. Such children will be able to contribute to the economic well-being of society, instead of being a "burden" on it.[11] Abortion for a fetal malformation, some feminists argued, is qualitatively different from an abortion for an unwanted pregnancy. The first they perceived as an unacceptable "quality control" of maternal productivity, while the second was seen as a legitimate exercise of woman's inalienable right to decide whether she wishes to be a mother in a given moment of her life.[12] Moderate opponents of abortion held an opposite view: they perceived abortion for "social reasons" as the totally unacceptable killing of a future child, but believed that in some cases a termination of pregnancy after a diagnosis of a severe fetal problem could be seen as a lesser evil. Feminists and moderate abortion opponents agreed nevertheless that an abortion for a fetal indication is qualitatively different from an abortion for refusal of maternity.

In the 1980s, abortion for a fetal indication acquired the label *eugenics*.[13] Some participants in the PND debate linked PND with positive eugenics and the parental dream of a "perfect child."[14] For others, the term *eugenics* pointed to links between PND and negative eugenics, especially the Nazi extermination of disabled people.[15] The linking of PND with negative eugenics, first advanced by pro-life activists and a small fraction of radical feminists, gained strength through its rapid adoption by disability rights activists, and its integration into their struggle against negative attitudes toward people with disabilities. However, the Danish sociologist Lene Koch points out the dangers of an indiscriminate use of the term *eugenics* to denounce tendencies one wishes to criticize:

> Eugenics is open to pejorative use because it is rarely, or only superficially, defined. We may begin by questioning the effects created by the rhetorical use of the term "eugenics." It seems that the reference to eugenics, perhaps precisely because it is poorly defined, serves the purpose of rendering the activity in question ethically unacceptable. . . . As long as we choose to remain ignorant of the history of eugenics, the term will remain a demon available to all sorts of abuse. The witless reference to "eugenics" with no further specification is empty and more often a function of our own projections and intentions than a reference to history. In addition to its problematic uses as a reference to the past it has the dubious advantage in the present to be able to absorb all sorts of worries and fears from both sides of the genetic negotiation table.[16]

The historian of eugenics Diane Paul has put it more concisely: "To assert that a policy with undesirable effects is also 'eugenic' does not add anything substantive to the accusation. What it does add is emotional charge."[17] The emotional charge of associating PND with Nazi extermination of disabled people steered debates about this diagnostic approach toward discussions about broad moral principles—and away from examining the technical aspects of PND, its contextualized uses, and the interests, including financial, involved in the dissemination of this biomedical technology.

Birth Defects and Disability

Birth defects include conditions defined mostly as a disability, and those defined mostly as a chronic disease. The term *chronic disease*, the historian of medicine George Weisz had shown, first appeared in the United States in the 1920s and 1930s. The rise of a specific category known as chronic disease was a consequence of attempts to plan and control health costs and address the growing health needs of aging populations.[18] In the United Kingdom and in France, this category was introduced only after World War II. Moreover, entities defined as chronic diseases in the United States, the United

Kingdom, and France were not identical: situated uses of this classificatory category reflected local, social, and political needs. The term *disability*—and its earlier variant, *handicap*—had also arisen as a situated answer to specific social and political problems, above all a need to manage the consequences of human-induced impairments: wars and workplace accidents. The term *disability*, the historian Beth Linder proposes, became increasingly popular throughout the twentieth century, largely because of its usage in the emerging social welfare state. In the early twentieth century, *disability* became a household word for American families with next of kin who had served in the Union army during the American Civil War. Upon their military discharge, injured soldiers would receive "disability ratings" according to a schedule used by the federal government to assign monetary worth to body parts lost in battle. The loss of each body part was correlated to the impact such a loss would have on a man's ability to perform manual labor.[19]

In the twentieth century, *disability* was linked with the provision of invalidity pensions and free or subsidized health care to veterans and people harmed by their work conditions. Disabled/handicapped people were expected to receive compensation that would make them more equal to able-bodied individuals. Like handicapping in sports, such compensation aimed to offset an unequal capacity to compete in the labor market.[20] Thus, from the early days of its introduction, the term *disability* had a bureaucratic implication: access to specific rights and privileges.[21] In the first half of the twentieth century, this term covered permanent impairments and chronic diseases. Some of the most commonly cited disabilities to warrant payment from the US Civil War–era Pension Bureau were chronic diarrhea, tuberculosis (consumption), asthma, epilepsy, hernia, rheumatism, and malaria.[22] The inclusion of infectious diseases in this list can be explained by the permanent physical, sensorial, and intellectual impairment caused by such diseases: for example, smallpox caused blindness, and tuberculosis caused irreversible malformations of the spine.[23] Moreover, pathologies such as tuberculosis or syphilis were seen as one of the main causes of hereditary birth defects.[24]

Chronic diseases were lumped with disability as late as the 1980s. One of the first peer-reviewed journals in the domain of disability, established in 1982, was *Disability Studies and Chronic Disease Quarterly*. In 1985, its editors decided to remove the words *and chronic disease* from the title.[25] This decision to dissociate disability from chronic disease was related to the rapid development of disability studies as a distinct area of scholarship, and a parallel rejection by the disability rights community of the medical model of disability, which was replaced by the social model of disability. The medical model of disability, scholars and activists maintained, conceptualizes disability as a long-term or permanent illness or injury and proposes to "fix" it, or at least to attenuate its negative effects at the level of individual functioning. It is a model based on the assumption that an impairment is a deficit that affects every aspect of the life of the person with the disability. Having a disability is associated with the need for medical treatment, financial help, psychological support, and adequate care.

By contrast, the social model of disability stresses that the problems of disabled people originate primarily in prejudices and physical barriers created collectively by able-bodied people, not in the presence of a specific impairment. Physiological impairment is a biological reality, but disability is above all the product of a disabling society, and can be eliminated through social interventions such as adapted education, living conditions, and care—and especially the disappearance of prejudices and discrimination. The medical model became anathema for those disability scholars reluctant to identify disability with disease. As a consequence, these scholars tend to focus on the "healthy disabled"—those who do not use medical services and who can best approximate the activities of nondisabled people.[26] People such as these, especially if they do not suffer from severe physical or intellectual limitations, are also those most likely to live on their own, hold jobs, and participate in social life. The focus on healthy disabled may stem from the importance of groups such as the Deaf community in developing the argument that one should speak about difference or different ways of being in the world rather than "disability." It may also stem from the important role of organizations of people with stable physical impairments, such as the powerful Association of French Paralyzed People (l'Association des Paralysés de France), within the disability movement. As Susan Wendell argues, by minimizing the struggles of illness, these disability activists led many disability scholars to neglect the realities of an impairment coupled with a disease. Consequently, the unhealthy disabled—who seek out medical interventions and live their lives frustrated and disheartened by pain, fatigue, depression, and chronic illness—were relegated to the margins of the disability movement.[27]

The opposition between advocates of the medical and the social models of disability is often not absolute. Disability activists who resist the medical model of disability may at the same time seek recognition from the medical establishment of their special needs, because such an "official" recognition initiates opportunities for receiving targeted help. Societies that tolerate high levels of inequality in other domains such as education or living conditions still aspire to provide equal, or at least not strikingly unequal, access to health care. Biological misfortune is often judged differently from economic or cultural misfortune.[28] In an era of "bureaucratized medicine," an official recognition of disability, which often includes a specific diagnosis, is an essential first step in gaining access to institutional advantages.[29] Parents of children with genetic impairments may resist a reductionist definition of their child's condition, yet at the same time eagerly use it to obtain access to special medical and educational services.[30] Nonetheless, people with certain inborn disabilities/diseases increasingly define their condition not as a dis/ease but as an important element of their identity.

Disability activism facilitated the rise of groups of individuals linked through a recognition of shared genetic heritage, a phenomenon named "biosociality" by the anthropologist Paul Rabinow."[31] In turn, the rise of disability-focused identity policies intensified opposition to PND, in the name of defending human diversity and rejecting a utilitarian view of human beings which measures people only according to their productivity. From the late 1980s onward, disability rights activists have become the most visible opponents of PND and selective abortion, and the most powerful advocates of the equation of pregnancy termination for a fetal indication with a "eugenic" extermination of disabled people. The grounding of disapproval of PND in broad moral principles in turn blurred distinctions between highly variable situations, settings, and interventions.

Disability Rights and PND

Early debates about PND were firmly situated within a public health agenda and were guided by an explicit aspiration to prevent the birth of children with disabilities.[32] In the 1960s, 1970s, and 1980s, this argument was also supported by mothers/parents of disabled children. Mothers of children with hereditary diseases such as hemophilia or thalassemia pressed researchers to find a way to diagnose this condition before birth to allow them to have healthy children, and were willing to go to great lengths to prevent the birth of another disabled child. Mothers of children harmed by rubella fought for better state services for their children, and also backed the development of anti-rubella vaccine.[33] Then from the 1990s onward, a positive view toward efforts to limit the birth of impaired children was replaced with a strong critique of such efforts by disability rights activists.[34] The feminist and disability rights activist Marsha Saxton, one of the most eloquent promoters of this opposition, explained:

> The message at the heart of widespread selective abortion on the basis of prenatal diagnosis is the greatest insult: some of us are "too flawed" at our very DNA core to exist, unworthy of being born. This message is painful to

confront. It seems tempting to take on easier battles or even just to give in. But fighting for this issue, our right and worthiness to be born, is the fundamental challenge to disability oppression; it underpins our most basic claim to justice and equality: we are indeed worthy of being born, we are worth the help and expense, and we know it! The great opportunity with this issue is to think and act and take leadership in the place where feminism, disability rights and human liberation meet.[35]

The German advocate of disability rights Theresia Degener, born without limbs because her mother took the drug thalidomide during pregnancy, has argued that the introduction of prenatal testing strongly resonates with eugenics aspirations and is contrary to a feminist ethos.[36] PND, Degener explains, "transformed pregnancy into a medical production process in which women, at most, constitute the means of production, with production management having long since passed into the hands of gynecologists and human geneticists."[37] A selective abortion of a malformed fetus is qualitatively different from a termination of an unwanted pregnancy. While the latter is a reaction to elements in the woman's life that are unrelated to the fetus—such as the woman's living conditions, family relationships, and how she wants to shape her life—termination of pregnancy for a fetal indication reflects a wish to opt for a so-called normal child and reject a disabled one. The widespread acceptance of the definition of what a normal child/human being is, Degener concludes, is very dangerous, because it can be implemented only within the framework of a politically motivated control program that potentially militates against the interests of all people—men and women, able-bodied and disabled.

Many disability rights activists have strongly protested against an automatic and unthinking description of the birth of a disabled child as a "tragedy," a description they regard as insulting to all impaired people. Some extended this protest to an argument that the widespread use of PND for a fetal malformation—whatever it may be—is an implicit statement that life with a disability is worthless.[38] This emotionally powerful argument was later renamed the "expressivist objection" to PND and selective abortion.[39] Such an objection, when not connected with opposition to abortion in general, is grounded in the assumption that a woman who discovers that she is pregnant always instantaneously knows whether this is an unwanted or a wanted pregnancy. In the first case, she has the right to decide whether she wants to be a mother; in the second, she has a duty to accept the fetus she is carrying, independently of the traits of her future child.[40] This view also assumes that a pregnant woman either immediately and unconditionally

accepts a disabled child, or swiftly rejects this child.[41] In summarizing the expressivist objection to PND, the disability scholar Claudia Malacrida explains that an abortion of "any" child should be a woman's right, while abortion of "this" (disabled) child should not.[42]

The development of the expressivist objection to PND was closely associated with the rise of a social model of disability. Not so long ago, advocates of this model argue, Black people were viewed as born with a "racial handicap," that is, endowed with lower intellectual capacities and an inferior moral sense compared with White people, while homosexuals were perceived as sick people who suffered from a psychiatric disorder. When homophobic and racist views became unacceptable, Blacks and homosexuals were "cured" of their ills. The same will happen, disability activists propose, when a "healthist" and "ableist" society recognizes that nobody is immune from accidents, sickness, and old age, and learns to perceive the variability of human shapes and abilities as an asset, not a problem. People will continue to have different bodies that will function in many ways, but these differences will be not seen anymore as obstacles to full participation in society. Prenatal testing for fetal anomalies will then become unnecessary.

Scholars such as Marsha Saxton, Theresia Degener, and Adrienne Asch, who promoted this point of view, developed fine-grained arguments and engaged in stimulating debates with colleagues who questioned their arguments. Other disability rights activists developed a less nuanced discourse grounded in an analogy between PND and Nazi-style aspirations to exterminate disabled people. Such a view may reflect a long-standing frustration with negative attitudes of many health professionals toward disability, the slow progress of disability rights, and the persistence of discrimination against disabled people. The growing acceptance of the expressivist objection to PND may also be associated with its effective promotion by opponents of abortion, who coupled their own strong objection to the classification of human beings along a scale of perfection with arguments developed by disability rights activists, such as a critique of a utilitarian, neoliberal society in which only "productive" people are seen as entitled to full human rights, while those unable to contribute to the collective well-being through the production of goods or supply of services are perceived as a "burden."

Advocates of the expressivist objection propose a radical distinction between an abortion of an unwanted pregnancy and an abortion of a wanted pregnancy following a diagnosis of a fetal anomaly. Those who reject this approach believe that a woman's decision whether to continue a pregnancy is always situated. A woman's refusal to give birth to a special-needs child, they argue, is not qualitatively different from a woman's refusal to become a mother in a specific moment of her life.[43] Arguments employed by abortion opponents against termination of pregnancy for social reasons—defending a life-enhancing attitude that glorifies love and care, and radically rejecting a shallow, materialistic "death culture" in which a woman feels free to kill her future child for self-centered reasons such as professional aspirations or considerations relative to the material well-being of her family—are not very different from those employed to criticize women who terminate a pregnancy for a fetal anomaly.[44] In both cases, women decide that they cannot cope—for whatever reason—with additional tasks of motherhood, or "special motherhood." And in both cases, those who condemn abortions put to the fore the woman's selfish attitude. When Pope Francis was asked in February 2016 whether a pregnant woman who learns that the fetus has a severe brain malformation caused by an infection with the Zika virus could consider an abortion, he answered that abortion is "what the mafia does," that is, "a crime, an absolute evil." He then explained, "You kill one person to save another, in the best case scenario. Or to live comfortably, no?"[45]

The objection of disability rights activists to the presentation of a birth of a severely impaired child in a negative light occasionally affected the discourse of pro-choice activists who defend women's right to interrupt a pregnancy. One of the main Brazilian militants in favor of abortion rights, Debora Diniz, has criticized an article proposing that Brazilian women who learn that the fetus they carry is severely malformed after infection with the Zika virus should be able to decide to terminate the pregnancy "in light of the severity of the malformations being identified (not just neurological but also of hearing and sight), with likely extreme negative consequences for the families affected."[46] Diniz strongly disagrees with this presentation of the dilemmas produced by Zika. She and her colleagues, she explains, are submitting a petition to the Brazilian supreme court to allow pregnant women infected with Zika to choose termination, because "women have the right to decide to be freed of psychological torture imposed by the epidemic. It is not the fetus's future impairments or the 'extreme negative consequences for the families affected' that moves our demand, but the urgency to protect women's rights in the epidemics."[47]

A woman's presumed wish to "live comfortably," criticized by Pope Francis, may include an apprehension that she may be obliged to radically change her life and dedicate herself to the care of a disabled child. The anthropologist Annemarie Mol decided to undergo amniocentesis because, she explains, "given where I am—I have a healthy child and work that fascinates me and it is difficult enough as it is to juggle between them—I follow the advice [to test for Down risk if the woman is over the age of thirty-five]."[48]

Mol's main points are "given where I am" and "juggle." Decisions about whether to continue a pregnancy, be it with a "normal" or an "impaired" fetus, may reveal "how close to the edge many parents feel when they imagine the juggling of work and family obligations should disability enter an already tight domestic economy."[49] Asked about the reasons for their decision to either continue or terminate a pregnancy after a diagnosis of a fetal impairment, French women focused above all on their capacity to welcome this child into their family and integrate her/him into the web of its social and affective relationships in a given moment of their lives. Another important argument was the (potential) suffering of the future child, even when the experts classified the predicted impairments as minor or moderate. As several prospective parents put it, life on this planet is difficult even without the additional problem of having to deal with a disability.[50]

Orthodox Jewish rabbis implicitly acknowledge the situated character of prenatal decisions. Orthodox Judaism opposes abortion for a fetal indication, but such an opposition is not absolute. The Israeli anthropologist Tsipy Ivry describes two ultra-Orthodox Jewish women who received a PND of Down syndrome. One explained to the rabbinic counselor that she will be fine, and her family can raise an impaired child if this is what God sends them. The community rabbi confirmed that this woman did not overestimate her and her family's strength. In a similar case, a woman diagnosed as having a Down syndrome fetus expressed a fear that a Down syndrome child would "ruin" her family. Her community rabbi confirmed that the second woman's family "are strong people; but not in this area, they're not going to withstand this." The rabbinic counselor ruled accordingly that she could have an abortion. He also told her that "although it is you undergoing the procedure [abortion], the one who shouldn't sleep at night is me, not you; I gave you the halachic [the Judaic religious law] permission." The woman, he added, should know that "she has acted according to the halacha and it's okay."[51]

A woman who feels that she cannot cope with the needs of a disabled child in a given moment in her life may feel differently if her life circumstances change.[52] This is especially true for moderate or "borderline" impairments. A French woman who decided to interrupt the pregnancy after finding out that the fetus lacked a limb explained why: she already had one impaired child, and her husband had just been diagnosed as having Hodgkin's lymphoma and begun intensive chemotherapy. She felt unable to cope with the additional stress of a disabled newborn. Another woman who faced the same decision decided to continue the pregnancy, because her family was fine and she already had two healthy children. Moreover, she was reminded by her mother that the mother's grandfather had lost a hand during World War I: it was possible to inscribe the child's missing limb in the family's history. Both women ended their interview by stating that under different circumstances, they could have made the opposite decision.[53]

Alison Piepmeier's study of the decisions of women who learned they were carrying a Down syndrome fetus led to a similar conclusion.[54] Piepmeier has a daughter with Down syndrome, and her study of decisions after a positive PND was informed by her personal experience. Her initial assumption was that in mapping the decision-making process, she would find important differences between women who elected to terminate the pregnancy and those who decided to continue it, especially as regards the timing of the assignation of personhood to the fetus. She assumed that women who terminated their pregnancy saw the fetus mainly as an abstract biological entity, while those who decided to continue their pregnancy strongly identified the fetus as their child. However, she did not find any qualitative difference between the decision-making processes of these two groups. All the interviewed women found the process extremely stressful and painful, and all explained that they had to decide about their child's (never the fetus's) fate without knowing how severe the child's disability would be and how it would affect their family's dynamics. As Piepmeier points out, however, similar considerations and feelings can nevertheless lead to diametrically opposed decisions. These decisions, she explains, are always situated, and many women among those who elected to continue the pregnancy and those who chose to terminate it stressed that there is no "right" choice.[55]

The expressivist objection to PND, summed up in the influential volume *Prenatal Testing and Disability Rights*, edited in 2000 by Erik Parens and Adrienne Asch, relies on two arguments: moral and epistemological.[56] PND, promoters of the expressivist objection explain, is morally wrong because it harms people with disabilities and their families, places a lower value on "imperfect lives," and implicitly condemns women who fail to prevent such imperfect lives.[57] It is also epistemologically wrong because it is grounded in false knowledge. Medical experts who promote the selective abortion of impaired fetuses have dramatically inaccurate views of the quality of life of people with disabilities. Physicians' opinions reflect their professional prejudice, and are radically divorced from the lived experience of disabled people.[58] The only reliable experts on a given inborn condition are the people who have firsthand experience of life with this condition: disabled people and those who care for them.

In his review of *Prenatal Testing and Disability Rights*, the bioethicist Paul Ford argues that while this volume raises many important questions, it is mostly about disability, not prenatal testing. The book provides an excellent description of the social injustice experienced by people with disabilities and their emotional reactions to their current needs. The book's discussion of PND is more problematic:

> The prenatal testing debate presented in this book is made muddy on at least three accounts: the use of an extremely broad and inclusive definition of disability, the treatment of the moral question of abortion, and the channeling of social injustices experienced by those with disabilities. Although a part of the debate is always about what is meant by disabilities, the term is used in very different ways in this book; at times it means something like trisomy 13 (a lethal mutation), and other times it means something like deafness. This leads to a great deal of miscommunication and misunderstanding.[59]

In his thoughtful comments on the expressivist objection and the diametrically opposed injunction of some bioethicists such as the philosopher Dan Brock that, given a choice, parents may have an obligation to give birth to a child who is free of impairment in order to increase the child's chances in life, the disability rights activist Tom Shakespeare similarly affirms that developing a general discourse on disability and PND is impossible. Disabilities are different, as are the families that make decisions about the fate of pregnancies. It does not make much sense to treat Tay-Sachs disease, Down syndrome, deafness, and cleft palate in the same way. Many of the problems produced today by the presence of an impaired child result from insufficient support for disabled people and their families. Such support, and not the opposition to PND, should be the focus of disability rights activists. The fear that PND will eliminate human diversity, Shakespeare adds, is an imaginary preoccupation. Less than 10% of all disabled individuals have inborn impairments. Even if the prenatal detection of such impairments was 100% effective—and this is very far from being the case today—it could not produce a disability-free society or even one with a much lower proportion of disabled people. Another argument employed by PND critics, that a selective abortion of a fetus with a given disability badly hurts the feelings of people with this disability, is understandable from an emotional point of view, but is insufficient justification for limiting parental choices. Eating meat, Shakespeare adds, may badly hurt vegetarians' feelings, but this is not an adequate reason for restricting other people's dietetic preferences.[60]

Some feminist scholars have proposed that the expressivist objection to PND adequately captures the difficulties of disabled people while sustaining the important concept of the unique worth of every human being. At the same time, the demand of unconditional devotion from mothers overlooks how such a demand compromises women's autonomy. It disconnects parenting from its social context and disregards major differences in the level of support provided to families of impaired children. It also disregards that in some circumstances, raising a child with a severe disability obliges the child's mother to spend the rest of her life as the main caregiver for this child, and may also deplete the family's financial resources. Our understanding of the ethics of maternity and parenting, these scholars believe, has been deeply influenced by an ideology of motherhood that privatizes child care and prescribes maternal self-sacrifice as part of the natural female role. When societies do not provide adequate care for all their disabled members, a victory for one discriminated group, people with disabilities, may be obtained through the subjugation of another discriminated group, women.[61]

Another problematic aspect of the expressivist objection is its implicit equation of a fetus with an already existing child, an attitude which strongly resonates with the views of radical opponents of abortion. Yet many future mothers/parents distinguish between a potential and an already existing child. In the last chapter of his book *Far from the Tree*, a compassionate and often moving description of parents' experiences with their "differently abled" children, Andrew Solomon talks about the birth of his and his husband's son, conceived with Solomon's sperm and a donor's egg and carried by a surrogate mother. The egg donor was chosen through a careful selection process, which, Solomon recognizes, was quasi-eugenic. He and his partner looked for intelligence, character, health, and appearance: "I did not want to devaluate the extraordinary lives I have come to respect, yet I could not deny that I wanted a child who will be familiar enough so that we could soothe him or her with our mutualities."[62] In the early stages of their "child project," the couple also asked the surrogate mother to undergo screening for Down syndrome risk and other fetal anomalies, implicitly considering an abortion in case of a positive result. By contrast, when after the birth of their child the physicians suspected (erroneously) that the newborn might have suffered bleeding to his brain, Solomon immediately and with great emotion identified himself with his son—that is, a child, not a fetus.[63]

Uncertain Prognosis and Intellectual Impairment

Debates about PND frequently focus on the moral dilemmas generated by this diagnostic approach. PND, the bioethicist Arthur Caplan explained, produces an irreducible opposition between individuals' right to make choices freely about their reproductive and procreative behaviors, including the right to as much information as possible, and the morally contentious option of terminating the pregnancy, which may result from obtaining certain information, and which includes an implicit assumption that information about the fetus produced by prenatal tests is always reliable and unambiguous.[64] His main concern is the use of this information, especially the risk of unintended outcomes, which may result from the rapidly growing capacity to identify specific fetal traits before birth, increasingly blurring the line between normality and disorder. This advanced technology may also lead to parental indulgence in whims, fancies, and biases: "To the extent to which the lines between choice and whim remain vague, the potential exists for prenatal testing to be enmeshed in the pursuit of the frivolous or to be put in the service of ignorance, prejudice and bigotry."[65]

Caplan, like many other bioethicists, is mainly preoccupied with events that may unfold in an unspecified future.[66] Such concerns are remote from the dilemmas of most pregnant women who undergo PND. In the early twenty-first century, not many prospective parents are so preoccupied with their future child's physical beauty or musical abilities that they would consider an abortion if the fetus does not have the desired traits or talents. By contrast, many prospective parents are being confronted with the diagnosis of a condition with an uncertain prognosis, or a small but significant risk of a severe impairment of the fetus. Fears of such outcomes may only be intensified by the rapid expansion of the scope of prenatal tests.[67] Many pregnant women and their partners are compelled to make a difficult decision about the management of risks for their future child.[68] Moreover, the definition of an acceptable risk of disability is not the same everywhere: the "tyranny of [prenatal] diagnosis" is often a situated phenomenon.[69] In France, many hospital ethics committees (CPDPNs) perceive a 10% risk of a severe mental impairment as enough to justify a woman's request to terminate a pregnancy, while in Germany genetic counselors attempt to dissuade women from having abortions for risk of intellectual impairment even when such risk greatly exceeds 10%.[70]

The risk of disability for the child is at the same time the risk that a disabled child's care will have negative consequences for the family.[71] Disability rights activists, who rightly protest against a systematic and unthinking presentation of the birth of an impaired child as a "tragedy," may be reluctant to recognize the great diversity of inborn disabilities and their consequences. In the early days of PND, promoters of this approach discussed the financial and emotional costs of care for severely impaired children, but this topic became less visible later, partly because of pressure from activists who insisted on the positive value of educating children with disabilities and the importance of public policies that support impaired people.[72] Few people object to the principle that society should provide sufficient help to disabled persons and their families. But acceptance of this generous principle is hampered by practical difficulties in fulfilling all the urgent and often competing societal needs. In the meantime, in the great majority of societies families continue to carry the main responsibility for caring for special-needs children and, not infrequently, impaired non-autonomous adults. The material and emotional costs of such care vary greatly: they depend on the family's socioeconomic status, relations between their members, their values and beliefs, their psychological makeup, and the precise nature of the disabled person's physical and emotional problems—a child with a partial visual impairment does not need the same investment in care as a child with severe cerebral palsy.

People with significant learning/intellectual disabilities may need an especially high level of intensive maternal/parental investment in their care. Many genetic anomalies detected by PND are linked to significant intellectual impairments, an issue frequently evoked in early debates about PND.[73] The same is true for many structural anomalies of the fetus detected by diagnostic ultrasound. For example, in 2015 scientists discovered that inborn heart defects are associated with a high risk of neurodevelopmental disabilities.[74] Today, scientists and activists often employ the term *learning difficulties* to describe inborn cognitive problems. This term may be misleading, because often the main problem of people with such problems is not their inability to undertake advanced studies, master complex topics, or even hold a full-time job, but their difficulty in developing meaningful interactions with others, living an independent life, and protecting themselves from abuse and self-harm. When the dominant discourse shifted from a uniformly negative picture of children with a disability, especially intellectual, to a focus on inclusion and progress, it became increasingly difficult to admit that some of these children make only limited progress and are violent, are poorly adjusted, or experience mental health or medical problems along with intellectual disability, and that some families break apart under the pressure of these issues.[75] According to the autism activist Mark Osteen,

> Disability studies' adherence to the social-constructionist model, with its heavy debt to Foucault, had helped to foster a set of biases and misrepresentations that, ironically, replicate those historically aimed at disabled people. First, disability studies has been unwilling or unable to theorize impairments, suffering and pain—somatic conditions that accompany or precede disability—or to theorize the body itself, perhaps because doing so would seem to yield the floor to medicine. Second, Foucauldian paradigms that view the subjects as

pure products of competition for power minimize human agency. Third, the breadth of its interdisciplinary research has swept together a wide array of different conditions and embodiments, many of which have little in common. Fourth, disability studies' focus on visible physical disabilities has blinded it to the existence of other, perhaps less obvious but equally significant, (and widespread) disabilities. Last, and most important . . . disability scholarship has ignored cognitive, intellectual or neurological disabilities therefore excluding the intellectually disabled, just as mainstream society has done.[76]

Ideally, the sociologist Gil Eyal and his collaborators propose, society should create a "prosthetic environment" that will allow people on the broad spectrum of autism, intellectual disability, and mental disease to be safe and thrive, but it also should extend the boundaries of such a prosthetic environment and promote a greater range of social environments and experiences. Society, they affirm, should adapt to the needs of people that are different and will not stop being different. All children and all adults with developmental delays should receive all the help they need to fully reach their potential and, later, all the collective help they need to be fully integrated in society.[77]

Yet helping people with intellectual disabilities to live in a safe and supportive environment, without a doubt a worthy goal, depends on the level of public investment provided to achieve that goal. Most industrialized countries, and all developing and intermediary ones, struggle with escalating health care costs and difficulties in providing the right level of care to aging populations, while the overall trend in many countries seems to be (in 2018) the dwindling of resources for the welfare state. In such a situation, increasing help to one group of needy people often leads to a decrease in the level of assistance to other groups of equally needy people. It does not seem very likely that in the near future societies will be able and/or willing to provide the ample resources necessary for the creation of a safe "prosthetic environment" for all the people with intellectual disabilities.[78] Absent such an environment, the care of people with intellectual impairments often depends on their parents' (often mainly mothers') skill in navigating bureaucratic mazes, persistence, and above all intense investment of energy and time. Middle-class mothers can delegate some of their care tasks; low-income mothers often do not have that opportunity. In addition, even affluent industrialized societies rarely relieve parents of impaired and dependent children from worrying about their child's future when they become incapacitated or die.

Societies have made major advances in providing help to people with intellectual disabilities. These advances are summarized in a journal article,

"I Am John," which recapitulates, through a synthetic narrative of a man with "mental retardation," the different stages of dealing with people with this condition in the United States.[79] John's narrative ends with an account of recent developments that enable people with intellectual disabilities to live autonomous or semiautonomous, accomplished lives:

> I live in an apartment. I'm six feet tall and have brown hair and brown eyes. I work hard in a restaurant. I bowl on Saturdays and swim on Sundays. In the summer, I play on a softball team. I like rock-and-roll music and pizza. I vote. I pay taxes. I have a girlfriend. We go to rock concerts and dances. I also have mental retardation. I am John.[80]

One sentence is missing from this description: "I'm the best-case scenario." "John" is mildly intellectually impaired. He is able to live a quasi-independent life, has friends and a partner, and holds a job. He also, one can assume, received effective help that enabled him to live such a life.[81] Not all people with intellectual disabilities are so lucky.

Andrew Solomon sums up the unique predicament of parents of intellectually impaired children:

> In typical circumstances, to have children who won't care for you in your dotage is to be a King Lear. Disability changes the reciprocity equation; severely disabled adults may still require attention in midlife, while other grown-up children are attending to their own parents. The most effortful stages of dealing with a child with special needs are generally held to be his first decade, when the situation is still novel and confusing, the second decade, because cognizant disabled adolescents, like most teenagers, feel the need to defy their parents, and the decade when the parents become too impaired to continue to provide care and worry acutely about what will happen to their child after they are gone. This account fails, however, to reflect that the first decade does not vary so much from the norm as the subsequent do. Taking care of the helpless disabled infant is similar to caring for a helpless nondisabled one, but continuing to tend to a dependent adult requires a special valor. . . . One mother of a twenty-year-old with severe disabilities said to me, "It's as if I'd had a baby every year for the past twenty years—and who would choose to do that?"[82]

The disability activist and mother of a severely disabled child Helen Featherstone provided a thoughtful and balanced view of the rewards but also the difficulties of raising a child with serious developmental delays.[83] She argued that in some cases, parents tend to exaggerate their child's abilities and character in order to survive psychologically in a difficult situation, a stance she classified as a necessary delusion.[84] Such cases, one may argue, are not "delusions" but examples of good adaptation. Parents adjusted their expectations about their child to a different reality and learned to enjoy this child as s/he is, an excellent coping strategy when it works.[85]

The care of a disabled child can be a source of deep satisfaction. It can also be very challenging. Only rarely, however, do parents of severely impaired children criticize the masking of some of the harsh realities of their children's lives:

> Where is the description of the months or years of grueling hospitalization with the associated gastrostomy tubes, jejunostomy tubes, and fundoplications; the tracheostomies, shunts, and orthopedic, eye, and brain surgeries; hyperalimentation, oxygen tanks, and ventilators? Similarly, there was no mention of bankruptcies, divorces, mental and physical breakdowns, deaths in late childhood, neglected siblings, and suicides caused by the extreme burdens of caring for severely medically and developmentally compromised children.[86]

The activist Helen Harrison explained that it may be difficult for parents of impaired children to speak openly about their daily struggles:

> Upon becoming parents of a disabled or "high-risk" child, one of the first things we learn to do is lie—to our friends and family, to the doctors, to our child, and to ourselves. We quickly learn that others do not want an honest answer when they ask, "How are you (or your child) doing?" and we oblige by giving the positive and politically correct answer. . . . We don't lie just to reassure others. An arguably more important motive is the need to comfort ourselves and give positive meaning to the immense physical and emotional difficulties of our lives. . . . We lie to deny, or at least postpone, unpleasant realities. We believe that our children's problems can be overcome with therapies, interventions, and, of course, the "right" parental attitude. . . .
>
> Our children's doctors and therapists instruct us in euphemisms: our children do not have cerebral palsy or autism, they have "tight muscles" or problems with "sensory integration." Our children are not retarded, but "developmentally delayed." Disability rights advocates caution us against using phrases that begin with the words "suffering from," even when our children's condition involves substantial pain. We learn to keep a straight face when we describe our children as "handi-capable," "not typically developing," or "severely differently-abled." . . . Parents who express unhappiness about the conditions of their children's lives (and their own) may find themselves barred from

support groups for "negativity." Physicians and therapists label such parents as "angry" and "embittered" and refer them for counseling and even for investigation as potential child abusers. . . . It is clearly dangerous to challenge other people's coping mechanisms.[87]

Today, disability activists focus on families' capacity to adapt and on the advantages and gratifications of raising disabled children, especially those with an intellectual impairment.[88] Researchers who are also parents of disabled children have provided thoughtful, fine-grained, and poignant descriptions of "special parenthood," its struggles and its blessings.[89] At the same time, many health experts continue to view the education of children with such impairments as a source of serious difficulties for their caregivers.[90] One can argue that both viewpoints are accurate but not for the same people, and sometimes for the same people but not at the same moment in their life. Some people are willing to take risks, thrive in risky situations, and derive immense satisfaction from overcoming obstacles and challenges; others are risk-averse and perform less well under stress. Some people find out that a situation they initially thought would be intolerable turned out to be not only acceptable but deeply fulfilling; others find that the situation is indeed very difficult. PND often cannot predict the future child's precise risk of impairment. It also cannot predict the consequences of the presence of an impaired child for the family. Disability activists usually point out that predicting the future of any child, or the family dynamic after the birth of a child, is impossible. This is an accurate statement, but since the "taming of chance" in the nineteenth century, our societies have calculated risks, and some risks have been perceived as higher than others.[91] The birth of a child with Down syndrome may be viewed as such a risky event.

The Down Syndrome Conundrum

The transformation of PND into population-based prenatal screening was to an important extent driven by the desire to prevent Down syndrome (DS).[92] "Screening for Down" also became a focus of struggles against the selective abortion of impaired fetuses. Parents of children with DS (or, to be more accurate, some parents of children with this condition) are often at the forefront of the opposition to abortion for this indication. Articles written by these parents often describe the child as lively, funny, bright, and happy.[93] Parents of DS children also complain, with excellent reason, about the excessively negative representation of DS and the persistence of prejudice against individuals with this condition.[94] Attitudes toward people with Down have partly changed thanks to public campaigns which present them as "different" rather than "retarded"; greater visibility of children and adults with this condition, especially those on the high end of DS-related intellectual impairment, in the public space and in the media; and accessible information about DS on the web.[95] Nevertheless, pregnant women who learn that they carry a DS fetus frequently opt for an abortion, including some who initially opposed termination of pregnancy for this indication.[96]

A meta-analysis of studies on women's decisions following DS diagnoses found out that when the participants were prospective parents recruited from the general population, 23–33% said that they would terminate; when the participants were pregnant women at increased risk of having a child with DS, 46–86% affirmed they would terminate; and in practice, 89–97% of women who received a positive diagnosis of fetal DS during the prenatal period chose termination. They justified this decision by their comprehension of increased difficulties for themselves as primary caregivers of a disabled child, additional problems for their other children, a potential strain on their marriage, fears that the child would never be independent, and worries about the fate of the child after their death.[97] A woman who aborted a DS fetus explains:

> I did do this thing for myself and my family. . . . I did not do it for him at all. Maybe he would have suffered. Maybe not. Maybe I did not want to imagine an isolated, lonely, depressed, retarded adult man. So maybe I did spare him that end. But life has no guarantees for any of us. We all face the possibilities of leading a sick, unhappy life. So I cannot insist I did it for him. I did it for me. I did it for my marriage. I did it for my other child. I did it for my home and job and way of life.[98]

A woman who terminated for DS and cardiac anomalies detected during an ultrasound reports:

> My husband and I thought we would keep the baby if it was "just Downs." I even decided against the triple screen test because I knew I would keep it. . . . But there is nothing like hearing those words to your face, and I am no longer confident what I would do in that situation. I never speak for myself anymore unless I'm actually in the situation because it's so, so hard to know how you would actually react. Reality of the fetal anomaly news hits so much harder than conceptual thinking takes into account.[99]

Another woman draws on her personal knowledge as a teacher of young people with DS to explain why she personally could not see herself as a mother of a child with this condition:

> I knew going in what it would mean to have a child with DS. I remember all too well how cute and sweet and loving the little ones could be. I also remembered how hard to manage the teen boys were. Not that all teen boys cannot be hard, but with DS kids it's a different story. I remember in my teaching that we had to make sure they did not get too bored. If they did, they would start masturbating or trying to touch other students. It was hard as a teacher. As a mother I could not imagine it . . . knowing what I knew about when they are no longer so "cute," well I chose not to have Leif.[100]

DS was and is at the very center of efforts to screen all pregnant women for fetal anomalies. In countries such as the Netherlands, a more positive societal attitude toward people with DS, a good level of support for people with this condition, and (probably) a valorization of women's care tasks are expressed in the low uptake of screening for this condition. If a woman does not consider the possibility of aborting a DS fetus, she will usually avoid potentially stressful and risky diagnostic tests. In such a context, professionals do not see testing for DS as a moral obligation for a pregnant woman, and value the advantages of not knowing. In Denmark or Israel, the norm promoted by professionals is that knowledge about the status of the fetus is a positive value, independent of the woman's attitude toward abortion for this condition, although the great majority of women terminates a pregnancy with a Down fetus.[101] In the United States, DS activists do not attempt to dissuade women from undergoing prenatal tests. Such tests are presented as important to help couples to prepare themselves for the arrival of a "special child." The activists' goal is to promote a positive image of people with DS and prevent selective abortion for this indication.[102]

In discussions about abortion for DS, both sides often implicitly assume the existence of a relatively homogenous entity called Down syndrome, and tend to present people with DS in either a uniformly positive or a uniformly negative light. However, experts who first studied DS (at that time called "mongolism" or "mongoloid idiocy," a term with strong racist undertones) were aware of the great variability in the health and capabilities of people with this disorder. People with DS frequently looked alike, but this superficial similarity masked important differences. In the first edition of his widely read *Textbook of Mental Deficiency* (1907), the British psychiatrist Alfred Tredgold pointed out that despite the great physical similarity of all the "mongols," people with this condition display a wide range of mental disabilities: "the milder members generally learn to read, write and perform simple duties with a fair amount of intelligence; the majority belong to medium grade of mental defect, a few are idiots" (that is, severely mentally impaired people).[103] After the redefinition of DS as trisomy 21, specialists stressed that while the presence of three copies of chromosome 21 invariably denotes DS, it fails to provide information about the severity of symptoms in a given individual. The British geneticist Lionel Penrose, one of the leading specialists on DS in the mid-twentieth century, attempted to uncover genetic markers that could be correlated with specific manifestations of DS.[104] He was unsuccessful, as were his followers. In 2018, too, PND of trisomy 21 does not indicate the level of impairment of a trisomic fetus and how much care the child will need.

Parents recruited for research on DS through Down syndrome associations emphasized the positive aspects of educating DS children.[105] Other parents of DS children tell more complex stories. Some who insist on their strong attachment to and unconditional love for their DS child are nevertheless critical of media stories about DS children that tell only how cute and sweet these children are, and fail to report the parents' difficulties: dealing with the severe health problems that occur in some DS children, confronting their children's tantrums and disruptive behavior, coping with their children's limited speech and incontinence, and facing the knowledge that their caregivers' obligations will never end.[106] Parents' objections to DS associations' focusing on high-performing DS people and disregarding children with more severe manifestations of this disorder are not new. In the 1960s, a British mother of a child with DS wrote to the British Down Syndrome Association: "Stuart is nearly six years old. He is not one of the more gifted Down children, in fact the blunt truth is that he is severely retarded. Despite our great efforts to help him to overcome his handicap, . . . he still cannot talk, dress or undress himself." Another mother complained that the association kept focusing on high achievers among children with Down, and that "they never write or say anything about low-graded Down. My daughter Kelly is 13 years old, but mentally only 2 years old, and I have never read anything from the Association that had anything to do with her, they only seem to like successes. I'm sorry if I sound very bitter, but it's how I feel very let down by the Association."[107]

Some parents think that the new rules about the autonomy and dignity of mentally impaired children are a double-edged sword. Despite the caregivers' good intentions, they can make the situation worse for these children. In the United Kingdom in 2012, a young woman with DS was placed by social services in a flat on her own. After six months, she had put on five stones (approximately seventy pounds or thirty-two kilograms). Her mother spoke to the social worker, who replied that it was her daughter's choice to eat all day long, then quoted the Human Rights Act. This young woman, her mother believes, is incapable of making such a choice and needs support and guidance, not another bag of potato chips. Alas, she added, individualized supervision and care are expensive, and the conservative British government reduced the funding of care for adults with disabilities. The replacement of the term *mental handicap* with *learning disability* was meant to destigmatize people with intellectual deficiencies, but in fact might have worsened their situation: "The impression created is that we are dealing with people who are just a little bit slow. . . . There is no allowance for the fact that a disabled child will become a disabled adult, seemingly no understanding that a lifelong learning disability is exactly that—lifelong."[108]

Many testimonies of parents and siblings of people with DS give a different, much more upbeat view. They convincingly show that trisomic people can lead happy, fulfilling lives, and that their parents and siblings perceive the experience of living with a DS person as an important source of personal enrichment.[109] On the other hand, the existence of serious educational and care challenges may increase the probability of a less than optimal outcome.[110] All the stakeholders in the polarized debates about PND of fetal anomalies agree to the principle that prospective parents should be able to receive objective and balanced information about the consequences of the detected anomaly for their future child. When dealing with highly variable conditions and with emotionally loaded subjects, the provision of "objective and balanced" information may become a very challenging task. This is nevertheless the goal of recent US laws.

The Prenatally and Postnatally Diagnosed Conditions Awareness Act (Public Law 110-34) was enacted into US federal law in 2008 to increase the provision of accurate, up-to-date, and balanced information about DS to women and families considering prenatal testing. This act, strongly supported by senators Edward Kennedy (Democrat, pro-choice) and Samuel Brownback (Republican, pro-life), is also known as the Kennedy-Brownback Act.[111] Several US states strengthened the federal law through passage of state laws conveying the same message. For example, a Massachusetts law (no. 03825) adopted in November 2011 explains that parents who receive prenatal or postnatal diagnosis of DS should receive at the same time up-to-date, evidence-based written information about DS that has been reviewed by medical experts and national DS associations.[112] A similar act promulgated by the State of Pennsylvania in June 2014, named Chloe's Act for a girl with DS who inspired the legislation, requires health care providers to make sure that women receive "a full range of factual and supportive information." The law was adopted by a rare bipartisan vote, with practically no opposition.

Laws that frame obligatory information about DS lump together two distinct situations: pregnant women who learn they are carrying a fetus with DS (today, increasingly during the first trimester of pregnancy) and may face difficult decisions about the pregnancy's fate, and parents who have just learned that their newborn child has DS and may need material and emotional help to cope with this potentially distressing news.[113] The message implicitly conveyed through the distribution of the same materials to these two different groups is that a DS fetus is identical to a DS child. These materials aim "to correct the incomplete information that leads many women to terminate their pregnancies after a diagnosis or screening."[114] Tool kits containing information for parents/prospective parents, such as the one produced by the National Down Syndrome Society, contain upbeat, and undoubtedly sincere, testimonies of parents and family members of DS children—those who are coping well with raising a child with this condition. Accordingly, they contain statements such as "The more I interact with someone who has Down syndrome, the more I think I am the one who has one chromosome less, instead of them having one extra. They tend to be loving, caring and forgiving—features we are missing a lot in general society."[115]

Recently, US anti-abortion activists attempted to bar abortions for DS. In 2013, such a bill was adopted by North Dakota, although no physicians were prosecuted under that law. In 2015, a similar bill was proposed by the Ohio National Right to Life Committee. Mothers of children with Down who testified in favor of passing this legislation spoke about the extinction and culling of DS people. The president of Right to Life explained:

> We all want to be born perfect, but none of us are, and everyone has a right to live, perfect or not. You go to any supermarket or mall and see these families who just happen to have a child with Down syndrome, and they will tell you how fortunate they are to have those children. Pretty soon, we're going to find the gene for autism. Are we going to abort for that, too?[116]

In 2015, the father of a DS girl—born after her parents had learned that the fetus was trisomic, decided to continue the pregnancy, and faced pressures from health professionals to reconsider their decision—wrote an op-ed in the *New York Times* criticizing the proposed Ohio law. Such conservative initiatives, he argues, are insensitive to the dilemmas of parents who had made a very difficult decision. Moreover, these are not accompanied by a proposal to extend services for special-needs children. The great majority of readers' comments on this piece agrees with the author's two main points: the need to allow women/parents to freely choose whether they want to give birth to a DS child, and the hypocrisy of some conservative politicians, who strongly oppose abortion for a fetal anomaly yet advocate for reducing state assistance for disabled children and adults.[117] These responses also offer contrasting views of DS. Some commentators, often parents or relatives of people with DS, present a very positive view of this condition:

> My son defies all the doom and gloom painted here. He is super healthy; works 28 hours a week at a job he has held sixteen years; makes his own meals except dinner, which he shares with us; pays rent and does his chores more reliably than any other family member; uses his iPad and iPhone to talk to his friends; and is generally a delight to be around.

> As the father of a 25-year-old with Down Syndrome, I want to assure all respondents offering a long list of rationalizations, that we're OK. Abortion is a personal choice, but the litany of catastrophic consequences attributable to Down Syndrome so many are offering are simultaneously sad, scientifically specious, and unnecessary.

> Naia is preparing to go to her high school (11th grade), where she is in an inclusive classroom. Her counselors and team tell us that she is on track for a full diploma.

> Twenty years ago my cousin chose to give birth to a child with Down's syndrome. That child is now doing well in college. She never has been a "terrible burden" on anyone, and given her track record, she likely never will be.[118]

Other people, often physicians and educators, give a more pessimistic view of DS:

> The poster child for Downs is a grinning little kid who does not reveal the spectrum of Downs disabilities. Yes, many can learn. But many are severely retarded and will never learn. Those who do learn to walk, talk, read a bit, and do simple jobs require a lifetime of extraordinarily hard work from parents. Inch by determined inch, progress is made. Then early-onset Alzheimer's turns it all around. . . . Parent after parent came to my neuropsychology office

(I'm now retired): "He used to be able to fold the laundry." "He used to talk to me." "She used to be able to get dressed on her own." . . . It would be cruel and immoral to force Down's on any pregnant woman. Instead, show them a poster of an adult "child" with Alzheimer's.

As a pro-choice woman and physician, there is no debate here for me. I have treated many children with Down syndrome. Some are happy and live fulfilling lives, but many have serious medical, neurodevelopmental and psychiatric issues. The kids we see on TV or hear human interest stories about are the lucky ones who did really well despite their diagnosis. No one is writing stories about the kids and families that I see.

I am a woman whose entire career was spent working with developmentally disabled children of all ages up to 21 years. . . . Children with Down Syndrome are sick more often and often more sick than their age mates without known genetic disorders. I know too that the incidence of divorce and abandonment of family is disproportionately high among families with children with handicapping disorders. . . . My evidence is not of one child but of hundreds of children and their families.[119]

A woman who decided to terminate a pregnancy with a DS fetus and describes the diagnosis and consecutive decision as the "darkest time in my life" makes a passionate plea:

Please don't pick on tri 21; this is a life-altering and life-threatening genetic condition and like in any other such condition, there are best-case and worst-case scenarios. I am often reminded and comforted by the poster I saw when I first entered the clinic—it said only, "TRUST WOMEN."[120]

Information packets provided as a consequence of Chloe's Law and similar legislation, the bioethicist Arthur Caplan argues, are strongly biased toward a positive view of DS. The message they convey is that DS children may have health and learning problems, but medical advances, devoted parenting, and societal resources will overcome the majority of such problems.[121] These laws, Caplan adds, are radically changing the ideal of value-neutral genetic counseling, because disability and pro-life groups want information that puts disability in a positive light and abortion in a negative light to become part of all counseling.[122] Although information about Down is theoretically expected to help the woman decide about the fate of her pregnancy, the voices of women who elected to terminate their pregnancy after a diagnosis of fetal DS are conspicuously absent from the information packets about this condition.[123]

Materials distributed to prospective parents in the United States rarely dwell on the severity of the medical problems found in approximately half the children with DS. They explain—accurately—that the health issues of these children are treatable, but not that in some cases the treatment may be long, difficult, and unsuccessful: about 10% of children with DS die before the age of five.[124] They seldom provide quantitative data about the severity of mental impairments among people with DS.[125] An additional, rarely discussed issue is the increased risk of sexual abuse for people with intellectual disabilities, especially women.[126] On the other end of the life spectrum, people with Down face a high risk of an early onset of dementia. Specialists estimate that 77% of people with DS aged sixty to sixty-nine years suffer from dementia, as compared with 2% of people of the same age group in the general population.[127] Genetic counselors trained in the United States in the 1990s were invited by the organizers of the training program to visit families with DS children, but not institutions for those with more severe impairments. They also were not invited to meet with parents whose DS child had died.[128]

DS activists often quote a modern fable, "Welcome to Holland," that compares the parenting of a DS child to people who planned a trip to Italy but unexpectedly land in the Netherlands. Bitterly disappointed at first, they gradually discover the Netherlands' quiet charm and hidden treasures.[129] Taking into account the variability of effects produced by the presence of three copies of chromosome 21, it might have been more accurate to tell a story about people who think they are going to Italy, then learn in midair that their original destination has changed and that it is impossible to know where their airplane will land: the less spectacular but enchanting and peaceful Netherlands, the beautiful but more challenging Albania, or possibly, a conflict-torn Syria.[130]

Differences in the capacity of people with DS to live autonomous or semiautonomous lives have important repercussions for their caregivers, who often are the mothers/parents. In one study, the lowest IQ found in DS people aged twenty-one to forty-two years was 8 and the highest 67. It also found that 40% percent of people with Down recruited for the study were able to read, about a quarter could stay by themselves during the whole day, and two-thirds were able to feed, dress, and bathe themselves. By contrast, approximately 25% were unable to accomplish any of these basic functions, and 17% could not be left alone even for a very short time.[131] Care of adults with severe Down may become more demanding as they age. A systematic survey of adults with DS in Rome revealed that nearly all went to mainstream primary schools, and the majority had secondary education. Later, however, they received very limited support from the community. Their overall level of autonomous functioning was low, and diminished greatly after the age of thirty. In the absence of adequate institutional support, nearly all the surveyed adults with DS lived with their aging parents. Only 10% had occupations outside their home, and very few had social contacts with people besides their immediate family.[132]

Parents who undergo prenatal counseling, the sociologists Aliza Kolker and Meredith Burke have argued, should be informed about all the aspects of DS, positive and negative, including a possibility of the child's early death: "We believe that a condition that entails morbidity and mortality of this magnitude cannot and must not be presented as, 'not a disease but . . . one way of being human.' This falsifies reality."[133]

PND of Down syndrome played a key role in the more extensive use of prenatal screening in the 1990s. From 2012 onward, it has been at the center of efforts to generalize non-invasive prenatal testing (NIPT) for DS. Some parents of children with the condition are hostile toward NIPT, perceived, not without reason, as yet another step in the extension of screening for DS. They emphasize the potential for people with DS to have happy and fulfilling lives, and the existence of high achievers with this condition. They also point out the very real risk of accelerating the transformation of screening for Down into a self-evident element of routine pregnancy monitoring.[134] UK parents of DS children have reported professionals' negative reactions to a woman's decision to continue a pregnancy with a DS fetus; pressures from family members and friends to have an abortion; and, after their child was born, disparaging and hostile reactions to this child, including in some cases remarks that children with DS should not allowed to be born. At the same time, Jane Fisher, the director of the British charity Antenatal Results and Choices, whose aim is to support all women's decisions about the fate of their pregnancy, explained in 2015 that women who decide to terminate a pregnancy after a diagnosis of DS may face strong disapproval for wanting a "perfect baby," or are perceived as selfish and unprepared to "put the extra work in." UK women who openly spoke about their termination of a pregnancy with a DS fetus faced a mind-boggling outpouring of vitriol and hate mail. The reality is, Fisher concludes, that mothers are criticized, whichever path they choose.[135]

A Rarely Mentioned Risk: Psychiatric Disease

The care of an intellectually disabled child may be made more challenging by the intertwining of intellectual disability (and sometimes other types of disability as well) with mental health problems. A moving article about Rosemary Kennedy, the intellectually impaired daughter of Rose and Joseph Kennedy, explained that while "mentally slow," she was able to participate in her family's social life and underwent training to become a teachers' aide. Then in her twenties she "regressed," had violent tantrums, and attacked people near her. Her desperate father allowed her physicians to perform a prefrontal lobotomy, hoping that this intervention would put an end to Rosemary's disruptive behavior. The surgery left her severely impaired. The article starts with the description of Rosemary Kennedy as an "intellectually disabled" person, and ends by describing her fate as a "mental health tragedy."[136] The two conditions are indeed frequently linked. The problems of educating children with intellectual disabilities may be amplified by a difficulty to cope with the child's psychiatric and behavioral problems.[137] Mothers of children who display disruptive behavior (with or without additional learning difficulties) are often blamed for their child's unruly conduct, including by other mothers.[138] In some cases, children with intellectual impairments who were easy to care for when young become more difficult to manage as teenagers. They may become angry, agitated, and aggressive, and their parents may find it difficult to cope with the new situation.[139]

Historically, mental impairment has been systematically linked with a higher-than-average frequency of psychiatric and behavioral problems. Such problems, much more than IQ level or educational achievements, were seen as the main obstacle to the successful social integration of people with mild and moderate intellectual disabilities.[140] Connections between "mental retardation" and behavioral/psychiatric problems were often exaggerated in earlier periods, an attitude that contributed to the mistreatment of people with intellectual impairments. Such connections tend to be downplayed today, a possible reaction to past abuses and the persistence of prejudice.[141] Experts continue, however, to report a higher-than-average frequency of psychiatric disorders among children and adults with intellectual impairments. DS children are more rarely affected by psychiatric problems than those with other intellectual difficulties, but this difference tends to diminish with age.[142] People with DS have a lower ability to read other people's feelings and to adjust to them, and may have difficulties in changing routines and paying attention; some may be prone to uncontrolled anger and other behavioral problems. Those with milder intellectual impairment, more aware of their limitations, may have a more pronounced tendency to become depressed and anxious.[143] Specialists estimate that the frequency of psychiatric and behavioral problems among people with DS is four to five times higher than the frequency of such problems in the general population. Between 20% and 40% of adults with DS display such problems—or, to put it otherwise, between 60% and 80% of people with DS do not have psychiatric difficulties. The risk of psychiatric problems in people with Down is in all probability amplified by their social isolation and the scarcity of community living opportunities for DS adults.[144]

Intellectual disabilities became increasingly coupled with manifestations on the autism spectrum. The recent "autism epidemics," the sociologist Gil Eyal and his collaborators propose, was a direct consequence of the deinstitutionalization of people with "mental retardation," a movement that started in the 1960s. Such deinstitutionalization, coupled with the rise of parents' activism and of networks of educational and behavioral expertise, led to a reclassification as autistic for many of the children previously defined as having "mental retardation." The result was a blurring of boundaries between intellectual impairment, psychiatric disease, and abnormal neuropsychological development.[145] Parents of children with autism maintain, more often than those of children with other kinds of intellectual difficulties, the hope of a significant improvement in their child's condition.[146] They may also be confronted more often with the behavioral difficulties of their child, such as explosive and destructive anger, an obstacle to the child's successful socialization.

Recent books about the history of autism—or rather, on the rapidly expanding category of the autism spectrum—stress the need for society to enlarge the definition of normal and accept and embrace behavioral differences. John Donvan and Caren Zuker end their book *In a Different Key* with a story about an autistic teenager bullied on a bus because of his unusual comportment until another person addressed the main bully: "He has autism; what is your problem?"[147] It is very important to protect people with a nonstandard behavior from being tormented in public spaces. It is also important, but more difficult, to promote the acceptance of such behavior in long-term relationships. And it is even more difficult to encourage tolerance of a truly disruptive behavior. According to some evaluations, a majority of children/teenagers with autism have violent outbursts of anger. In one large study, 68% of autistic children demonstrated aggressiveness to a caregiver, and 48% to non-caregiver.[148] The promotion of a greater tolerance of an unusual conduct is an excellent thing. However, an exclusive focus on people with mild behavioral impairments who display, for example, repetitive movements may paradoxically increase the invisibility and exclusion of those with less easily manageable difficulties.[149]

The aunt of an intellectually impaired child with serious behavioral problems, in protesting the inadequate support of parents of "difficult" children, explains:

Raising a developmentally disabled child is often a thousand times harder than raising a non-disabled child. And the parents of disabled kids love them, so that's not the problem. Imagine a child who screams, bites and kicks through every day. And then becomes an adult who does the same thing. Imagine a child who is never fully continent and weighs 300 pounds at 35. . . . Pregnant women who find out they are carrying developmentally disabled kids must think past the cute stage. They must think about adulthood. Because the disabled child is always the center of everybody's life in a family. And the responsibility never, ever ends. I know that not all developmentally disabled kids are like what I've described above, but many are.[150]

The father of another intellectually impaired child states he would never trade his disabled child for anything, but at the same time criticizes well-meaning but misguided "pro-lifers" who condemn selective abortion for a fetal impairment:

They will never be there for the child when they need care, will never gasp at the size of the medical bills greater than my annual salary, will never look at their normal child and consider whether bankruptcy is an option, they will never bite their tongues while enduring wild behavioral challenges, changing bandages or dealing with the intimately human issues of disabled kids that are very different in public than they are in private.[151]

The father of an autistic child describes his and his wife's plight when their teenage son started to use uncontrolled urination as a way to express his unhappiness, frustration, and anger, an attitude which led to the son's social isolation and a parallel isolation of his parents.[152] A woman whose five-year-old son is on the autism spectrum speaks out, displeased with what she sees as sugarcoated euphemisms employed to describe this condition:

"Agitated?" Here's what agitation looks like at my house . . . when our son's rage and frustration spin out of control he'll lash out by hitting, biting, kicking and spitting. He'll throw his toys, chairs, table and easel; run around breaking things; and become a threat to us and to himself. . . . "Crying." The study talks of the crying. The word pales in the face of our son's dissolutions into tears. These days, if he hears a simple "no" or learns of some change in plans, he might launch into a 10-minute jag, where he argues fiercely with us in between the sobs. Then he can quickly escalate to ear-piercing screams lasting another 15 minutes or more. . . . There's also the isolation. I'm a very social person and before my son developed his "bad reputation," I worked hard to cultivate friends in the building. But then the play dates and birthday party invitations dried up. It's a terrible thing to be ostracized.[153]

In extreme cases, some psychiatrists employ electroconvulsive therapy to make life with severely disturbed autistic children bearable for their families and for the children themselves.[154] In commenting on the especially tragic—and extremely rare—case of Trudy Steuernagel, a Kent State University professor killed by her autistic son, to whom she was fiercely devoted, Stacy Clifford Simplican, who has a brother with autism, argues that it is important to acknowledge the "complex dependency" of some disabled people. People with developmental disabilities and/or mental pathologies, Simplican explains, are vulnerable, but so are the people around them, family members and paid attendants. Narratives about disability may mask the existence of complex dynamics of power within families that care for disabled people, and an increased risk of abuse of impaired individuals and their caregivers.[155]

Autism is seen today as a condition with a high heritability: it is more frequently found in identical twins; families with one autistic child have a higher-than-average chance of having another child with this condition; such a chance rises dramatically in families with two autistic children; and several genes are linked to this condition. There are no (in 2018) prenatal tests for autism. There are, however, prenatal tests for the rapidly increasing number of genetic conditions—for instance fragile X and DiGeorge syndrome (22q11.2 del)—connected with a higher risk of autism and other psychiatric problems, such as severe attention deficit disorder.[156]

Decisions following a PND of DiGeorge syndrome, a condition linked with many health problems but also a high probability of psychiatric disorders, may be especially difficult because of this condition's great variability of expression. Some people are only minimally affected by this mutation, but nearly half of them will have serious psychiatric and behavioral issues. Care for people with this condition may be especially challenging. Their caregivers—nearly always their parents—explain that a high probability of a psychiatric disorder was their greatest source of anxiety, much more than the health problems associated with DiGeorge syndrome.[157] Many among those who took care of affected adolescents and adults complain about an insufficient level of support for families, and worry about what will happen to their child when their health fails or they die. Typical statements:

In general care-managers have been disappointing, particularly in the area of matching needs to programs.

It is much more difficult to advocate for an adult than a child. Doctors, social workers, teachers etc., want the adult to ask/explain things themselves which can be next to impossible for adults with learning disabilities or with anxiety issues.

Families fall apart under the severity of this illness. You don't know what you're dealing with or how to help—or where to go. It is terrible.

The family provides much support that is at times beyond what we can bear. The added stress that this places on everyone makes life very difficult and strains familial relationships.[158]

A woman facing a diagnosis of chromosomal anomaly linked to a higher risk of mental problems and who decided to terminate the pregnancy explains:

There is a lot about this [chromosomal] deletion that looks like a mental illness. I grew up with a sister who is schizoaffective. I have a lot of fear about that. . . . And lo and behold, those very characteristics are part of the behavioral phenotype that has been identified for the chromosome deletion: it was terrifying. . . . When I began reading about this deletion, I went from being absolutely sure I would not terminate to shaking with terror. How could I have a child like this being the way that I am? . . . If I were a different kind of person, looser maybe, I could do this, I could have had this child, I could have parented this child. But here I am, with my own tremendous limitations.[159]

A positive PND is frequently the notification of a risk: a possibility of the best- or worst-case scenario, and all the options in between. Judgmental, simplified pronouncements, be they in favor of a woman's decision to continue a pregnancy or in favor of a termination, rarely take into account the great variety of ways that people deal with risks: for themselves, their future child, and their families. The writer Ayelet Waldman learned that her ten-week-old fetus had a small but non-negligible chance of having severe physical problems and intellectual delays. Waldman, a self-described pessimist, was convinced that her child would indeed suffer from severe anomalies and immediately decided to terminate the pregnancy. Her husband, the writer Michael Chabon, presented by Waldman as an optimist, was convinced of the opposite, and initially opposed an abortion. Later, however, he accepted his wife's point of view:

I think, really, that we have no choice. If we do what you want, if we have an abortion, and it turned out that Rocketship [the fetus] would have been healthy after all, I can live with your mistake. I can love you, no matter what. But if we do what I want, if we have the baby and it turns out that he is not okay, it's too massive of an error. The ramifications are too lasting, not only for us but for Sophie and Zeke [their children]. My mistake will burden them for the rest of their lives with the care of their brother, and burden us so much that our relationships might be in danger.[160]

PND and other approaches such as preimplantation genetic diagnosis, destined to prevent the birth of "flawed" children, have produced new knowledge, new areas of medical intervention, and new ways of preventing consequences that may be seen as negative by prospective parents. They also have produced new dilemmas and, in some cases, maternal/parental stress and guilt. As the anthropologists Tine Gammeltoft and Ayo Wahlberg explain in their discussion about selective reproduction, such technologies

promise to provide new knowledge and enhanced control of reproductive processes, offering novel pathways to intervene in the making of new children. Yet as practiced and experienced, ethnographic evidence indicates, these strivings for control tend to generate new doubts and unknowns. Rather than producing a brave new world of reproductive mastery, selective reproductive technologies throw their users into social worlds of contingency, ambivalence, and disorientation, worlds in which they must grapple with new and perhaps intensified reproductive anxieties and uncertainties.[161]

SIX

PND, Reproductive Choices, and Care

Families We Live With and Families We Live By

PND was developed to prevent disability. According to some disability activists, the "prevention" goal, besides being morally questionable, is rooted in a deeply erroneous view of life with disabilities. Families of disabled children indeed face more difficulties than those of able-bodied children, but these difficulties are caused by inadequate support for disabled people and their families and the persistence of prejudice and discrimination. To the parents of a disabled child, s/he is just their child, and raising her/him is a source of gratifications and problems similar to raising an able-bodied child.[1] The passionate—and deeply felt—argument about the lack of difference between raising "special-needs" and "normal" children is partly grounded, one may propose, in idealized images of the family as an infallible source of support and affection. The historian John Gillis investigated the role of such images in Western societies.

Gillis distinguished between families we live with and families we live by.[2] The first reflect observable behaviors and describe how real-life families behave in specific circumstances. The second are constructed from family myths, memories, and symbols, and reflect how families give meaning to people's lives through the ritualization and stylization of reality. In families we live with, parents may be violent to each other and to their children; children can be cruel, manipulative, sly, and jealous; and disabled family members may be patronized, disrespected, ostracized, and exploited, and may react to their difficult situation by showing less than stellar qualities. Families we live by are an infallible source of unconditional love, care, and support; these qualities are especially apparent in mothers. Families we live with have existed at least as long as recorded history, but families we live by appeared in Europe in the late eighteenth and early nineteenth centuries with the rise of the bourgeois culture. Before that time, members of the same family had specific obligations. Love among family members, including maternal love, though not invented by the Western bourgeois culture, was for a long time regarded as optional.[3] Then in the late eighteenth century, there appeared a very different—and highly idealized—image of the family as a haven in a cruel world and a major source of support, care, and affection. This new image of the family was expressed through rituals such as shared meals, family time, holidays, and celebrations: birthdays, weddings, anniversaries, and funerals.

The two images of the family, the existing and the idealized, are coextensive and consubstantial. They are inseparably intertwined: it is impossible to study one without the other. Experience influences the construction of family stories, images, and rituals, while the latter shapes the experience of families. The combination of families we live with and families we live by produced a highly resilient structure. In the 1960s and 1970s, the counterculture movement and then the feminist movement rebelled against the idealization of the heterosexual, middle-class, white family. Feminists analyzed the subjugation of the weak—often women and children—by a male-dominated social order. The feminist critique of the oppressive aspects of family life affected many aspects of relationships between the sexes, but it did not dismantle the powerful image of families we live by as an infallible source of support, care, and unlimited love.[4] Debates about PND, especially those linking this issue with disability rights, are frequently colonized by idealized images of families we live by, in which women are never ambivalent about their pregnancies and maternal role; parents always unconditionally love and support their children; there is no room for tension, competition, and jealousy between family members, even less for abusive behavior; and the presence of a disabled child provides additional stimulus for intrafamily cohesion.[5]

Families we live by is a situated concept: the ideal of family life is not the same in all countries and social groups. The US debate about abortion, the sociologist Kristin Luker proposed, is mainly a debate about women's role in the family and society (or, to follow Gillis's terms, their perceived role in the families we live by).[6] For the opponents of abortion, all the women, including those who elect to work outside the home, are above all mothers and caregivers, and families are grounded in essential differences between male and female tasks. A woman can successfully fulfill many responsibilities, but her prime duty and vocation is to give and sustain life. People who hold such a view categorically reject the legitimation of aborting a malformed fetus by arguing that the care of a severely impaired child will limit the life options of the child's mother, because being a mother and a caregiver is much more important and rewarding than any other female activity. Pro-choice activists hold an opposite view. For them, woman's role in society is not predetermined by her gender. Being a mother and a caregiver is but one option among many others, and a woman should be free to decide at each time of her life whether she elects this option. The debate about abortion, Luker argued, is not about the status of the fetus as an autonomous individual (a developing child, a potential patient) or as a part of the pregnant woman's body; nor is it mainly about the definition of abortion as a murder or an elective surgery. It is mainly about a contrast between these two radically different worldviews of women and families.

Luker's conceptual framework firmly situates PND within a debate about gender roles and women's place in society. This framework, I believe, is as valid today as it was in 1984, when Luker published her book. I interviewed two French (male) physicians about the care of intellectually disabled people. Both explained that mothers of intellectually impaired children/people usually have the primary responsibility for their care, a responsibility which lasts as long as the mother lives. They had, however, very different views of this maternal role. The first physician proposed that for some women at least, an intellectually disabled child is an ideal child, one who will never grow up, become autonomous, separate from the mother, and leave her with an empty nest—hinting that these women, although undoubtedly loving and devoted mothers, may also be driven by a selfish satisfaction rooted in their quasi-absolute control over an eternally dependent child. The second physician expressed his deep admiration for women who often put their own aspirations on hold in order to take care of an intellectually disabled child, a task which will last all their lives, hinting that such selfless devotion to the well-being of an impaired individual represents the highest ideal of maternal and human virtue. To be sure, discussions about PND and selective abortions, like those on abortion in general, cannot be dissociated from those about families, care, and gender.

Hereditary Conditions and PND

PND was developed in the 1960s in an effort to prevent the transmission of hereditary diseases such as hemophilia, Tay-Sachs disease, Duchenne muscular dystrophy, or thalassemia. Women with a known hereditary condition in their family and hence a high probability of having a child with this condition aspired to escape the "family malediction" by giving birth to healthy children.[7] In the 1970s and early 1980s, their aspiration was perceived as self-evident and unproblematic; it was also presented as one of the main elements that stimulated physicians to extend the scope of PND.[8] The aspiration to give birth to a healthy child ceased to be perceived as self-evident with the rise of the claim (the expressivist objection, discussed in chapter 5) that a selective elimination of malformed fetuses sends a strong message that life with disabilities is not worth living. This argument can have a great emotional impact in families with hereditary diseases.[9]

Advocates of the expressivist objection speak about a generic "disability," and do not differentiate between fetal anomalies that stem (mainly) from accidents of pregnancy and those that stem from hereditary conditions. However, in the absence of a hereditary condition in a family, women's situated attitudes toward PND reflect mainly their and their partners' values, beliefs, and experience.[10] In families with hereditary conditions, such attitudes may also reflect intrafamily dynamics. Hereditary diseases produce a "genetic kinship," mediated by the transmission of abnormal mutations. The notion of genetic kinship originated in a juxtaposition of two different regimes of knowledge: one is produced by geneticists, and the other is generated by family members. The construction of pedigrees, a central element in clinical geneticists' work, is grounded in genealogical data provided by affected families. Clinical geneticists rely on cutting-edge molecular biology/sequencing technologies, and at the same time are entirely dependent on subjective knowledge about kinship transmitted in families, the accuracy of family members' memories, and their unconscious and sometimes conscious decisions to include specific individuals in the family tree. The definition of genetic kinship that emerges in this context is a hybrid one, inseparably macromolecular and sociocultural, as is the reception of prenatal testing in families with hereditary diseases.[11]

Attitudes toward PND for hereditary disorders are frequently shaped by the dominant image of a given disorder. Selected hereditary conditions such as phenylketonuria (PKU) are seen as curable, drastically limiting the demand for their diagnosis during pregnancy.[12] Other pathologies, such as polycystic kidney disease (in the majority of cases, a dominant hereditary disorder) and heterochromatosis (a recessive hereditary disorder), are seen mainly as "diseases," not mainly as "hereditary." These conditions escape geneticization and mutation-centered biosociality, because they have a highly variable expression (some of the people who inherit the disease-related gene are sick and others not), there are great differences in the severity of the disease among affected people, and, above all, treatment is possible.[13] People with polycystic kidney disease and heterochromatosis are more worried about prevention, diagnosis, and therapy than about the transmission of a hereditary trait to offspring.[14] In some ways, their attitudes may be closer to those of nineteenth-century patients: more inclined to be aware of a familiar predisposition (diathesis) for a weak heart or digestive troubles than to be part of a group which constructs its "biosociality" around shared mutations.

When a condition is clearly perceived as hereditary, the acceptability of medical interventions for preventing the birth of children with this condition especially depends on whether it is seen as a disability—that is, a deeply ingrained identity that should be respected and cherished—or as a disease, a physiological dysfunction which does not modify the identity of an affected person. The paradigmatic example of the first category is deafness, or rather belonging to the Deaf community, and of the second, hereditary diseases which have become the focus of public health campaigns, such as Tay-Sachs disease (a metabolic disease that invariably causes early death), targeted for elimination in Ashkenazi Jewish communities, or thalassemia (a blood disorder which today is not incompatible with life but often produces a severe impairment), targeted for elimination in, for example, Cyprus.[15] In Israel, the ministry of health is actively promoting the identification of the carriers of hereditary diseases such as Tay-Sachs, cystic fibrosis, or familiar dysautonomia with the explicit aim of preventing the birth of children with these conditions through a combination of preconceptional, prenatal, and preimplantation diagnosis.[16] A comparison of the views of adults with inborn impairments confirmed that those with a condition usually perceived as a disease—sickle-cell anemia, cystic fibrosis, and thalassemia—were often open to the possibility of selective prevention of the birth of children with their pathology. By contrast, people with a condition usually defined as a disability—Down syndrome and spina bifida—were often distressed or saddened by the possibility of prevention of the birth of children with their condition.[17]

Cancer is invariably seen as a disease, and as a cruel enemy that blindly strikes healthy people.[18] A small minority of cancers are transmitted in families. Women carrying mutations in the BRCA gene, which greatly increases their chances of developing breast or ovarian cancer, frequently think hard about their reproductive options. The majority finally decide to have children without prenatal or preimplanatory testing, often with the hope than when their children are adults and face an increased cancer risk, there will be an effective treatment which will limit such risk. However, a few women choose preimplantation genetic diagnosis in order to prevent the transmission of the mutated gene to their offspring. Women's wish not to transmit the BRCA mutation has been linked to their conviction that they will pass to their children (especially daughters) the risk of going through the same painful experiences that they did: fighting their own illness, encountering difficulties in finding a partner, negotiating their plans for a family, or witnessing the death of a close relative. One BRCA-positive woman explains, "I can overcome what concerns me, but it is difficult to bear the fact that persons I love are ill, not well in their minds and in their bodies. You can imagine how much stronger this would be if it were my daughter, my own child."[19]

When a hereditary condition is seen above all as a disability, reproductive decisions of women at risk for giving birth to an affected child are often influenced by their intimate acquaintance with people with that condition. Physicians are frequently convinced that "responsible mothers"—responsible, that is, to society at large but also to their future child, since good health is often seen by health professionals as an absolute value—should use PND to prevent the birth of a child with a hereditary disease. Some women from affected families disagree. They believe that "responsible mothers"—responsible, that is, to the affected members of their family with whom they have strong emotional links—should reject PND because it can hurt people to whom they are deeply attached, often a previously born child or a sibling.

The attitudes of UK women from families with hereditary disorders which attack muscles—Duchenne muscular dystrophy (DMD), an X-chromosome-linked condition, and spinal muscular atrophy (SMA), a recessive autosomal condition—illustrate the complex tension between women's wish to give birth to a healthy child and their strong identifications with impaired family members. Both DMD and SMA are usually seen as a disability, perhaps because an affected individual often uses a wheelchair. DMD, which leads to a progressive weakening of muscles, is (as for now) always fatal, but the rapidity of the progression of the muscular impairment is not uniform. Moreover, affected children may have additional problems, such as neurological disorders and learning disabilities. Until recently, the majority of boys with DMD did not survive beyond adolescence, but recent progress in care allows some of these boys to live to their twenties and thirties. When interviewed about their reproductive decisions, mothers of boys with DMD provided contrasting views of "maternal responsibility." Some believed that they should not have children anymore, even if they could make sure that the next child would be healthy, because s/he would be saddled with the responsibility for the care of a sick brother. Others were strongly in favor of the use of PND to ascertain that their next child would not be affected. Still others felt that they could not face PND and termination for DMD, and rejected PND when they decided to be pregnant again.[20] One woman with an affected firstborn child elected PND in her second pregnancy (and gave birth to a healthy child), but decided to forgo prenatal testing in her third pregnancy. She felt that the burden of undergoing prenatal testing which might lead to an abortion was worse than the possibility of having another affected son.[21]

Transmission of SMA is more complicated than DMD. A prenatal detection of the mutation that induces this disease does not predict its severity. It is impossible to foresee whether the affected child will develop type I SMA, which leads to death in early childhood; type III SMA, which produces a progressive muscular impairment that may be relatively mild in young adults and become more pronounced with age; or an intermediary type II SMA. Physicians assume that the gravity of the disease is correlated with manifestations of SMA in a given family, but such a correlation is far from being absolute.[22] Facing a high level of uncertainty, family members often evaluate the expert's knowledge in the context of their experience of this condition.[23] Since SMA's expression is variable, the experience of life with this pathology is variable, too.[24] The diversity of individual experiences with SMA was reflected in the multiplicity of ways that women justified their attitude toward PND and selective abortion. Many felt obliged to take into consideration the negative view of PND of family members with SMA. These women saw disabled people's objection to PND as oppressive yet irrefutable. They wanted their children to be healthy, but they did not wish to negate the value of the lives of affected relatives or hurt them.[25] Consequently, many decided, like mothers of children with DMD, to "choose not to choose": either not to test the fetus to avoid a difficult decision or, more often, not to have children. The choice itself has become a condition to be avoided, severely limiting these women's reproductive options.[26]

The presentation of some hereditary impairments as diseases and thus a legitimate target for elimination, and others as disabilities whose continuous presence enriches the human race, is partly arbitrary. One can argue that the frequently progressive SMA can be classified as a disease, and the impairment-inducing thalassemia as a disability, or that there is no difference in kind between a perturbation of motor neurons by the SMN1 gene, linked with SMA, and a perturbation of hemoglobin production by the HBB gene, linked with beta-thalassemia. If one accepts this argument, an aspiration to "eradicate" SMA is as legitimate as an effort to eliminate thalassemia from affected populations. Such a view was adopted by some physicians, especially those who were active in the promotion of PND in the early days of its spread, and who perceive efforts to reduce the number of children with inborn impairments as an unquestionably desirable goal. History cannot, however, be erased at will. For various reasons—sometimes because they are readily visible; sometimes because of institutional and legal reasons;

sometimes as a result of the intervention of a dedicated group of activists—selected inborn conditions became perceived as a part of a given person's identity rather than an external affliction. Disability activists' objections to PND cannot be dissociated from the dramatic history of oppression of disabled people which culminated in the murderous Nazi regime. This history does not belong only to a faraway past. Despite the important gains of disability rights movements, many disabled people continue to face negative attitudes, discrimination, and prejudice.[27] Strong objection to PND may be one way some disability activists react to an unfair society, and perhaps also an unjust fate.

The opposition to PND is not shared by all the disability activists. A majority of such activists in Israel are not opposed to PND, and do not consider this approach a negation of the value of their lives. The same is true—for very different reasons—for disability activists in Vietnam.[28] In other countries, too, not all the people with disabilities are hostile to PND and selective abortion. Twelve out of fifteen disabled people in Chicago interviewed in depth about their attitudes toward PND have a positive view of this approach. They reasoned mainly in terms of a possibility of improving people's lives, but one interviewee explains:

> I see what disability does to other children who may already be born and then have a disabled brother or sister. As with my family, if I could prevent that in any way, I would want to do that. . . . I've seen some people who are activists to the point that they're annoying and I don't think that does the disabled community any good. . . . Sometimes it seems like they're promoting the birth of disabled children. I don't think that's appropriate. I think that having people with all different problems can make a society richer, but in a sadder sort of way.[29]

Transmission Dilemmas

One of the most important reproductive decisions mutation carriers face is whether to have children. PND was presented in the 1970s and 1980s as a means of enabling women/couples to overcome an unfortunate genetic legacy. Some women at risk of having a child with a hereditary disease elect not to have biological children, or, as Annabel Stenzler, a woman with cystic fibrosis, puts it, "to leave no child behind."[30] "Carl Anderson," a Marfan syndrome patient who, according to the geneticist Robert Marion, suffered because of his unusual looks (people with Marfan syndrome are very tall and thin, with long limbs) and his syndrome-related physical problems, such as limited eyesight, made the opposite choice. He chose to marry a near-blind woman, literally unable to see his many physical oddities, and did not tell her that he was affected by a hereditary disease. After the birth of a daughter with Marfan syndrome, when his distressed wife asked him why he did not tell her that he had this condition, he answered:

> Sure, and what would you have done? I tell you I've got Marfan syndrome and I'm going to be dead within ten years, and the first thing you'd do is take off. And more than anything else in life, I wanted to have a kid. If you found out that this disease was inherited, and that there was a 50-50 chance that any kid I had would have it too, you'd never have gotten pregnant. And if, by some accident, you did happen to get pregnant, you'd have had an abortion. By not saying anything, at least now we have this baby.[31]

Deciding whether to test for the presence of a hereditary trait in a fetus may reflect complex intrafamily relationships. A female carrier of the hemophilia gene, interviewed for the European interdisciplinary study Ethical Dilemmas Due to Prenatal and Genetic Diagnostics (EDIG) about her decision to undergo PND, explained that she did not want a child with hemophilia because she had witnessed the difficulties of her older brother. The brother, she reported, failed to receive appropriate treatment for the condition when he was young, and was often in severe pain. His intense physical suffering made him "very difficult," and jealous of his sister's good health. When he learned that his sister had decided to undergo PND for a risk of hemophilia, he interpreted this decision as a statement that she did not want people like him to live. The sister was deeply pained by his attitude, and by her failure to convince him that her choice to undergo PND was directed not against him as a person but against his illness and attendant suffering. Both siblings seemed to be badly hurt by the availability of PND, although the sister did not regret her decision: she believed that her duty was to spare her child suffering.[32]

Carriers of genetic traits may have complex attitudes toward genetic testing. Mrs. D., a physician who also was interviewed for the EDIG study, had a daughter with congenital adrenal hyperplasia (CAH), an autosomal recessive condition caused by excessive exposure to male hormones during pregnancy. Among other things, CAH produces a "virilization" of girls' sexual organs.[33] Mrs. D., who was pregnant when interviewed, claimed she would have never considered a termination of pregnancy for CAH, because such an act would have been akin to "terminating" her daughter. On the other hand, she was extremely and quasi-obsessively preoccupied by fetal well-being during her pregnancy, and had many ultrasounds and other prenatal tests. Mrs. D.'s strong desire to do everything possible to have a healthy child, coupled with an equally strong rejection of abortion for the hereditary condition of her daughter but not for other inborn anomalies, may be connected with possible guilt for transmitting a "defective" gene to her daughter.[34] Yet not all mothers of children with a chronic condition have the same reaction as Mrs. D. Sometimes a parent with a genetic disease is determined to end its chain of transmission.[35] During my observations of fetopathologists in France, I encountered the case of a woman with a mild form of Larsen syndrome—a hereditary bone and joint anomaly—who had a daughter with a more severe form of the syndrome. When in her next pregnancy she found out that the fetus carried the same condition, she immediately decided to abort this fetus.

Women with mitochondrial diseases, a family of progressive conditions characterized by highly variable expression and unpredictable effects in offspring, not only do not see their condition as a part of their identity, but some perceive the possibility of preventing its transmission and giving birth to a healthy biological child as an essential part of "compensating" for the misery linked with their disease. These women—not infrequently already sick, and aware of the high probability that they will gradually become more impaired—are willing to undergo a highly experimental procedure: the production of a "two-mothers egg" with a nucleus from the original mother and cytoplasm from an egg donor, in order to produce a disease-free child.[36] During the 2015 parliamentary debate in the United Kingdom about the implementation of this technology, some women argued that the birth of a healthy biological child would alleviate the trauma of the death of a previous child or, alternatively, decrease the suffering produced by their own progressive disease. Hopes to attenuate the often inexorable progression of the disease through the use of cutting-edge biological technologies to produce healthy children have become deeply ingrained in how these women cope with their condition.[37]

Most of the mothers of children with birth defects (two-thirds of the forty investigated families) interviewed by the sociologist Susan Kelly decided to "leave no child behind."[38] One of the most frequent reasons they gave for their decision were doubts about experts' capacity to prevent the birth of another impaired child. The birth of the first child was perceived as a catastrophic event that radically undermined these women's faith in the medical profession and made them apprehensive about the future. Specialists' affirmations that the odds of having a second child with the condition were "small" were not at all reassuring. As one woman puts it, "They told me that if I ever got pregnant that I would never have another one like him. But you know, there's always a possibility it might happen here." A woman whose daughter was diagnosed as having a complex neurological disorder explains:

> When we found out something was wrong, that knocked any idea of another child out for me, and my husband agreed. We just could not take the chance on another one, you know, being something wrong with it. . . . And then to even have a normal, which they tried to tell me the odds, but it wasn't great enough for me. I mean, if it had been zero then I might have considered it. But to think of having another one was just too much.[39]

Another woman had a first child with a bone disease, and a second child with a kidney disorder. She then decided that another pregnancy was a much too risky enterprise:

> I tied my tubes, there was no way. These two are enough. . . . I know this is going to sound pretty bad, but I have got one child with [bone development disorder] and my next child I have has kidney problems and has to have surgery, why in the world would I want to bring another child to have to go through something else.[40]

A woman who underwent screening for fetal malformation, was told that everything was fine, and then gave birth to a Down syndrome child summarizes the reason women distrust the probabilistic evaluations proposed by the experts:

> When he was born I was looking for everything because we had the AFP [measure of the level of alpha-fetoprotein in the pregnant woman's blood], we had the ultrasound and everything. It looked okay. . . . Sunday when the pediatrician came in . . . he just walks in and said, well, as you probably know your child has Down's. . . . I'm done with statistics. You know, you said this or that, statistically this or whatever. Well, they didn't play in our favor. . . . So with the whole thing I thought if everything is okay I'm not having any kids, this is it, you know.[41]

The case of the writer Emily Rapp was even more dramatic. Rapp, herself a disability rights activist (she has only one leg), wanted to have a healthy child:

> Although my disability is not genetic, and my life has been always rich and full, it has also involved a great deal of suffering, both physical and mental, and

it was this prospect I wanted to spare any future child of mine. Even as I felt a miracle of creation—that stunning pause, a suspension of disbelief—the first time I saw my child's heartbeat on the ultrasound, I knew that if a compromising genetic anomaly were discovered, I would have an abortion.[42]

Rapp underwent a chorionic villus sampling (CVS), and was told that the test did not display any genetic anomaly of the fetus. Her son was born, however, with a rare variant of Tay-Sachs disease, a hereditary condition she and her husband had elected not to include in the panel of genetic tests for their future child, because they were told that in their case—Rapp is not Jewish, although her partner is—odds that the child would have Tay-Sachs were astronomically small. But, she adds, "odds, it turned out, did not matter."[43] Her experience echoes the one of the woman who explains that she is "done with statistics." Probability calculus is not very useful if the rare exception is your child.

Mothers of children with fragile-X syndrome (an X-chromosome-linked condition that causes intellectual disability), like mothers of children with Duchenne muscular dystrophy, have provided widely divergent interpretations of the term *responsible motherhood*. Some believe that a responsible mother would not risk giving birth to another disabled child. As one mother of two children with fragile-X syndrome puts it, "I love my children and I probably would love to have a whole houseful of them but I said, 'You know what? I'm not going to do that. I'm going to be responsible. Take care of the two I have.'" Another woman explains that having another biological child after learning her carrier status would have been irresponsible. She elected to adopt a second child.[44] Other women internalized the expressivist objection to PND, refusing to promote a culture that accepts only perfect children:

> I have a couple friends who—I think they have like a 20 per cent chance of passing on a gene that has something to do with the nervous system or muscular system—and they decided not to have kids. They're going to adopt, and I'm like, "Hey, adoption's great and everything." But it's going to get to a point where if people are predisposed to a certain gene and they know about it, they're not going to have those kids, or they're going to go in vitro, and so in a way a lot of these special kids like our kids I think will be sort of taken out of Western culture countries.[45]

The mother of a child with fragile-X who refused PND in the following pregnancy and had a second child with the condition felt strong disapproval of her decision: "He's a beautiful kid. They're beautiful and I mean that's why I went on to have more kids. We get a lot of criticism for that." Moreover, women's views can change. A couple with one child with fragile-X explains, "We decided to have another child. Went through the whole PGD [preimplantation genetic diagnosis] process. That didn't work. So we just decided that it was important for [our son] to have a sibling regardless, and if we had two kids with fragile X, then at least they would have each other."[46]

People with a rare hereditary disease, hypohidrotic ectodermal dysplasia (HED), similarly display a wide spectrum of reactions to the transmission of this hereditary disorder. HED is found mainly in men, since the most frequent mutation linked with this condition is X-linked. People with HED have sparse hair, few teeth, and problems with perspiring, which lead to overheating during exercise and in hot weather and a greater risk of developing a dangerously high fever. Children with this condition can die of overheating, failure to thrive, and infection, but a majority survive to adulthood. People with HED usually cope well physically, but many are stigmatized because they have unusual facial features and teeth (the latter problem can be corrected today with dental implants, if the affected person can afford them). In addition, people with HED are frequently concerned about the transmission of this condition to offspring. Unsurprisingly, their attitude toward transmitting their mutation is strongly colored by their own experience of life with HED.

Some affected men categorically reject the possibility of transmitting HED: "I never wanted children to go through what I've gone through. That has always been at the back of my mind. . . . I would not wish this on anyone at all." One young man with HED has an intermediary position: he is in favor of PND and selective abortion for this condition, then adds that he would not discourage his sister from having an affected child: "A child's a child at the end of the day." Other men are convinced that an abortion for HED is wrong, or believe that the presence of the same condition in their children would produce stronger links with these children. Others still have a strong emotional rejection of PND for this condition. One man explains that if his daughter would be unwilling to have a son like him, he would consider such a decision to be a disparaging judgment of the quality of his own life.[47]

HED may be seen as a "borderline" genetic anomaly: people with this condition have health issues and problems with their atypical appearance, but can live a quasi-normal, autonomous life. This makes reproductive decisions especially challenging. Bonnie Rough's grandfather suffered from HED, but also from psychiatric problems and addiction, both probably indirectly linked with his impairment. On the other hand, Rough's brother, who inherited HED, coped well with problems associated with his genetic anomaly. Rough decided to undergo testing to learn whether she is an HED carrier, and upon learning that this was the case, was for a long time ambivalent about transmitting HED to her children. In observing her brother, she was aware of the difficulties of living with this condition: "As we grew up together, I saw what it was like for him to navigate the physical, emotional, and social challenges of being different." She also knew that her brother had overcome most of these challenges. Rough hesitated for a long time about whether she should undergo in vitro fertilization and PGD, undergo PND if pregnant, or choose to remain unaware about the genetic status of future children. Finally, she decided that she was more comfortable with natural conception and, if necessary, a selective abortion. She underwent a CVS and reported that the technician who made the test deliberately turned the fetal heart monitor to its highest volume, telling her, "These are heartbeats of your child." Genetic tests revealed that she was pregnant with a healthy girl. When pregnant again a few years later, Rough learned that the fetus was male and carried the HED mutation. She decided to abort the fetus. For her and her husband, "this wasn't necessarily the hardest thing we had ever done. But it was absolutely the saddest." In her next pregnancy, Rough elected again to have CVS, and gave birth to a healthy girl. She then decided to write a book, to share her experience of facing a series of complex and profound medical decisions with no obvious "right" answers.[48]

Rough tells a relatively straightforward story: her refusal to transmit a mutation to offspring. Anne Martin Powell's history of hereditary disease in her family is more convoluted. Powell, who comes from a medical family, had a severely disabled brother, Bob. At first, Powell's family kept Bob at home, where, despite his severe limitations, he was much loved by his sister. When Bob was nine years old and his care at home became too challenging, the family moved him into a specialized institution. His health gradually deteriorated, and he died as a young adult. Bob's problems were originally attributed to a nongenetic cause—the consequences of encephalitis as a baby. In 1966, Powell's other brother, David, a physician interested in hereditary anomalies, learned that Bob had in fact a newly described X-linked hereditary disorder, Lesch-Nyhan syndrome. Powell, worried about her own chances of transmitting the disease, was tested and told that she was not a carrier. She gave birth to a healthy son. A few years later, Powell's brother told her that early tests for Lesch-Nyhan syndrome might have been inaccurate. She took the test again, and this time tested positive.[49]

In the meantime, Powell became pregnant again. At that time, the early 1970s, it was not possible to find out whether the fetus had Lesch-Nyhan syndrome. Already unsure about the continuation of her pregnancy—it was accidental, and very close to the birth of her son—Powell considered termination, but at the same time felt ambivalent about it. She vacillated between her positive memories of her brother Bob as a child, and her awareness of later devastation brought to him by his illness. She was "haunted by the horror of what happened to him . . . images of his devastation lunge at me out of nowhere, again and again, making me gasp." She finally decided that she "could never intend another child to face the fate that awaited Bob. . . . I will stop this damn predator in its tracks." In the early 1970s, it was still difficult to obtain a second-trimester abortion for a fetal malformation. At first, Powell's gynecologist told her that her only option was a hysterectomy. Finally, with the help of her brother, she found a hospital willing to induce an early expulsion of the fetus after an administration of prostaglandin—and marveled about the perfection of the tiny aborted fetus.[50]

Powell says she was at peace with her decision after her abortion, and even elated by it:

> I believe that the possibility of a child's suffering has been averted and I have escaped the difficult, life-long psychological burden that I see my mother bearing due to Bob's condition and demise. It occurs to me that I have a strong drive to protect not only my two children, but myself. I feel but a modicum of guilt about my selfishness convinced that what Mother taught me was correct: each of us is responsible for our own happiness.[51]

She later adopted another child. Nearly forty years later, Powell decided to write a memoir about her feelings for her brother Bob, and about her experience as a carrier of his genetic condition. In it, she explains that she wanted her children and grandchildren to understand how someone as physically incapacitated as Bob could enrich his family's life. Dr. William Nyhan, one of the physicians who described Lesch-Nyhan syndrome, heard about her project and proposed to test her blood to find out what exactly the family mutation was. No such mutation was found: the earlier result, which had led to an abortion, was a false positive.[52] Yet Powell does not regret that abortion: she believes that she had made the best possible decision in light of the medical knowledge available at that time. And while she wants badly to make her children and grandchildren understand how a severely impaired sibling can make an important contribution to a family, she also wants to convey her understanding that her brother's disease, that "damn predator," should not be allowed to be propagated in future generations.[53] This may be perceived as an illogical position, but as the pro-choice activist Jessica Valenti put it, "Nothing is truly consistent about being pregnant or having a baby. Nothing is simple, nothing is clear-cut. And it doesn't have to be."[54]

Situated Choices

Many of the dilemmas produced by testing for a hereditary condition in the family—the uncertainty of the results and their interpretations, the stressful medical odyssey, the difficulty in deciding the right course of action, and the fear of regret—are produced by other diagnostic tests, too. Hereditary conditions are, however, unique, because they transcend the individual and extend to the family. The diagnosis of a hereditary disease reshapes the family history—its past, present, and future. It produces, according to the sociologists Kelly Raspberry and Debra Skinner, "phantom ancestors" who live on through their genetic legacy.[55] The diagnosis of such a condition also produces "phantom descendants." People aware of the presence of an inherited condition in their family often project themselves into the future. A man with HED imagined himself with an HED grandson and explains that he'll have a special bond with such a grandson, "because you are going to be able to identify with them a lot more, especially when they get older."[56] A woman interviewed by Raspberry and Skinner explains, "I think it's a parent's responsibility to keep the well-being of your family for even generations to come. So even if you're going to have a child that's just a carrier, are you going to be able to deal with the fact that your grandkids are going to possibly have it?"[57]

People, especially women, who face difficult reproductive choices perceive an extended temporality of their physical body, which includes multilevel connections with past and present kin. Such connections do not simply mirror the degree of genetic proximity as represented in pedigrees and family trees but incorporate dense networks of relationships within the family. The temporality of genetic conditions as perceived by families with such conditions is perhaps best captured by the notion of "topological folds in time," elaborated by the French philosopher Michel Serres. Time, especially when dealing with individual/family history, Serres has noted, is not linear but "kneaded" or crumpled, and it produces unpredictable effects of proximity and distance, continuity and discontinuity. Often, the proper image for understanding temporal developments is neither time's arrow (the idea that time is linear, like a digital clock or a calendar date) nor time's circle (the idea that time is circular, like an analog clock or a sequence of seasons), but an origami—unpredictable spatiotemporal associations between surfaces.[58] Individual responses to a family history of hereditary diseases mirror the uncertainty of the phenotypic effect of a given mutation, the unforeseeable effect of the intrafamily dynamics it can produce, and how such dynamics can unfold in time.

Decisions whether to test or not to test for a genetic condition in the family—and what to do with the results of such testing—display the complexity of individual dilemmas linked to PND, and also their interactions with institutional, cultural, religious, economic, and political variables. The latter increasingly include specific forms of disability activism. In domains where such activism is present, the views of a small number of deeply committed—and often charismatic—people, be they individuals with a specific inborn impairment or, especially when this impairment induces a severe intellectual disability, parent-activists, may shape the perception of that impairment. The existence of a collectively elaborated view that stresses the value and dignity of life with a given condition can be immensely helpful for some of the people so impaired and their families. It can also generate problems for others with that condition, because a dominant view of a given disease/disability makes the development of other views difficult. Where the predominant understanding of an inborn condition (for example, polycystic kidney disease) is not shaped by "biosocial" activism, affected people and their families may have more options to construct their own narratives about the meaning of life with this condition.

Health crises, the US writer and columnist Meghan Daum argues, are by definition chaotic. They don't always impart lessons, and contrary to what we like to tell ourselves, they're just as likely to bring out the worst in people as the best. These crises tend, however, to be reformulated as redemption narratives. Such narratives help the affected people cope with the crisis and attenuate the guilt and the fears of family and friends. Daum adds that they also display a cultural preference for sentimentality and neat endings over honesty and authenticity.[59] The arbitrariness and messiness of life—including life with an inborn impairment—is seldom a theme of uplifting stories. It is also rarely present in debates about reproductive decisions. Discussions about the expressivist objection to PND, and those holding the diametrically opposite view, the parental "duty" to strive to give birth to healthy children, are often couched in absolute terms.[60] But women's reproductive decisions are rarely guided by rigid views. According to the disability rights activist Tom Shakespeare,

> Logic and consistency and resolution are required of academic bioethics. In practice, pragmatism, feelings, and lived experience may lead individuals to

decide in ways that are not intellectually rigorous, and this is not always a bad thing, especially if the particular questions are not definitively resolvable anyway.[61]

PNDs of genetic conditions transmitted in families are but a small fraction of positive PND results. A majority of women who receive a diagnosis of severe fetal malformation are not aware of the presence of a hereditary condition in their or their partner's family. Today (in 2017), most fetal anomalies found through PND, chromosomal anomalies (especially Down syndrome) and structural anomalies of the fetus, are seen as accidents of pregnancy. A woman who receives a positive PND result is often unexpectedly confronted with a need to balance multiple risks: for the future child, herself, her family, sometimes the larger community.

Chloe Atkins, a lesbian academic who is a wheelchair user as a result of a rare neurological disease, discusses the irreducible complexity of dilemmas produced by the possibility "to see what is about to be born." Atkins and her pregnant partner long hesitated whether to undergo PND, weighing their reluctance to terminate a pregnancy for a fetal impairment against their apprehension over the need to care for a disabled child.[62] They were unwilling to put the pregnancy at risk, and were haunted by the possibility of a second-trimester abortion (they did not have access to first-trimester testing for Down). Atkins was torn between her identity as a feminist disability activist and the one as a future parent. She believes that she has a rich and fulfilling life despite her physical limitation, is strongly committed to the social model of disability, and is sensitive to the argument that the "elimination" of flawed fetuses is a negative statement about people with disabilities. On the other hand, she was also aware of the fact that the costs associated with a child's disability have to be borne primarily by the child's parents, and was not sure whether she had the psychological reserves needed to deal with a child whose care would require great patience. Atkins knew that people with Down syndrome can lead happy and fulfilling lives, but also that some among them have severe health problems, and that some suffer because of their impairment. She acknowledged that her hesitations about whether she was willing to parent a disabled child were motivated, not only by an aspiration to protect the future child, but also by a wish to protect herself and her family from a risk of material and psychological hardship:

> For me, it was all about looking after my own interests and those of my family members who were already here. I didn't like this characterization of myself, but my own logic wouldn't support any other conclusion. I could not comfort

myself with a belief that in considering such screening I was behaving altru-
istically. . . . In the end, I cannot say whether I would choose to abort a fetus
or not. I have yet to confront that situation. What I do know is that such de-
terminations are highly contextual and not always rationally based—even for
the most well informed and well educated.[63]

Atkins's recognition of the complex motivations of women who decide
whether to use PND contrasts with a simplified representation of such decisions. As Michael Bérubé, the father of a Down syndrome child and a dedicated advocate for people with this syndrome has argued, the "choice" discourse is meaningful only if pregnant women and their partners can elect
whether they wish to continue a pregnancy after a positive PND knowing that
they will not be criticized or ostracized for their preference—whatever it is.[64]

Gendered Care

Women in industrialized countries are told that they have either already
achieved equality with men or are very near achieving such equality. Women
indeed have achieved legal and political equality, obtained access to education and the professions, and are able to control their fertility. Despite these
crucial attainments, masculinity continues to be defined (mainly) through
a capacity to act effectively in the external world, while femininity continues
to defined (mainly) through women's (hetero) sexual attractiveness, and
their maternal and caregiving qualities.[65] In all the current societies, including the industrialized ones, women continue to be seen as "natural" caregivers and to carry the main responsibility for care tasks.

The modern image of a woman as, above all, mother and caregiver came
into being two and a half centuries ago. In 1762, in the fifth chapter of his
Émile; or, Treatise on Education, which deals with the education of women,
Jean-Jacques Rousseau explained that women are fully equal to men: they
have the same organs, the same needs, the same faculties, the same physiology. With, however, one crucial difference:

> The consequences of sex are wholly unlike for man and woman. The male is
> only a male now and again, the female is always a female, or at least all her
> youth; everything reminds her of her sex; the performance of her functions
> requires a special constitution. She needs care during pregnancy and freedom
> from work when her child is born; she must have a quiet, easy life while she
> nurses her children; their education calls for patience and gentleness, for a
> zeal and love which nothing can dismay; she forms a bond between father

and child, she alone can win the father's love for his children and convince him that they are indeed his own. What loving care is required to preserve a united family! And there should be no question of virtue in all this, it must be a labor of love, without which the human race would be doomed to extinction.[66]

To put it in a nutshell, men *have* a sex, women *are* a sex. The French word for "sex," *le sexe*, was in the eighteenth century a synonym of the word for "women." A man's interest in sexuality and reproductive function is only a small part of his life ("The male is only a male now and again"), but a woman's entire life is governed by her reproductive functions. The essence of womanhood is mother's care work.

Two hundred and fifty years later, women in industrialized countries occupy a very different position in society, but Rousseau's injunction that women's chief task is "loving care" is still valid. In both conservative and some variants of feminist discourse, women are expected to dedicate themselves to nurturance and the care of others. Nurturance is opposed to the capitalist marketplace, and stands for unconditional, self-sacrificing support and love. Such a nurturance discourse, the anthropologist Faye Ginsburg has explained, carries contradictory values: it is a source of female power, a counterdiscourse for capitalist values, and a factor in women's subordination.[67]

Women's care tasks are not limited to the care of children, but for many people a woman's achievements as a mother continue to be seen as the central tenet of female identity. In many important ways, women continue to be judged by their ability to be the right kind of mother that produces the right kind of children. The definition of what exactly a "good mother" is varies greatly in different societies, but not the principle that women can be judged harshly if they fail to conform to their maternal role. Recent debates about breast-feeding illustrate this point. The dominant point of view is that a woman who does not breast-feed her baby or does it for too short a period, presumably for selfish reasons, deprives this baby of the chance to be healthy and intelligent. Critics of this view argue that the importance of breast-feeding in affluent Western societies has been greatly exaggerated, and the observed differences between breast- and bottle-fed babies mainly reflect differences of class and social status, since lower-class mothers often cannot take the time necessary for breast-feeding. The real focus of the breast-feeding debate, such critics argue, is not babies' health but the definition of a good mother as a woman who gives an absolute priority to her child's needs.[68] In comments on a 2015 *New York Times* article, "Overselling Breast-Feeding," some readers argue that "the pro-breastfeeding initiative has taken yet another very personal choice for women and transformed it into a public forum for exercising judgment over women's bodies," while others claim that women should breast-feed even if this is difficult for them, "because having a baby isn't about you, it's about doing what you can to give your baby the best start in life."[69]

The breast-feeding controversy illustrates the new focus on "intensive motherhood." The improvement of women's status in society also increased the pressures on women to be perfect mothers who produce smart, accomplished, and successful children. In societies that proclaim gender equality, the success of the woman as a mother often continues to define her success as a human being: "The whole culture supports a mother in the opinion that her children are what she has made them . . . even if she has a good job, her happiness and her sense of achievement rest far more on her children's physical and emotional perfection than does her husband's."[70]

Paradoxically, a child's "minor" impairment can occasionally produce more severe maternal distress than a severe one. Disabilities described by health professionals as "minor," the activist Helen Harrison noted, can keep children from living independently, or from ever having a social life or being able to function in society.[71] Moreover, a child's problems may be attributed to a faulty maternal attitude. When the child merely falls behind the achievements of other children, the mother may feel ashamed and guilty. Her feelings of inadequacy as a mother may be amplified, especially when the child has mild or borderline learning difficulties, by the child's despair upon realizing that s/he cannot perform as well as his/her peers.[72] Similarly, if the child develops psychological or behavioral problems, the mother is blamed for the child's troubles, and not infrequently internalizes such blame.[73]

Because of their key role in the shaping of the next generation, mothers tend to be depicted in simplified, black-and-white terms. Two research groups in my department investigate motherhood and child care, yet rarely interact with each other. One, connected with the French National Disability Institute, is strongly committed to the progress of disability rights. The image of motherhood presented by researchers in this group draws on an idealized view of families we live by. Mothers of disabled children can make mistakes or buckle under excessive strain, but members of this group implicitly assume that motherhood is always benevolent, and that mothers' main motivation is an unconditional, selfless, and unambivalent love for their children, especially impaired ones. The second research group studies child abuse by parents, with a special focus on mothers who kill their young child, then present the child's death as an accident or a crib death. The image of motherhood elaborated by members of this group is grounded in a pessimistic view of families we live with. They view motherhood as a risky and unstable condition, are convinced that mothers often harbor ambivalent or negative feelings about their children, and believe that some are capable of extreme violence. Accordingly, researchers in this group assume that when professionals encounter unexplained child abuse, they should consider the mother as a potential culprit until the contrary is proved.

The increasing responsibility and stress associated with motherhood in the highly competitive Western society of today, the sociologists Rosemary Hopcroft and Julie McLaughlin argue, can explain the paradoxical finding that in countries with high gender equality, the gap in rates of depression between men and women is greater than in countries with lower gender equality. This difference is attributed mainly to the greater stress involved with parenting, which has a disproportionate influence on women. Maternity increases the frequency of depression in women in high-gender-equality societies, while it reduces the frequency of depression in women without paid employment in low-gender-equality societies. The overall levels of depression are lower in countries with high gender equality, which, it so happens, are also more affluent than those with low gender equality, but equality between men and women benefits men more.[74]

When a woman gives birth to a disabled child needing a significant investment in care, the probability she will carry out most of the care-related tasks is high. As the anthropologist Ayo Wahlberg proposes, it is not possible to discuss decisions about the birth of a child with a genetic disease without taking into consideration the financial, psychological, social, or emotional effect of his/her care on the caregivers.[75] Physicians who participated in early debates about selective abortion for a fetal impairment openly discussed the costs of care for mothers. The Australian pediatric ophthalmologist Norman Gregg, who first described (in 1941) inborn malformations induced by maternal infection with the rubella virus, explained that "in some cases the strain imposed on the mother in caring for the child is more than she should be expected, or permitted to endure."[76] Commenting in 1959 about termination of pregnancy in women infected with rubella—a frequent practice, despite the illegality of abortion at that time—the British pioneer of clinical genetics Julia Bell argued, "There are three main aspects of this problem concerned with (a) the risks of severe handicap to the unborn child, (b) the risks of acute distress and difficulty for the potential parent, perhaps for the rest of *her* life, (c) the burden likely to rest upon the Welfare State."[77]

In 1959, Bell's expression "her life" was unlikely to be a manifestation of politically correct language for that time; it is much more probable that it was, like Gregg's concern for woman's plight, a recognition of women's role as the main caregivers of impaired children. Fathers can become dedicated advocates of their children's cause—and write moving books about them—but they tend to be less involved in the physical aspects of caring for disabled children and less inclined to give up their professional aspirations to dedicate themselves to such care.[78]

Mothers' devotion to children with inborn impairments may in some cases reflect a deeply felt conviction that they failed this child. Their feelings of guilt and shame, however irrational, may compel them to compensate through a dedication of all their time and energy to the care of the impaired child. Some women return to a traditional homemaker role, while others find ways to fit work commitments around their heavy care burden by reducing their hours, working altered schedules, passing up promotions, or working part-time. Many bear in silence the costs of insufficient support from family, community, and the state.[79] Moreover, economic restrictions may aggravate the caregivers' difficulties. In the United Kingdom in January 2017, David Mowat, Parliamentary Under Secretary of State for community health and care, evoked "the rapidly rising life expectancy among people with learning disabilities whose care was very expensive" as one of the main reasons for escalating costs of care and the need to rely more on services provided by the family.[80] Becoming the primary caregiver of a disabled child/person can be seen as a more meaningful and fulfilling occupation than a full-time job—be it in a supermarket or a law firm. It can also come at a significant cost.

A 2014 study has shown that training in psychodynamic approaches to emotion regulation helps mothers who care for adult children with intellectual disabilities or psychiatric diseases cope with the difficulties generated by care tasks. At the start of the study, 85% of the mothers participating reported significantly elevated stress; 48% said they were clinically depressed, and 41% reported anxiety disorders. After they had received training in psychodynamic methods, many participants stated that the methods helped them deal with their stress and take better care of themselves and their children. The authors of the study explain that it is important to teach these mothers how to quell distress and anxiety, because they often have a lifetime caretaking commitment to their child and therefore need a long-term solution. Training in psychological methods will make their difficult task more bearable. With advances in pediatric care, the authors conclude, more children with neurological, cognitive, and psychiatric disabilities are living well into adulthood, and most continue to reside with their aging parents. Care for adults with intellectual disabilities often compromises the physical and mental health of their caregivers, nearly always their mothers, increasing the need for approaches which will protect the mothers' health.[81]

Care tasks may be presented as a "natural" part of the maternal role. Some opponents of PND have promoted a traditional model of the self-scarifying, entirely selfless mother.[82] Women from less-privileged social strata who are the main caregivers of their impaired child may face additional problems such as poor health, difficult family circumstances, and precarious economic status, which make their task especially challenging. Women from more privileged social strata who care for an impaired child also face many practical and emotional difficulties, but their plight may be partly alleviated by their greater capacity to purchase goods and services, and better skills in securing for their children the advantages provided by their community. Paradoxically, an increase in the level of help for disabled children in many industrialized countries might have contributed to the widening of the gap between middle-class educated mothers and major segments of lower-class ones.[83]

Many women internalize the idea that their first obligation is to their disabled child. Mothers of severely disabled children fear being criticized by people around them if they decide to place their child in institutional care, and many believe that doing so would be a severe transgression of their maternal duties. Helen Featherstone—whose son Jody was severely brain damaged following a toxoplasmosis infection during pregnancy, and who wrote a deeply moving and lucid memoir about caring for a severely disabled child—explained:

> I was afraid that the world will judge me as a rejecting, inadequate mother if I let Jody live somewhere else. If there was one person in all the world who opposed placement, I did not want to do it. I paid no attention to a phalanx of friends and relatives advising institutional care; I figured I was doing more, not less, than they advise.[84]

Women choosing to delegate the care of their disabled child may indeed be condemned for this decision. The musician and orchestra director Julia Hollander, overwhelmed by the care needs of her daughter Imogene, who suffers from a severe form of cerebral palsy and needs constant attention, ultimately placed her in foster care on weekdays, explaining that she felt unable to be a full-time mother for her. She then wrote a book about her experience. Hollander was harshly criticized by some of the book's readers, who saw her as narcissistic, self-centered, and career obsessed. These critics rejected Hollander's claim that she was not the right person to become a full-time caregiver for a severely disabled child, arguing that this is not a question of personal choice. A woman who gives birth has a nonnegotiable obligation to become a good mother and put her child's interests before her own. Women who fail to live up to this standard may be accused of being selfish, cold, and unfeeling.[85]

The good mother of a disabled child is described by the clinical geneticist Natasha Shur as a "real tiger mother." Such a mother

> fights illness, schools, hospital policies, and insurance companies in order to advocate for her child. She follows her child's cues, accepting that the future is not always in her control. At the same time, she works to maximize her child's potential. . . . The genetics clinic mother does not always accept medical advice or conventional parenting approaches. With brilliance and resilience, she creates a supportive and child-centered world. She battles not against her child but for her child. This mother's song and prayer is universal: happiness and well-being. Her love is unconditional. In times of adversity, she demonstrates the strength of a tiger, and the heart and soul of a mother. . . . And, with hard work and constant efforts, her child can make more progress than was expected by the initially pessimistic expert.[86]

"Tiger mothers" of nondisabled children typically end their task when their adequately coached child enters an Ivy League college or its equivalent. The task of "tiger mothers" of children with severe inborn impairments never ends. Judging from the diagnoses of children described by Shur as making remarkable progress thanks to their mother's dedication, it is probable that many among them will need a significant level of care as adults, and will continue to receive such care from their mother.

According to the Norwegian sociologist Halvor Hanisch, one of the ways in which parents, especially mothers, confronted with the difficulties of caring for a disabled child, cope with the strains of such care is the production of a "hyperbolic discourse of love."[87] Parents of disabled children express grief and distress:

> With Jesper's birth we were forced to say goodbye to the life we had been living including all those dreams, hopes, and wishes for his life—and say hello to a new and unknown life of the sorrow, the shattered dreams and hopes, the difficulties and the work with him and all the time it takes.[88]

At the same time, they strongly affirm the positive aspects of dealing with a special-needs child:

We have no doubt that there's a reason Jesper is here. He helps put our lives in perspective. He has helped teach many classmates, children of friends, and cousins that even though someone is different they can still have a valuable place in this world.

For me, it has been a huge gift to have a disabled child. It is, of course, always a gift to have a child, but life with Markus has opened so many doors to many other sides of life, aspects which otherwise we never would have known of.

I am convinced that it, all in all, has made me stronger and more whole as a person. . . . What life with Solvej has taught me is that there is strength, richness and community in recognizing our dependence on and need for each other.[89]

Norwegian parents' powerful discourse of love echoes statements of disability rights activists such as Gail Landsman, Eva Feder Kittay, Emily Rapp, and Michael Bérubé, who provide rich, nuanced descriptions of the substantial rewards of "special parenting."[90] The tension between two discourses on the care of the special-needs child—one focused on the difficulties of dealing with the child's impairment, the fragile nature of the existing material arrangements, and the permanent struggle with acceptance of the child's condition, and the other focused on the gratifications and joys of parenting this child—is partly resolved, Hanisch argues, through the language of love. The narrative of an "unconditional acceptance" is not a factual description of a situation. It expresses parents feeling how they "can" or "hope" to feel. This misinterpretation is a powerful reflection of parental love.[91] In a hyperbolic discourse in which the narration itself is an act of love, disabled children are described as "angels," "goodness personified," and a source of "pure and whole-hearted love." Idealized images of such children in present-day narratives recall some of the images of them in nineteenth-century literature, such as the archetypal personage of Tiny Tim in Dickens's *A Christmas Carol*.[92]

Hanisch proposes that an idealized image of a disabled child in the parental "hyperbolic discourse of love" strongly resonates with an idealized view of one's partner in romanticized images of heterosexual love.[93] This is a felicitous comparison. It attracts attention to the role of fictional narratives in helping couples create an idealized image of their life, the "couple we live by." This fictional image performs an essential task: it sustains human bonds and promotes desirable behavior patterns such as cooperation, selflessness, understanding, and compassion. It provides the "couples we live with" an ideal to which they strive, but also a blueprint for daily behavior. On the other hand, feminists have pointed to the role of idealized images of heterosexual love in the reproduction of unequal gender relationships and in keeping many aspects of women's discrimination private and thus invisible. The feminist slogan "The private is political" aims, among other things, to highlight oppressive aspects of the "politics of normalized heterosexual love" when it is translated into an obligation for women to provide nonreciprocal attention and care to men.[94]

The opposite of love is hatred, and it, too, can exist. The disability activist Tom Shakespeare evokes the thorny issue of less than happy outcomes for disabled people and their families. Some families, Shakespeare argues, are an essential source of loving support for their disabled members, but others can become a locus of abuse, inequality, and oppression for them.[95] Families can also harbor complex relationships, and be divided over support for the disabled people in their midst. Helen Featherstone described an older woman who in her sixties still cared for her forty-year-old severely disabled son: "At sixty her strength was failing; she could no longer get him out of bed. The result of her life work: her other children hated her for sacrificing the family to their disabled brother."[96]

An imaginary letter to a "beloved older sister" with Down syndrome (now deceased) displays the tensions between the discourse of love and the costs of care. For many years, the younger sister did not realize that her sibling had a special problem. As young children they felt close; later they drifted apart. Her mother was pained by her unfolding destiny as a full-time, lifetime caregiver of her older child. The younger sister left home, became a teacher, married, had children, and distanced herself from her sibling's problems. As she relates, the older sister developed serious health problems in addition to her developmental delays, and their mother became increasingly overwhelmed by her caretaking duties:

> I would have you both to stay for a week at a time so I could properly see you but I can't deny how exhausted I would feel by the end of it. . . . I tried to cheer you both up but when I had my own children it became really hard to keep everyone happy at all times, and I realized I had to put them first. . . . As our mother became depressed I told her she needed a break but she wouldn't listen. The phone calls were hard and listening to a suicidal mum before putting tea on the table for your nephew and niece began to take its toll to the point where I'd sometimes avoid any phone calls if I had things to do.[97]

After the death of her sister, the writer remembers her with great sorrow, and recalls the important moments they had shared:

Your presence has graced me with an appreciation and zest for life and a tolerance of others. I will always remember how it felt when you took my hand and kissed me when I felt sad. Rest in peace, big sister. Your humor and kindness will stay with me for ever.[98]

Down syndrome is at the very heart of PND debates, especially the thorny problem of long-term care of people with more severe variants of this condition. The solution for such a conundrum of care, disability rights activists argue, is to socialize a significant proportion of the care tasks. Women become depressed and suicidal because they are obliged to cope alone with a difficult situation. The problem is not the care tasks by themselves but the caregiver's isolation. Mothers of disabled children should be able to rely on effective and compassionate care provided by the community and on networks of solidarity and support. Everybody agrees with this proposal—in principle. Yet shifting the caregiving for disabled children and adults from their families to society does not seem to be a very realistic option, at least in the short term. Even in affluent and disability-friendly countries such as Norway, parents of disabled children speak openly about their struggles, tensions, and fears about the future.[99] The generous discourse on the importance of accepting children/people with diverse bodies and capabilities relies to an important extent on the willingness of parents, especially mothers, to carry on the majority of material and emotional care tasks.

In one strand of discourse on PND and disabilities, women/mothers are seen as selfless Madonnas, "long-suffering mothers whose nurturance is unconditional and ever present," expected to be willing to dedicate a substantial portion of their lives to care for a disabled child. Those who refuse to accept this role and elect a "eugenic" abortion are seen as unnatural women, selfish and narcissistic, who, moreover, contribute to the "genetic genocide" of differently abled people. In another kind of discourse on PND, women are expected to fulfill the role of rational "agents of quality control on the reproductive production line."[100] Such a discourse criticizes the "irresponsible" and "backward" women who reject PND and/or those who decide to continue a pregnancy after a positive PND, a decision that harms their families and increases the burden on the already struggling health care system. Enthusiastic promoters and dedicated opponents of PND and selective abortion hold diametrically opposed points of view, but share a deep distrust in women's capacity to make their own reproductive decisions and an unwavering conviction that they have a moral right to harshly judge women's "deviant" behavior. As the British literary critic Jacqueline Rose puts it, "Society continues to believe it has the right to trample over [the] mental life of mothers."[101]

In 1992, the HBO television film *A Private Matter* told the story of the Arizona TV presenter Sherri Finkbine, who in the summer of 1962 attempted to have a legal abortion after she discovered that thalidomide, the drug she had used early in pregnancy, induces fetal malformations. She went public with her decision, lost her job, and encountered violent reactions and threats. In the movie, Finkbine, played by the actress Sissy Spacek, declares, "I'll have the baby. Spend the rest of my life taking care of him like everybody wants. And everybody'll say, 'Isn't she wonderful?'" Commenting on this movie, the US author Anna Quindlen wrote: "So many people wanted to tell her what was best for her life, then walk away and leave her to live it. So many want to do that today: insist that they know best what makes us good mothers, women, people, and then leave us to live with the consequences of their convictions."[102]

A Nonscrutinized Diagnosis

A Total Social Conflict with Layers of Opacity

The detection of a fetal malformation "[makes] things more complicated, more and more complicated—as if it could get any more complicated."[1] PND is indeed an irreducibly complex and difficult subject. Women's decisions whether to undergo prenatal testing and their reactions to a positive result are molded by a wide range of personal, institutional, and structural variables that shape their attitude toward the risk of giving birth to a special-needs child. As a woman who underwent an abortion after diagnosis of a fetal malformation puts it, "Nothing is ever simply black and white. And laws too often are not even able to cover all the areas of gray."[2] However, debates about the consequences of the generalization of the PND *dispositif* are rarely dominated by multiple shades of gray. They are frequently entangled in "affective economies" mobilized for political goals.[3] Debates about selective reproduction have become highly contrasted studies in black-and-white, or, to follow the sociologist and bioethicist Charles Bosk, "essentially-contested total social conflicts" (ECTSC).[4] Such conflicts,

> very quickly, turn three-dimensional individuals, who once lived within dense social networks with unique biographies, into two-dimensional iconic figures; they become symbolic proxies for this or that group's interests. As ECTSCs, the concrete cases that dramatize and legitimate the need for bioethics are transformed into platonic essences. The complexities, nuances, and subtleties that are embodied in private clinical conflicts are erased once those conflicts become collective social spectacles. What is lost in ECTSCs is the "hereness" of place, the "this personness" of the parties in conflict, and the intense "nowness" of a private dilemma that has escalated into a public conflict. The loss of these three distinctive local elements results in all the subtleties

and shadings that create the decision-making paralysis being washed out by the klieg lights and culturally resonant dramatic framings that mark the discussion of private troubles as public issues receiving media attention.[5]

PND has become an "essentially-contested total social conflict" as a result of its entanglement with a highly polarizing issue: selective abortion of impaired fetuses. Two powerful constituencies joined forces in an opposition to selective abortion: the "pro-life" movement, and an influential segment of disability rights activists.[6] Despite their major differences, these two groups implicitly agreed to equate a fetus with a child, and an abortion of an impaired fetus with a "eugenic" elimination of an impaired individual—a view strongly rejected by advocates of a woman's right to decide whether she wishes to continue a pregnancy.

The polarization of debates about PND led to the masking of some of the key problems linked with the broadened use of PND. Like a classic detective story in which each protagonist attempts to hide something, nearly all the stakeholders in debates about PND are reluctant to discuss specific issues. Feminists, who struggle, especially in the United States, against a rising tide of opposition to abortion rights, are reluctant to discuss the liminal and unstable status of the fetus; the possibility that a woman can be ambivalent about her reproductive decisions; and, if she decides to terminate a pregnancy, can be deeply distressed about the loss of her "baby" as opposed to a clump of cells. Health professionals are reluctant to discuss how the broadened use of PND may advance their professional interests and strengthen their power over specific jurisdictions. They are also reluctant to talk about PND as a profit-generating endeavor and—especially before the rise of non-invasive prenatal testing, a PND technology entirely driven by commercial interests—about the key role of manufacturers of instruments, reagents, and software in the circulation of the PND *dispositif*. Public health experts, who in the 1960s and early 1970s openly compared the costs of the broadened use of PND with the savings achieved through preventing the birth of disabled children, do not mention this issue anymore, despite the fact that in all probability, cost-efficiency considerations continue to play an important role in shaping health policies.[7]

A woman who elects to terminate the pregnancy more often justifies this decision, perhaps above all to herself, by an undoubtedly sincere wish to prevent the suffering of her future child and, if applicable, to protect the well-being of her other children (the "good mother" posture). Her decision is less often explicitly justified by the fear that the birth of an impaired child will seriously limit her own life options and may endanger her marriage (the "bad mother" standpoint). Advocates of a woman's right to end a pregnancy of an affected fetus carefully avoid discussions about the material aspects of late-term abortions. Opponents of pregnancy termination for a fetal indication who do not reject abortion in general are reluctant to discuss the great diversity of inborn impairments; the predicament of some "unhealthy disabled"; the material and emotional difficulties of mothers/parents/siblings of severely impaired children; the specific problems of children/people with an intellectual disability or a psychiatric disorder; and the difficulty of predicting how the birth of a special-needs child will affect the child's family. Often, advocates of PND either evade discussions about problems produced by a rapid increase in the number of prenatally diagnosed conditions with a variable expression and an uncertain prognosis, or they argue that these problems will be solved when all the pregnant women and their partners have access to appropriate guidance and counseling.

The existence of a great number of unmentionable topics had led to the rise of wide zones of silence around PND and the termination of pregnancy for fetal indications. In discussing the decision to introduce screening for Down syndrome in France, a French public health expert explains that politicians could not openly admit that their aim was to reduce the number of children born with chromosomal anomalies in order to decrease the costs of care for these children. It was necessary for them to "wrap up" this goal, and promote the new measure using the uncontroversial argument that screening for Down syndrome will decrease the number of amniocenteses, and thus will reduce health care expenses and limit the risk of abortion of an unaffected fetus.[8]

Debates about PND often eschew—or mask—many uncomfortable aspects of the use of this technology. In Poland, as in Brazil, abortion is illegal but very common. Affluent Polish women can terminate a pregnancy in safe conditions, while vulnerable women—young, poor, isolated, uneducated, marginalized—suffer disproportionately from the consequences of unsafe abortions. To protest against this situation, Polish feminists published a book, *A Like Hypocrisy: An Anthology of Texts on Abortion, Power, Money and Justice*.[9] One can imagine a study of PND called *P Like Murkiness: On Prenatal Diagnosis, Uncertainty, Risk, Gender, and Politics*.

Situated Choices

The term *prenatal diagnosis* encompasses two distinct sets of practices: universally accepted tests destined to improve the health of the mother and her future child; and more controversial tests, which make fetal anomalies visible and are destined to allow pregnant women and their partners to decide the pregnancy's future.[10] In countries where abortion is legal, women who receive a positive PND are told that they have the right to decide their pregnancy's fate.[11] Pregnant women who are not opposed to abortion believe that since they will have the main responsibility for the care of a disabled child, they will strongly defend their right to choose the pregnancy's fate.[12] At the same time, the choice discourse often masks the paucity of a woman's real-life options and her limited control over specific situations.[13] Pregnant women cannot choose the kind of society they live in or their socioeconomic status. They also cannot select society's attitudes toward people with disabilities, especially those who diverge from the iconic image of a cute, smiling (and presumably smart, cheerful, and well-behaved) child in a wheelchair.[14]

Disability rights activists often praise parental, and especially maternal, investment in the education and training of special-needs children, which enables these children to reach their full potential. Paradoxically, stories about exceptional families who raise disabled children may increase pregnant women's apprehension. Some women may fear that they and their unexceptional families will not be able to provide intensive care and unfailing support to their disabled children, and will be forced to deal with the guilt of failing these children. Women living in industrialized countries often make a realistic assessment of the support they would be entitled to receive for an impaired child; the amount of energy they would need to expend to secure their child's access to such help; the difficulty of maintaining this help over time; the danger of depleting the family's financial and emotional resources; and the degree of uncertainty about the fate of disabled people in economically and politically unstable times. The focus on women's autonomy often masks the constraints on people's individual choices in a state or market system.[15] Women who undergo PND frequently have limited choices and a heavy responsibility. They are loudly told that it's "up to them" to decide whether to continue their pregnancy and then, not infrequently, are harshly judged for their decision.[16]

Choices of women who live in developing countries are even more restricted. Brazilian women who use the national health service and cannot afford private genetic counseling can either accept "God's choice"—and their physicians' explanations that life is risky and unpredictable—or live with a feeling of frustration and injustice.[17] Tine Gammeltoft's study of the use of ultrasound in Vietnam similarly displays pregnant women's limited choices in a resource-poor environment. In Vietnam, unlike Brazil, abortion for a fetal impairment is not only legal but often encouraged. Vietnamese physicians who diagnose fetal malformations affirm that they always respect women's autonomy and their right to decide what the fate of their pregnancy will be, but in practice are often paternalistic and authoritarian. Vietnamese women who undergo a late-term abortion for a fetal indication seldom learn precisely what the diagnosis was, and do not receive counseling or psychological support. They are also painfully aware of the limited resources available to help disabled people in their country, and the opprobrium these people and their families incur. Moreover, in some cases the abortion itself is performed in difficult conditions, and physicians' attitudes toward women undergoing the procedure are related to their capacity to pay for medical services. Consequently, Gammeltoft's study of PND in Vietnam illustrates, among other things, the brutality of a neoliberal medical system in which physicians' interventions can occasionally be guided by greed, not their obligations to patients. It also displays the paucity of choices offered to women in disadvantaged settings who were diagnosed as having severe fetal defects.[18]

Situated Dilemmas of PND

The more extensive use of PND allows a pregnant woman to decide what level of her future child's risk of disability she is willing to tolerate, and what she perceives as an "acceptable" human being.[19] Such decisions are never simple. PND of a fetal anomaly associated with an option to terminate a pregnancy will, in all probability, remain a complex, controversial, and divisive issue. The main argument developed in this book is, however, that studies of the material and social technologies that enable us "to see what is about to be born" and a fine-grained analysis of situated uses of PND yield a better understanding of what precisely can be seen in a specific time and place, how it can be seen, and what the consequences of seeing what is about to be born are.

Many of the dilemmas produced by PND, such as the need to deal with uncertainty, fear, and sorrow caused by physicians' "bad news," the blurring of the boundaries between normal and pathological, and the dangers of overdiagnosis, are not qualitatively different from those produced by other diagnostic technologies. The diagnosis of a fetal malformation is very distressing, as is the diagnosis of a malignant tumor. Either finding may bring the certainty of a bad outcome (metastatic lung cancer, anencephaly) or indicate the existence of a health risk (localized breast cancer, Turner syndrome). Yet PND is also a unique diagnostic approach, because in the first half century of its existence, the main practical consequence of its use was the prevention of the birth of human beings possessing specific traits.

A main goal of the traditional practice of carefully selecting a prospective mate was the promotion of the birth of children with desirable traits. PND proposed a radically new way to achieve this goal: decisions grounded in technologies that make possible a direct scrutiny of the living fetus by health professionals. This unique feature led to the transformation of debates about PND into an "essentially-contested total social conflict" and a collective social spectacle. Such a transformation was made possible through a "purification" process that reduced complicated, multilevel situations to a series of simplified, one-dimensional "exemplary cases."[20]

Debates about PND are frequently shaped by abstract notions, affirmation of general philosophical and moral principles, and discussion of hypothetical cases and futuristic projections. It is not possible to eliminate the dilemmas produced by the diagnosis of fetal anomalies, or put an end to controversies on selective reproduction. Precisely for this reason, it may be especially important to have a discussion about current and future uses of this diagnostic technology that is as accurate and nuanced as possible, and avoid schematic views, excessive generalizations, shortcuts, and above all the "wrapping up" of heated or difficult issues.[21] Replacing broad and not infrequently vague statements about PND with those reflecting a real effort to understand the concrete, situated dilemmas produced by this diagnostic technology can open spaces for creative thinking, solidarity, and compassion.

To conclude, here is a short—and entirely subjective—list of PND's features that are often neglected, marginalized, or made invisible in present-day debates about this biomedical innovation.

PND Is a Mass-Distributed Diagnostic Technology

Examining PND as a routine diagnostic technology will facilitate associating discussions about the implementation and generalization of this diagnostic approach with discussions about the consequences of detection of other pathological conditions and health risks. It will link PND to issues such as newborn screening, preconception screening, dealing with incidental diagnoses, or testing for late-onset diseases. Focus on routine uses of PND will also draw attention to the role of socioeconomic variables and professional interests in the dissemination of this technology. For example, PND experts often react to criticism of their approach by claiming that present-day problems will be solved with more research, more specialization, and more counseling—that is, through the extension of their professional power. This strategy is, however, not unique to PND experts; it is shared by many professional groups within and outside medicine.[22]

PND Is a Feminist Issue

Feminism entered PND debates mainly through the proposal, advanced by some feminist activists, that PND favors the transformation of women into "mother machines," that is, the means of providing flawless offspring. This proposal was adopted, in a somewhat different form, by disability rights activists. Yet feminists discussed many other topics relevant to the understanding of PND and selective abortion: the nature of medical power, the role of experts, the social role of family dynamics in heteronormative societies, the ambivalence of the maternal role, the possibility of intrafamily oppression and abuse, and the relegation of unpaid care work within families to women. Insights produced by the feminist movement from the 1960s through the 1980s were, however, lost when discussions of PND became focused exclusively on abortion rights and the risks of a "eugenic drift," and were disconnected from discussions about family structure, mothers' tasks, and the gendering of care.

Debates about PND Should Not Focus Exclusively on Genetics

Debates about PND focus nearly exclusively on genetic tests. They tend to neglect the importance of the diagnosis of morphological anomalies of the fetus, and the key role of medical imaging technologies in their detection. The definition of PND as a risk management technology refers not only to genetic conditions with a highly variable expression but also to structural anomalies detected with medical imaging technologies. Routine ultrasounds, smoothly integrated into the monitoring of pregnancy and the preparation for childbirth, are at the origin of at least as many abortions for fetal indications as genetic tests. Moreover, many structural anomalies of the fetus with uncertain outcomes, such as brain anomalies linked to uncertain but potentially severely disabling outcomes, become visible only late in pregnancy, increasing the difficulty of deciding the pregnancy's fate.

PND Imperils People's Right to Ignore Their "Anomalies"

The growing scope of screening for fetal defects at the same time increases the probability of incidental findings. It may also lead to the detection of a hereditary anomaly; the subsequent finding of this anomaly in other family members may have dramatic effect on their lives. Many inborn conditions have a variable expression and produce "minimal" impairment in some affected individuals, and/or are expressed only in older individuals. People with such conditions may be unaware of their atypical makeup. Their right to maintain their ignorance, particularly if their condition cannot be treated, may be endangered by the widespread use of genetic testing of the fetus. Such testing may reveal hidden anomalies in parents and other family members. It may also rob children of their right to an open future. Children diagnosed before birth with an anomaly, especially one that usually will not be spotted at birth and does not need to be treated rapidly (for example, some sex chromosome anomalies), or those diagnosed as having an inborn condition that produces an impairment only later in life may lose the opportunity to be perceived by their parents as "normal" and "healthy," especially during their early formative years. They also may lose the possibility of living free from the shadow of their "anomaly."

PND Is Frequently a Diagnosis of a Risk and a Risky Endeavor

A positive PND diagnosis is often the diagnosis of a risk. Even when the diagnosis is seen as certain, in only a small proportion of cases is the prognosis certain, too. Paradoxically, an absolute prognostic certainty is frequently the certainty that the child will not survive. Debates about PND seldom discuss the material and emotional costs of "suspicious" prenatal findings, or the consequences of a diagnostic odyssey which may end with an uncertain result. They also seldom discuss the long-term consequences of stress that may not end with the birth of an (apparently) healthy child: today "birth defects" are not limited to events observed at birth. Moreover, despite the growing interest in epigenetics, discussions about PND rarely evoke the possibility that stress produced by prenatal testing may affect fetal development. It is important to protect women's right to know and their right not to know, and to perceive both an enthusiasm for screening and testing and a reluctance to enter a medicalized trajectory as valid, rational attitudes. It is also important to respect equally the decisions of women who feel unwilling or unable to risk raising a disabled child and those who refuse to terminate a pregnancy with a severely impaired fetus.

Women Who Undergo PND May Be Ambivalent about Their Pregnancy

Ambivalence is an inseparable part of the experience of health and disease.[23] In human reproduction, the indeterminacy of the biomedical encounter is amplified by the fluid and indeterminate status of the fetus, one of the main sources of ambivalence in women's reproductive decisions. The stereotyped view of PND is grounded in a radical distinction between women who decidedly did not want to be pregnant (refusal of *a* child) and those who decidedly want to be pregnant and then, following a positive PND diagnosis, elect to terminate the pregnancy (refusal of *this* child). In real life, many pregnant women have ambivalent feelings about their pregnancy and future motherhood, a stance that may be reinforced by the liminal and unstable status of the fetus. A selective termination of pregnancy may intensify the mixed feelings—of love, grief, guilt, and relief—that often suffuse an "ordinary" abortion experience.[24] Women's ambivalence about their reproductive decisions is often presented as irrational behavior, a view that hampers the understanding of their attitudes toward PND and termination of pregnancy.[25]

The Broad Term *Disability* Fails to Adequately Describe PND's Dilemmas

Activists often use the umbrella term *disability* to describe a wide range of conditions that produce very different impairments, different ways of living with an impairment, and different societal reactions to impairments. An indiscriminate lumping together of very diverse conditions and situations can harm people with disabilities and their caregivers. The popular media images of a disabled individual—a person with limited mobility, a member of the Deaf community, and, more recently, a high-functioning person with Down syndrome—inadequately represent the lives of people with less easily manageable conditions: a progressive disease or a severe intellectual impairment. Focusing on autonomous persons with nonprogressive conditions overlooks the complex issues of dependence and care, and the role of families, particularly mothers, in providing such care. It also provides an inaccurate view of inborn impairments, especially those responsible for the majority of pregnancy terminations for fetal indications.[26]

Debates about PND and Disability Avoid Difficult Questions

People with disabilities and their caregivers struggle with insufficient (to put it mildly) societal support, discrimination, and prejudice. It is not surprising that they evade discussions about issues that may shed a negative light on life with disability and may worsen their plight. Similarly, families of disabled people are usually described by activists as inspirational families we live by, not as actually existing families we live with. Such idealized descriptions do not acknowledge the possibility of ambivalent feelings of mothers and other family members, and those of people with disabilities themselves. They mask complex family dynamics and the existence of un-resolved conflicts. They also leave little room for investigating the difficult task of protecting impaired people from bullying, abuse, and exploitation, sometimes by family members (and occasionally also the abuse of family members by an impaired person), or investigating difficulties in the mate-rial and emotional care of severely disabled children and adults, especially those with intellectual impairment and psychiatric conditions.[27] An ide-alization of mothers and families of disabled people and disabled people themselves may compromise the trustworthiness of testimonies about sub-stantial gratifications and unexpected joys of raising a special-needs child. It is difficult to achieve credibility through the telling of partial truths.

Debates about PND Focus on
Individual Management of "Disability Risk"

PND frequently diagnoses a risk of unknown magnitude. This is true also for the paradigmatic positive result of PND, the diagnosis of Down syn-drome. Moreover, even when it is possible to accurately predict the level of impairment of the future child, it is difficult to predict the consequences of the child's impairment for her/his family: the outcome can be much better or much worse than expected. The pragmatic way to deal with unknown health risks is through health insurance, that is, shifting an important part of the risk to the collectivity. Such collectivization of a risk may be espe-cially important when the pregnant woman is facing a small but significant chance of a truly bad outcome. Women told about the 10% probability that their future child will be severely impaired often do not think about the 90% chance that this will not happen, but dwell on the one-in-ten chance that they will spend the rest of their lives as the main caregiver of this child. Access to high-quality, flexible, compassionate, and imaginative care solu-tions for all the impaired children and adults, coupled with a true respect for mothers'/families' decisions about the kind of care they want for their children could help pregnant women decide to eschew PND or, when faced with a positive PND, to accept the risk of giving birth to a impaired child. Today (in 2018), the availability of high-quality, collectively funded care for all disabled people is not a very realistic perspective in the great majority of industrialized countries, even less in intermediary and developing ones. When such care does not exist, heated debates about pregnant women's "choices" are mostly rhetorical exercises.

Debates about PND Promote a Restrictive Understanding of Human Diversity

The promotion of human diversity is not only an abstract acceptance of different bodies. It includes all the labor needed to help people with different bodies have satisfying and dignified lives. People who have only one hand and those with quadriplegia are equally entitled to full participation in human society, but the latter group needs a much greater investment in care to be able to do so. The same is true when one compares people on the high and low ends of the autism or Down syndrome spectrum. The generous discourse on the necessity of supporting human diversity relies in practice on free labor supplied by families, who are expected to be a source of unlimited support and love. Today, industrialized, democratic societies embrace—theoretically, at least—a great variety of sexualities and couple/marital arrangements, multiple sex/gender identities, and many medically enhanced approaches to reproduction. But in discussions about impaired children and their mothers/families, these same societies mainly propose sentimentalized images, some of which would have made the Victorians proud. The "human diversity" discourse within the PND debate usually focuses on the diversity of human bodies, not the diversity of affective links between human beings and the variety of societal arrangements. Such a discourse frequently assumes the absolute superiority of blood ties over other kinds of social bonds, and rarely takes into account the possibility of multiple and atypical models of parenting, kinship, and care.

Staying with the Trouble

The role of the academic, the British classics scholar Mary Beard explains, is to make issues more complicated.[28] This may be easier when dealing with ancient Rome than with recent developments. Scholars who deal with difficult topics and fundamentally unresolved, profound moral and material questions are caught between an aspiration to be policy relevant, and thus to simplify the debated issues, and the wish to be faithful to their material, and therefore to be especially attentive to context, contingency, and complexity.[29] Faced with a temptation to tell pacified histories, the feminist scholar Michelle Murphy argues, we should keep in mind the importance of working through discomfort, worry, anger, and pain. A politics of unsettling is cracking open the smooth into accounts of the messy and the partial, and strives to stir up and put into motion what is sedimented. At the same time, it embraces the generativity of discomfort, critique, and non-innocence.[30]

Attracting attention to neglected, marginalized, and occulted aspects of PND, the contradictory trends in its present-day development, and the tangled situations generated by its spread promotes reflexivity and critical thinking, a modest but nevertheless essential task. As the sociologist of science Olga Amsterdamska explained when she became the editor of the journal *Science, Technology and Human Values*, "All too often, political, philosophical and ethical positions rely on an assortment of 'illustrative examples' or deeply held convictions about what the world is like that are scantily supported and weakly argued. This is not enough."[31] At a memorial service for the historian of public health Barbara Rosenkranz, her granddaughter said that a few days before her death, Rosenkranz, already very feeble, had said to her daughter, "I'm confused." The daughter answered, "You'll always be in my heart." Rosenkranz, who suffered from dementia in her final years but in her lucid moments continued to display her great sense of humor, retorted, "That's so amorphous."[32] Sweeping statements such as "disabilities and disorders are also characteristic of the way in which members of the human species function. Human normality encompasses—or could encompass—disability and disease" are not false but amorphous.[33] In discussing whether and how to implement new biomedical technologies that have major social, cultural, moral, economic, and political ramifications, an affirmation of generous principles is not enough. When facing multifaceted, contentious practices, it is vital to carefully examine all their aspects, including those which generate discomfort, promote unsettling views, confront difficulty, and stay with the trouble.[34]

Minerva's Owl and Apollo's Lyre

PND, this book argues, is a path-dependent, situated, gendered risk management technology. In order to understand how this technology works, one should study the elements that shape medical practice, such as the medical division of labor, the organization of health care, the structure of medical institutions, professional and economic interests, activism networks, and legal frameworks. One also should study sociocultural variables, such as views of pregnancy, the fetus, and "responsible motherhood"; attitudes toward disabled children and adults; and caretaking duties of mothers and families. Yet all this is still only part of the story.

As a researcher and teacher, I strongly believe in the virtues of scholarly analysis. Yet, as a woman and mother who has mixed and not infrequently contradictory feelings about many aspects of the PND/selective abortion/care/disability rights tangle, I'm acutely aware of the limitations of the academic enterprise, especially when dealing with complex and emotionally loaded issues. Another way of promoting an understanding of such issues is through art.[1] Art speaks in a distinct language, illuminates issues that are often invisible in scholarly works, opens new vistas, disturbs, and unsettles. The academic endeavor strives to maximize rationality and clarity and has a low tolerance for tensions and contradictions. Art works through different channels and produces a different kind of knowledge. As the film director Eugène Green explains:

> The key notion of baroque was oxymoron, this rhetoric entity which holds together two terms seen as contradictory by reason to express truth that goes beyond reason. One of the world's big problems from the eighteenth century on is the refusal to embrace the oxymoron: the wish to have a single truth

that excludes everything that opposes it. The oxymoron, not the reason, leads to truth.[2]

A Polish poet, Justyna Bargielska (born in 1977), dedicated her 2010 collection of short stories, *Obsoletki*, to "sterile pregnancies." Bargielska's stories show how the experience of losing a wanted fetus/child is intermingled with other elements that shape women's lives.[3] She coined the term *obsoletki*—derived from the medical concept of *gravid obsoleta*, that is, death in the womb—as a collective designation for women whose pregnancy ended with fetal or newborn death. She became interested in this topic when her first pregnancy ended with a late miscarriage.

Bargielska writes about her own and other women's experience with compassion, brutal honesty, self-irony, black humor, and a sharp eye for the grotesque and the absurd. She also describes her attempts to introduce to Poland approaches that originated in the United States, such as establishing support groups for pregnancy loss and photographing fetuses and stillborn babies to allow parents to have mementos of their lost child. At first, Bargielska had hoped to find a professional photographer for the latter task, but when she failed to find one who was willing, she started to take pictures of dead fetuses herself. After a while she stopped, discouraged by the emotional difficulty of the task, but even more by the total lack of institutional support and the distressing conditions in which fetuses are stored in most Polish hospitals. Her experience was nevertheless one of the sources of inspiration for her book.[4]

In the story "On Throwing It Out from One's Mouth," Bargielska describes the burial of a very small fetus. When a fetus is smaller than thirteen centimeters, the Roman Catholic Church provides a special ceremony of "sprinkling" (with holy water) instead of a funeral. In Bargielska's story, the "sprinkled" fetus is placed in a small cardboard box that was decorated partly by the father, who painted cars, trains, and people, and partly by the mother, who painted an abstract violet and blue pattern. A good-looking Dominican priest makes the appropriate religious statement, and everybody cries in a very distinguished way. Then the undertakers roll away a stone that was covering the grave, put the cardboard box in the grave, cover it with earth, and roll the stone back. Bargielska adds—in all probability in an allusion to the frequent representation of miscarried/aborted fetuses as butterflies or angels—"I did not notice that my 'sprinkled' nephew has flown somewhere."[5]

In "How I Thought about What Is in That Jar," Bargielska describes her own miscarriage. When she arrives at the hospital's reception area, a physician, who happens to be Black, places her on a gynecological chair, informs her that she has had a miscarriage, scrapes a little, and puts something into a glass jar. Then Bargielska walks with a midwife to the gynecology ward. The midwife carries the jar. Once at the ward, Bargielska undergoes a dilation and evacuation procedure to empty the uterus. Her physicians tell her that during the latter intervention she will hear many noises, but none of these is her child's heartbeat. "Of course not," Bargielska answers. "The heart of my baby is in the jar." Later, however, she is less sure that this is the case. She phones friends and acquaintances to ask them whether it is possible that her child is in that jar—but, she explains, none among them had an experience with glass jars related to children or children related to glass jars. Bargielska still does not know what exactly is in that jar. But when she gets the autopsy results, she learns that the fetus had been dead long before it was expulsed: "For some time, I was a tomb."[6]

In "How to Make Photographs of Dead Children," Bargielska tells about photographing an eight-week-old miscarried embryo/fetus.[7] At the hospital, the midwife takes a small package from the refrigerator, and unfolds it slowly. She then asks Bargielska whether she had breakfast, worrying that she will feel sick if she had not. The midwife says next that some mothers have truly bizarre ideas. For example, one woman who in the nineteenth week of pregnancy miscarried a black and macerated fetus that had died much earlier insisted on pouring rose petals onto the small coffin brought by the undertakers who took care of the fetus's burial. While telling this story, the midwife completely unwraps the small package. Inside is something that looks like a piece of dried liver. Bargielska takes photos of this tiny dried "liver slice," then notices that she can distinguish some fetal parts within it. The woman who miscarried calls Bargielska to ask her how the photography session went. Bargielska answers that she is not sure what the photographs will show. "I know," the woman says. "I held Hanutka [little Hannah] in my hand." Later, sitting in front of her computer screen, Bargielska uncovers more parts of Hanutka: "little feet, little hands, place of tiny heart, and a head, the easiest part to distinguish, since it was the biggest."[8]

The story "Tammy Likes Grief" introduces an American woman who works with a charity that produces photographs of dead newborns.[9] Tammy, Bargielska tells us, likes grief. For example, she planted oaks to commemorate her son, who was stillborn and suffered from severe inborn malformations. The story describes Tammy's and Bargielska's trip to a hospital morgue to photograph a stillborn baby. A man who works there takes a package from the refrigerator and throws it onto the dissection table:

Together with the mother of the package we dress the contents in a green baby pajama. The contents' seams on head and torso are giving way. Doctors have no doubt that the baby died suffocated by its umbilical cord during childbirth, but why not do a postmortem if one can. One picture, a second picture. The mother kisses the contents of the package, the father cries loudly, and also kisses the package. Kitsch, super-kitsch. And then a long session of Photoshop at home.[10]

The title story of the collection, "Obsoletki," describes how Bargielska, inspired by advice given by her US friend Tammy, tries to bring US-style commemoration of fetuses and stillborn children to Poland. Bargielska and her friends decide to organize a rally in a park, during which grieving parents would receive free balloons: pink for girls and blue for boys. Tammy also told her that they could release butterflies, but they find out that this would be too expensive.

The event goes off without a hitch. The weather is excellent, and there is the right mix of grieving parents and TV and radio journalists. Everybody is very kind, and the parents write messages to their dead child on the balloons: "Dear son, we will remember you forever"; "Dear sweet pea, we love you. Dad and Mom." All is very uplifting. TV cameras record the takeoff of the balloons and the tears in the parents' eyes. After the ceremony, participants tell Bargielska how grateful they are: "It was magnificent; it was like the funeral that we were unable to organize"; "Look how wonderful it was; we lived through a disaster, but now we met so many extraordinary people, and these encounters changed our life, often for better." Surrounded by female admirers, Bargielska tells them that she always carries in her handbag an ugly, cheap plastic doll the size of the child she had lost, and adds:

You know what, go to hell, bitches, all of you. The only thing I really want to do now is to stay alone with my doll for a thousand years on a bottom of a dirty river. They answered: Go to hell yourself, you bitch! Why do you think that we do not want to do exactly the same?[11]

Bargielska describes *Obsoletki* as a "funny book about a loss." For a pregnant woman, a PND of a serious fetal malformation is indeed and above all the announcement of a loss—a loss of the hope to have an uneventful pregnancy and to give birth to a "normal" child, that is, one who does not have serious health and/or developmental problems, and whose care does not mobilize more than the usual share of parental resources. It is also the loss of the hope to avoid very difficult reproductive decisions.[12] Women described by Bargielska usually had a miscarriage or stillbirth; nevertheless, one of the stories tells of a woman who had an induced abortion of a malformed (probably anencephalic) fetus and later described herself as a "mother of two children with heads, and one without a head."[13] Asked whether having a photograph of a "lost" child is important for the mother, Bargielska answered, "I'm not here to tell a mother what is or is not important for her. Some mothers need it, others do not. One can only inform a woman that she might want such a photograph, and give her the opportunity to have it. It is not about telling people how they are supposed to feel." She explained, "We [women] hurt ourselves through a use of language which produces a radical division between 'pro-life' and 'pro-choice,' 'an angel' and 'a zygote.' . . . Women are alienated by language, that is, by something that, in principle, we should use to create bonds among us."[14]

The title of Bargielska's book is *Obsoletki*, but only some of the stories in this collection deal with pregnancy, miscarriage, fetuses, and stillborn children. Others are about life that goes on before, after, at the same time as, and around *gravid obsoleta*. Bargielska writes about her children's linguistic inventions, the funeral of her grandmother, going to a swimming pool, her neighbors and their troubles, apprehension of unwanted pregnancy, broken toys, childhood memories, fear of death, holiday plans, nightmares, ex-boyfriends, religion, a dispute with her sister, taking photographs, the death of her cat, a drunken woman, a rainy day. The persistent grief over a lost pregnancy is intermingled with a funny, crazy, curious, sad, boring, uplifting, confusing stream of everyday existence. Bargielska's book—with its unique mixture of despair and hope, the bizarre and the familiar, and seasoned with a hefty dose of grotesque, absurd, and black humor—provides a glimpse of an alternative way of looking at messy encounters between reproductive accidents and other life events. There are glass jars with bits of children, glass jars with homemade jam, and children who splash in a pool.[15]

Not many people, especially outside Poland, have read Bargielska's book. *Obsoletki* received very good reviews and in 2011 won the prestigious Gdynia Literary Prize for short fiction, but even in Poland it is far from being a best seller. By contrast, Frida Kahlo's 1932 painting *Henry Ford Hospital*, which depicts Kahlo's pregnancy loss, is a world-famous work of art. In the late twentieth century, Kahlo achieved iconic status, and this painting is one of the main reasons for this status. Besides being reproduced in nearly all the studies dedicated to Kahlo, *Henry Ford Hospital* is also present in books about women painters, representations of women's bodies in art, or feminist art.[16]

The story of the *Henry Ford Hospital* painting was dramatically staged in the 2002 movie *Frida*, directed by Julie Taymor. Kahlo, played by Salma Hayek, is shown as she wakes up in a hospital after a bloody and painful pregnancy loss. She becomes extremely agitated, jumps from her hospital bed, and starts to cry: "I want to see him! What did you do with him! I want to see my son!" Her husband, the well-known Mexican painter Diego Rivera, attempts to calm her down. In the next scene, we see Kahlo sitting in her bed and painting. On a shelf across from the bed stands a glass receptacle that holds a very large, near-term fetus.

Kahlo biographers tell a more complicated story, or rather several complicated and contradictory stories, about her pregnancy loss. Kahlo contracted polio when she was six years old and at eighteen was a victim of a bus accident that left her with a severe anomaly of the spine. All her life she suffered from many health problems and was often in pain. Her disability in all probability affected her attitude toward motherhood; another important factor was probably her stormy relationship with Diego Rivera. Rivera was an enthusiastic supporter and promoter of Kahlo's art, but he was also a notoriously unfaithful husband: among others, he became the lover of Kahlo's beloved sister. Kahlo, too, had many lovers, male and female.

In 1932, Kahlo and Rivera were visiting the United States. In late April they arrived in Detroit, where Rivera was invited to paint *Detroit Industry*, a series of mural panels for the Detroit Institute of Arts; Kahlo was approximately two months' pregnant at the time. This was probably her second pregnancy; the first, in 1929, ended by an induced surgical abortion at Kahlo's request. She explained later that her physician told her that with her spinal and pelvic anomalies, she could not carry a child to term; he therefore recommended an abortion. This explanation was probably inaccurate. With her atypical pelvis, Kahlo could give birth only by cesarean section, but her impairment was not necessarily an obstacle to carrying a child to term. Her reaction to her pregnancy in Detroit seems to indicate that she believed she had a reasonable chance to give birth to a healthy child—if she wanted.[17]

According to Kahlo's best-known biographer, Hayden Herrera, in 1932 Kahlo was initially not sure whether she wanted to have a baby. In a long letter to her Mexican physician, she explained that her US physician agreed at first that it would be better for her not to have a child, mainly because of her serious health problems. Consequently, he gave her "quinine, and a very strong purge of castor oil."[18] Kahlo took these medications probably in late May. Some bleeding occurred next, but not a full abortion. According to Herrera, her US physician then told her that she had a good chance of keeping the pregnancy, and Kahlo started to look forward to having a baby. However, he also strongly recommended that she be very careful, avoid strenuous exercise, and rest as much as possible. Kahlo, lonely and bored in Detroit while

Rivera worked all day on the institute frescoes, did not follow this advice. She wrote to a friend that the physician "tells me I can't do this, I can't do that, and that's a lot of bunk."[19] Kahlo lost the fetus in early July 1932, probably at the end of the fourth month of pregnancy. The expulsion of the fetus was painful and slow, and Kahlo was weakened by loss of blood.[20]

Kahlo's biographers disagree as to whether she suffered a miscarriage, or whether the fetal demise was a belated consequence of her earlier attempts at abortion. According to Herrera, Kahlo had three induced abortions and two miscarriages, including the one in Detroit. Herrera's claim is based on Kahlo's medical history, written in 1946 by one of her physicians, Henriette Begun.[21] But another Kahlo biographer, Martha Zamora, stated that Begun's report is unreliable, as are Kahlo's own accounts of her reproductive history. Kahlo, Zamora proposed, modified and dramatized stories about her reproductive life, including her account that she had miscarried while she probably was not pregnant.[22]

More recently, Salomon Grimberg argues that Kahlo had no miscarriages, only induced abortions. The ambivalence about the precise nature of her pregnancy loss in 1932 was reinforced by the proximity between the Mexican expression *hacerse un aborto* (to get an abortion) and *tener un aborto* (to have an abortion, that is, to miscarry). Grimberg presents quotations from Kahlo's correspondence with her sisters that seem to indicate that while in Detroit, she wanted to eliminate her pregnancy. Thus, after she had lost the fetus, she assured her sisters that "everything turned very well." Her sister Adriana responded that this is true. Children come from suffering; moreover, having a child would have prevented her from being free and traveling with Diego.[23]

Kahlo indeed made contradictory statements about her reproductive history. She claimed that she suffered greatly following each pregnancy loss, but also expressed a fear that children would negatively affect her creativity. She talked about children as a source of pain in a woman's life, and explained that they "would have filled my life horribly."[24] She also cited her inability to have children, together with the bus accident that shattered her vertebral column, as tragic events that have molded her life and which she tried to overcome through her art. Her disability and life with chronic pain probably played an important role in shaping her reproductive decisions.[25]

The direct cause of Kahlo's pregnancy loss in July 1932, Grimberg proposes, was her earlier ingestion of quinine. Herrera's statement that Kahlo was told in the hospital that the fetus had disintegrated in her womb makes the link between an earlier use of abortive substances and her pregnancy loss more plausible, because it may indicate that the pregnancy had ended well before the expulsion of the dead fetus.[26] On the other hand, quinine, if indeed this was the drug Kahlo had ingested, is an ineffective abortive substance: it can intoxicate the pregnant woman, but it rarely kills the fetus.[27]

Immediately upon her arrival at the Henry Ford Hospital, Kahlo was very depressed, but the next day she asked for pencils and paper and started to draw again. She then asked her physicians to lend her medical textbooks; she wanted see how her fetus should have looked.[28] Kahlo was told that the hospital did not allow patients to have these books, because the illustrations might upset them. Finally, Rivera brought her one, and Kahlo made a careful study of a male fetus. In July, the month of her abortion, she produced a drawing, *My Abortion*, which shows her standing up, blood (or fetal remains) streaming from her vagina and pooling around her legs. She is surrounded by a very big, floating fetus whose umbilical cord wraps around Kahlo's leg, and drawings of large dividing cells and small plants.[29]

Henry Ford Hospital, painted in the fall of 1932, has a different composition. In it, Kahlo lies naked on a bed, hemorrhaging and crying. Her body clearly shows the physical imprints of the lost pregnancy: swollen belly, engorged breasts, and vaginal bleeding.[30] She holds six ribbons to which are attached symbolic objects linked with her pregnancy loss: one is a fetus, nearly as big as Kahlo herself, which floats above her belly. Other objects are a broken spine; a salmon-pink torso, which symbolized, Kahlo explained, her idea of a woman's insides; a snail, which symbolized for her the slowness of the miscarriage; a huge pink-purple orchid, which resembles a uterus; and a complicated machine, probably an autoclave (an instrument used to sterilize surgical tools). Kahlo did not explain what the machine signified. Her friend the art critic Bertrand Wolfe thought that it may symbolize Kahlo's struggle with the iron grip of pain. A sterilizer can also be interpreted as a symbol of Kahlo's sterility—be it natural or provoked. It is also possible that this mysterious machine resonated with the modern medical technology Kahlo saw in the hospital.[31]

It may be impossible to know for sure whether in the summer of 1932 Kahlo "got an abortion" or "had an abortion"; whether she was depressed because her initial ambivalence toward the pregnancy ultimately changed to a desire to continue it, or because the prolonged suffering that produced her abortion, a very different physical experience from surgical termination of pregnancy under general anesthesia, exacerbated her distress at giving up maternity; and whether she felt sad, guilty, ashamed, confused, relieved, or a mixture of all these feelings. On the other hand, her actions and state of mind are probably less important than the power of her art. *Henry Ford Hospital* is an eloquent testimony to the raw intensity of Kahlo's emotions, and her unique capacity to transform these emotions into a painting which, despite its modest size (31.8 × 38.5 cm), literally grips the viewer. This picture is often presented as a turning point in Kahlo's development as an artist: the precise moment when she found her unique capacity to depict women's intimate experiences. *Henry Ford Hospital* is a totally unsentimental, powerful, and deeply moving painting. Its capacity to interrogate and disturb its viewer may be partly rooted in the ambivalent emotions its artist was experiencing at the time it was produced.

Kahlo's mighty visual art and Bargielska's powerful prose display the irreducible complexity and inescapable ambivalence of women's experiences of procreation, pregnancy, and pregnancy loss. Forcing debates about these experiences into a Procrustean bed of simplified and polarized opinions can seriously harm women, children, and men. Risks of schematization and oversimplification are, however, rarely visible to people caught up in the heat of the argument and convinced that their opinions and actions are right. Scholars warn about the dangers of reductionist views, but they are seldom listened to. Perhaps this is because the last thing people who are engaged in an activity they consider truly important—for the society, their domain, their career, their self-image, all of the above—want to contemplate are the "unknown knowns": things known to be problematic but deliberately avoided because they can impede advances in a given area.[32] Only when it is too late to weigh in on the events as they unfold will people be willing to listen to scholars and let them explain what has happened.[33] As Hegel famously said, "The owl of Minerva begins its flight only with the onset of dusk."[34] Artists do things differently: they express, not explain, and propose radically different ways to deal with loss, ambivalence, and complexity. At times, Apollo's lyre may succeed where Minerva's owl fails.

ACKNOWLEDGMENTS

My first major debt is to the research group at the origin of my interest in tangled diagnoses, the participants in the collaborative project Prenatal Diagnosis and the Prevention of Disability: Biomedical Techniques, Clinical Practices and Public Action. This project was conducted between 2009 and 2013 in my department, the Centre de recherche, médecine, sciences, santé, santé mentale, société 3 (CERMES 3), which is in the Institut nationale de la santé et de la recherche médicale of the Centre nationale de la recherche scientifique at the École des hautes études en sciences sociales, Paris V University. It was supported by Sciences, technologies et savoirs en société, a program of the Agence nationale de la recherche, ANR-09-SSOC-026-01. For those four years, a multidisciplinary group led by a disabilities studies scholar, Isabelle Ville, had regular exchanges on the history and present-day practice of PND. A close collaboration with colleagues whose main focus was disability rather than biomedicine forced me to venture outside my comfort zone and, not infrequently, to change my point of view. In a collective endeavor, it is sometimes difficult to tell who was at the origin of a specific idea. All the members of the group Prenatal Diagnosis and the Prevention of Disability—Emannuelle Fillon, Lynda Lotte, Veronique Mirlesse, Sophie Rosman, Benedicte Rousseau, Carine Vassy, and Isabelle Ville—enriched my thinking on this subject, while Veronique Mirlesse, an ultrasound expert, also improved my understanding of technical aspects of prenatal testing; they have no responsibility whatsoever for any shortcomings of my study.

My second major debt is to the fetopathologists, geneticists, and laboratory technicians in the two hospitals where I conducted my fieldwork: Andral Hospital in France and the Maternal Health Center in Brazil. In both facilities I received amazingly generous help and support, and I was impressed by the dedication and skill of professionals who often deal with difficult situations, animated by a true spirit of public service. This is especially true for Brazilian specialists, who often intervene in a very complex legal and social environment.

I'm very lucky to work in a department, the CERMES 3, where I can discuss my work with colleagues from different disciplinary backgrounds who share my interest in biomedical innovations, their clinical uses, and the societal consequences. Many thanks to Isabelle Baszanger, Simone Bateman, Luc Berlivet, Soraya Boudia, Catherine Bourgain, Martine Bungener, Maurice Cassier, Jean Paul Gaudilliere, Catherine LeGalles, Anne Lovell, Laurent Pordie, Jean François Ravaud, and Myriam Winance, who, each in her/his unique way, refined my understanding of science and medicine in context.

During the years I studied PND, I exchanged ideas about reproduction, pregnancy, women, fetuses, and health risks with a great number of colleagues who frequently are also friends. A very incomplete list of scholars from four continents who—to follow Ludwik Fleck's felicitous expression—constitute my "thought collective" on these subjects includes Christiane Aicardi, Madeleine Akrich, Salim Al-Gailani, Armelle Andro, Robert Aronowitz, Nathalie Bajos, Christiana Bastos, Dominique Behague, Anne-Emanualle Birn, Allan Brandt, Christine Brandt, Silvia Camporesi, Lilian Chazan, Adele Clarke, Nathaniel Confort, Angela Creager, Soraya De Charadevian, Debora Diniz, Michelle Ferrand, Evelyne Fox-Keller, Sahra Gibbon, Snait Gissis, Jeremy Greene, Yael Hashiloni-Dolev, Cathy Herbrant, Dagmar Herzog, Helena Hirata, Nick Hopwood, Tsipi Ivry, David Jones, Lene Koch, Cynthia Kraus, Joanna Latimer, Sandra Laugier, Sabina Lionelli, Veronika Lipphardt, Claire Marris, Catherine Marry, Pascale Molinier, Ornella Moscucci, Jesse Olszynko Gryn, Nelly Oudshoorn, Katherine Park, Naomi Pfeffer, Barbara Prainsack, Christelle Rabier, Emilia Sanabria, Maria Jesus Santestmeses, Helga Satzinger, Robert Sayre, Martina Schlunder, Lisa Schwartz, Alexandra Stern, Elisabeth Toon, Sezin Topcu, Joanna Radin, Robert Resta, Vololona Rhaberisoa, Hans Joerg Rheinberger, Charles Rosenberg, George Weisz, Claire Williams, and Anna Zielinska. Olga Amsterdamska, Harry Marks, Barbara Rosenkrantz, and Hélène Rouch are no longer with us, but I continue to talk to them in my head.

Several colleagues stand out among the members of my "thought collective." Rayna Rapp introduced me to PND and set a stellar example by her high standards of scholarship in this area. Many discussions with Rayna, and her unfailing support, guided my research on tangled diagnoses. Diane Paul generously shared with me her impressive understanding of the history of eugenics, genetics, and inborn disorders, and helped me to better grasp the differences between the health policies and practices of western Europe and the United States. In Brazil, Claudia Bonan, Marilena Correa, and Luiz

Antonio Teixeira familiarized me with the distinctive traits of the nation's health system and its politics and society. Christiane Aicardi, Anne Lovell, Veronique Mirlesse, and Barbara Prainsack read and commented on parts of the manuscript, Joele Bajoule improved my English, and two anonymous readers for the University of Chicago Press provided insightful remarks on the first version of this text.

At the University of Chicago Press, Karen Darling provided tireless support for this project, as did Erin DeWitt, Benjamin Balskus, and all the other people involved in the production of the book, while Sandra Hazel did an impressive job of improving my English and saving me from embarrassing typos. I am very fortunate to have worked with such a competent and dedicated group of professionals.

Reproduction and transmission are inseparably entangled with families. I am grateful to my children and their partners for helping me think, making me laugh, and setting my priorities straight, and for my grandchildren for being such fun. For nearly four decades now, Woody continues to be an exemplary supportive companion, one who provides literally everything, from stimulating intellectual discussions and a view from a different disciplinary perspective, to constant encouragement of all my endeavors, to more than a fair share of domestic tasks and effective solutions for mundane housekeeping crises. Finally, my very special thanks to Sylvie for demonstrating the ability of art to transcend academic discourse and provide an alternative view of embodied differences, and to Nicolas for showing that even the best narratives produced by scholars and artists have but a limited capacity to grasp the complexity and richness of individual lives.

NOTES

INTRODUCTION

1. This expression appears in chapter 5 of *Pirkei Avot* (פרקי אבות, *Chapters of the Fathers*), a compilation of the ethical teachings and maxims of the rabbis of the Mishnaic period. According to Jewish tradition, the text was gathered by Jehuda HaNassi in the third century AD, but includes teachings of rabbis collected from the third century BC onward.

2. We can argue that selective breeding—an old agricultural method but also the practice, especially among upper-class individuals, of carefully selecting a mate—was an indirect way to anticipate "what is about to be born." Selective breeding did not involve a direct scrutiny of the unborn, however.

3. Loudon, *Death in Childbirth* (1992).

4. Lopez-Beltran, "The Medical Origins of Heredity" (2007); Wright, *Downs: The History of a Disability* (2011).

5. Many historical studies of eugenics are available. In the context of prenatal diagnosis, especially useful ones may be Kevles, *In the Name of Eugenics* (1985); D. Paul, *The Politics of Heredity* (1998); Stern, *Eugenic Nation* (2005); and Comfort, *The Science of Human Perfection* (2012).

6. Al-Gailani, "The Making of Antenatal Life" (2010).

7. Oakley, *The Captured Womb* (1986).

8. Zallen, Christie, and Tansey, eds., *The Rhesus Factor and Disease Prevention* (2004).

9. In the post–World War II era, physicians usually spoke about antenatal care and then, from the 1970s onward, antenatal diagnosis. However, in the twenty-first century the term *prenatal diagnosis* is employed more frequently than *antenatal diagnosis* (e.g., in April 2016, a Medline search uncovered 9,807 articles with the term *prenatal diagnosis* in the title and 938 articles with the term *antenatal diagnosis* in the title). I elected therefore to use *prenatal diagnosis* in this study.

10. The project, The Implication of Prenatal Diagnosis in the Prevention of Handicap: The Use of Technologies between Scientific Progress and Public Action, was funded by the French Agence nationale de la recherche (ANR), or National Research Agency, between 2009 and 2013. It included scholars linked with the National Institute of Handicap, who work with the disability rights community. The French term *handicap* is equivalent to *disability* (or, rather, handicap-cum-disability).

11. Lindee, "Genetic Disease in the 1960s" (2002), 79. Lindee recalls that some editors of *American Journal of Medical Genetics* were uneasy with this statement. See also Lindee, *Moments of Truth in Genetic Medicine* (2005), 202.

12. PND of fetal impairments can also help the mother/parents to be better prepared for the child's difficulties after birth. This may be especially true in the case of visible deformities (cleft palate, missing limb, misshapen face, atypical genital organs), which may lead to a rejection of the newborn by unprepared parents. Moreover, in some cases (e.g., twin-to-twin transfusion, hypoplastic left heart syndrome), physicians attempt to treat the problem in the womb through fetal surgery; in other cases, physicians who detect a fetal anomaly requiring immediate surgical treatment after birth make sure that the birth takes place in a hospital equipped to provide this surgery.

13. Bianchi, "From Prenatal Genomic Diagnosis" (2012), 1049.

14. E.g., Julsingha, Tesh, and Fara, eds., *Advances in the Detection of Congenital Malformations* (1976); Fletcher, "Ethics and Trends in Applied Human Genetics" (1983).

15. Children with phenylketonuria cannot metabolize the amino acid phenylalanine. The accumulation of this amino acid in their body causes a severe mental impairment, which can be prevented if immediately after birth these children are put on a diet that limits the uptake of phenylalanine. D. Paul and Brosco, *The PKU Paradox* (2013).

16. On such hopes linked with fetal surgery, see Casper, *The Making of the Unborn Patient* (1998). Deborah Blizzard describes the development of fetoscopy, another approach to medical intervention on the fetus in the womb. Blizzard, *Looking Within* (2007).

17. Lindee, *Moments of Truth in Genetic Medicine* (2005). See also Comfort, introduction to *The Science of Human Perfection* (2012). On the other hand, as early as 1971 leading geneticists warned against unrealistic hopes associated with the predicted advent of genetic therapies. Fox and Littlefield, "Editorial" (1971).

18. Quoted in Finely, Finley, and Flowers, eds., *Birth Defects* (1983), 157.

19. Löwy, *Preventive Strikes* (2009).

20. Some researchers, the most famous among them Thomas McKeown, argued that bacteriology made a limited contribution to the decrease of mortality from transmissible diseases, and that such decrease should be attributed mainly to socioeconomic factors, above all better nutrition. McKeown, *The Modern Rise of Population* (1976). Today, most scholars reject the more extreme version of the McKeown thesis and argue that the decline of mortality from infectious diseases reflects a combination of economic factors and interventions grounded in bacteriological knowledge. Scherter, "The Importance of Social Intervention" (1988); Colgrove, "The McKeown Thesis" (2002).

21. The relevant texts are Riis and Fuchs, "Antenatal Determination of Foetal Sex" (1960); Nadler, "Antenatal Detection of Hereditary Disorders" (1968).

22. I loosely borrow the distinction between epistemological and sociological problems from Hon, Schickore, and Steinle's analysis of the malfunction of scientific tools. Such malfunction is not very troubling from an epistemological point of view, because the standard against which instruments are calibrated is known, at least in principle. But from a sociological point of view it is very disturbing, because social and material disorders are interdependent. Hon, Schickore, and Steinle, "Introduction" (2009), 5.

23. Fassin, "A Case for Critical Ethnography" (2013), 125.

24. Buchbinder and Timmermans, "Affective Economies" (2013). On oncologists' tendency to extrapolate from one "paradigmatic" condition to a very different one, see Löwy, *Preventive Strikes* (2009).

25. Foucault defined *dispositif* as "a thoroughly heterogeneous ensemble consisting of discourses, institutions, architectural forms, regulatory decisions, laws, administrative measures, scientific statements, philosophical, moral and philanthropic propositions." Foucault, "Le jeu de Michel Foucault" (1977), 299.

26. I summarize the history of PND in two papers: Löwy, "Prenatal Diagnosis" (2014): 154–62; Löwy, "How Genetics Came to the Unborn" (2014).

27. Bell, "On Rubella in Pregnancy" (1959); Coventry, "The Dynamics of Medical Genetics" (2000), chapter 7; Reagan, *Dangerous Pregnancies* (2010); Ville and Lotte, "Évolution des politiques publiques" (2013).

28. Sunday Times Insight Team, *Suffer the Children* (1979); Reagan, *Dangerous Pregnancies* (2010).

29. Riis and Fuchs, "Antenatal Determination of Fœtal Sex" (1960); Cowan, *Heredity and Hope* (2008), 91–95.

30. Harper, *First Years of Human Chromosomes* (2006). The detected aneuploidies were very different: trisomies 13 and 18 nearly always lead to the child's death shortly after birth; those children who survive for several years are severely disabled. As a rule, sex chromosome aneuploidy produces mild to moderate impairments, and many people, especially those with Klinefelter syndrome and triple X (women with chromosome formula 47, XXX), are not aware of their "anomaly." Down syndrome (trisomy 21) always produces an intellectual impairment, but its severity varies greatly. Some but not all individuals with trisomy 21 have significant health problems, such as heart defects.

31. Christie and Tansey, eds., *Genetic Testing* (2001); Harper, *First Years of Human Chromosomes* (2006).

32. Nadler, "Antenatal Detection of Hereditary Disorders" (1968).

33. R. Davidson and Rattazzi, "Review: Prenatal Diagnosis of Genetic Disorders" (1972).

34. Rodeck, "Sampling Pure Foetal Blood" (1978).

35. Penrose, "The Relative Effects of Paternal and Maternal Age" (1933); Hook and Chambers, "Estimated Rates of Down Syndrome" (1977).

36. Bermel, "Update on Genetic Screening" (1983).

37. See, e.g., the declaration of Keith Russell, President of the American College of Obstetricians and Gynecologists, in November 1973, quoted in Powledge and Sollitto, "Prenatal Diagnosis" (1974), 13.

38. Bekker et al., "Uptake of Cystic Fibrosis Testing" (1993); C. Williams, Alderson, and Farsides, "Too Many Choices?" (2002); Vassy, "How Prenatal Diagnosis Became Acceptable" (2005).

39. Brock and Sutcliffe, "Alpha-Fetoprotein" (1972).

40. Cuckle, Wald, and Lindenbaum, "Maternal Serum Alpha-Fetoprotein" (1984); Harris and Andrews, "Prenatal Screening for Down's Syndrome" (1988); Brock and Sutcliffe, "Alpha-Fetoprotein" (1972).

41. Wald et al., "Serum Screening for Down's Syndrome" (1996).

42. Blume, *Insight and Industry* (1992), chapter 3; Tansey and Christie, eds., *Looking at the Unborn* (2000), 23–25; Nicolson and Fleming, *Imaging and Imagining the Fetus* (2013).

43. Donald, "Ultrasonic in Diagnosis" (1969).

44. Bang and Northeved, "A New Ultrasonic Method" (1972); Tansey and Christie, eds., *Looking at the Unborn* (2000), 54–56.

45. Nicolaides et al., "Fetal Nuchal Translucency" (1992); Cuckle, "Rational Down Syndrome Screening" (1998).

46. The exact proportion of miscarriages induced by amniocentesis is difficult to evaluate, because it depends on the operator's skill. According to some specialists, when the operator is well trained the risk of miscarriage is much lower than 1%.

47. Sheldon and Simpson, "Appraisal of a New Scheme for Prenatal Screening" (1991), 133–34.

48. Wald and Hackshaw, "Combining Ultrasound and Biochemistry" (1997); Chitty, "Prenatal Screening for Chromosome Abnormalities" (1998).

49. Vassy, "From a Genetic Innovation to Mass Health Programmes" (2006); Schwennesen, Svendsen, and Koch, "Beyond Informed Choice" (2010).

50. Taylor, *The Public Life of the Fetal Sonogram* (2008).

51. Van Dijck, *The Transparent Body* (2005), 100–117; Sänger, "Obstetrical Care as a Matter of Time" (2015).

52. Samerski, "Genetic Counseling and the Fiction of Choice" (2009).

53. Vassy, Rosman, and Rousseau, "From Policy Making to Service Use" (2014).

54. Khoshnood et al., "Advances in Medical Technology" (2006).

55. Schwennesen and Koch, "Representing and Intervening" (2012); Vassy, Rosman, and Rousseau, "From Policy Making to Service Use" (2014); Crombag et al., "Explaining Variation" (2014).

56. I borrow the term *technoscientific entity* from Anne Marie Mol's description of blood as being not a stable, pre-existing entity but something that comes into existence as a separate, distinct "thing" entirely in relation to its specific sociotechnical milieu. Mol, *The Body Multiple* (2002).

57. Palomaki et al., "DNA Sequencing of Maternal Plasma" (2012).

58. Dondorp et al., "Non-invasive Prenatal Testing" (2015).

59. Ravitsky, "Non Invasive Prenatal Testing" (2015).

60. Besseau-Ayasse et al., "A French Collaborative Survey of 272 Fetuses" (2014).

61. E.g., Shakespeare, "The Content of Individual Choice" (2005), 225.

62. Hayden, "Prenatal Screening Companies" (2014). Autosomal trisomies are the triplication of nonsex chromosomes, which produce severe anomalies.

63. Wapner et al., "Chromosomal Microarray versus Karyotyping" (2012).

64. Greely, "Get Ready for the Flood of Fetal Gene Screening" (2011); Yurkiewicz, Korf, and Soleymani Lehmann, "Prenatal Whole-Genome Sequencing" (2014); Dondorp, Page-Christiaens, and de Wert, "Genomic Futures of Prenatal Screening" (2015).

65. Many historical, sociological, and anthropological studies on the rise of risk management in medicine are available. My research would not have been possible without studies of scholars such as Lupton, Rothstein, or Aronowitz. Lupton, *Risk* (1999); Rothstein, *Public Health and the Risk Factor* (2003); Aronowitz, *Risky Medicine* (2015). See also Welch, Schwartz, and Woloshin, *Overdiagnosed* (2011).

66. This study extensively employs testimonies of health care providers and consumers collected by sociologists and anthropologists of medicine.

67. Wertz and Fletcher, "Ethics and Genetics" (1989), 20–24; Wertz and Fletcher, "Geneticists Approach Ethics" (1993); Wertz and Fletcher, "Ethical and Social Issues in Prenatal Sex Selection" (1998).

68. E.g., Vassy, Rosman, and Rousseau, "From Policy Making to Service Use" (2014); Crombag et al., "Explaining Variation" (2014).

69. A partial list of studies on fetal conditions and PND that have played an important role in shaping my own thought includes R. Rapp, *Testing Women, Testing the Fetus* (1999); C. Williams, "Framing the Fetus in Clinical Work" (2005); Morgan, *Icons of Life* (2009);

Gammeltoft, *Haunting Images* (2014). There are many other important studies on this topic. My work stands on the shoulders of an impressive crowd.

70. Rothschild, *The Dream of the Perfect Child* (2005).

71. Kerr, "Reproductive Genetics" (2009).

CHAPTER ONE

1. Legalization of abortion facilitated the termination of a pregnancy for a fetal indication, but was not an absolute precondition for such a termination. In France and the United Kingdom, physicians have pointed to a severe harm to the pregnant woman's health, including her mental health, to justify termination of pregnancy for a risk of severe fetal malformation, for example after the woman's infection with rubella early in pregnancy. Coventry, "The Dynamics of Medical Genetics" (2000), chapter 7; Ville and Lotte, "Évolution des politiques publiques" (2013).

2. Lindee, "Genetic Disease in the 1960s" (2002), 79.

3. McLaren, *A History of Contraception* (1990).

4. Testimonies of US women undergoing abortions in the 1970s, immediately after the legalization of this procedure, already displayed a great diversity of attitudes toward the fetus. J. Nelson, *More than Medicine* (2015).

5. Morgan, "Strange Anatomy" (2006), 15.

6. This statement may be nuanced today (2017). Some Internet memorial sites for pregnancy loss reproduce realistic images of miscarried fetuses, including small and misshapen ones.

7. Morgan, "Strange Anatomy" (2006), 17.

8. McCoyd, "Pregnancy Interrupted" (2007).

9. Ibid., 43.

10. Bernadette Modell's testimony. In E. Jones and Tansey, eds., *Clinical and Molecular Genetics in the UK* (2014), 25. In the 1980s, Modell explains, thalassemia was diagnosed by the sampling of umbilical cord blood, a test which can be performed only relatively late in pregnancy. In the 1990s, with the development of new genetic techniques, it became possible to perform chorionic villus sampling (CVS) at eleven to twelve weeks, then test fetal cells for the presence of the thalassemia gene. If the test was positive, the woman could have a much earlier abortion. Ibid, 27.

11. Quoted in Stillwell, "Pretty Pioneering-Spirited People" (2015). Alexandra Stern has studied the early days of genetic counseling. Stern, *Telling Genes* (2012).

12. Ariss, "Theorizing Waste" (2003).

13. Pfeffer and Kent, "Framing Women, Framing Fetuses" (2007); Kent, "The Fetal Tissue Economy" (2008); Pfeffer, "What British Women Say" (2008).

14. Mirlesse et al., "Women's Experience of Termination of Pregnancy" (2011).

15. The absence of a clear-cut distinction between natural and induced termination of pregnancy in France may be related to the near complete absence of "abortion wars." Conservative Catholics in France are usually silent on abortion.

16. Dommergues et al., "Termination of Pregnancy" (2010). With a better resolution of obstetrical ultrasound and more frequent use of CVS, more French women receive a diagnosis of a fetal malformation during the first trimester of pregnancy. Mangionea et al., "Devenir des foetus" (2008).

17. France has lower C-section rates (20.2% in 2014) than the United States (32.8%), and much lower rates than Brazil (40% in the public sector, and 84% in the private sector). Molina et al., "Relationship between Cesarean Delivery Rate" (2015). However, the percentage of C-sections in France (20% in 2014) is somewhat higher

than the one recommended by the World Health Organization (10 to 15%). WHO's data on C-sections worldwide, 2015, accessed July 25, 2017, http://www.who.int /mediacentre/news/releases/2015/caesarean-sections/en/.

18. Information leaflet, Hospitals of the Languedoc-Roussilon region, *Livret d'inormation à l'usage des parents: Interruption médicale de la grossesse*, 6. Although this opinion is presented as having originated in the parents' experience, it is reasonable to assume that it is promoted by midwives.

19. Esplat, "Les sage femmes face à l'interruption médicale de la grossesse" (2012), 66; Garel et al., "French Midwives' Practice" (2007), 623.

20. British Pregnancy Advisory Service (BPAS), information on second-trimester surgical abortions, accessed July 25, 2017, https://www.bpas.org/abortion-care/abortion -treatments/surgical-abortion/dilatation-and-evacuation/. BPAS is the major provider of abortions in the United Kingdom.

21. Brazilian activists who fought for the right to a legal abortion for anacephaly (won in 2012) produced a striking poster that shows medical professionals clad in black gowns and masks, and a woman in labor also dressed in black, with a caption that reads: "When the birth is of an anacephalic child, you do not get a birth certificate, you get a death certificate." https://www.google.com/search?q=aborto+anencefalia+brasil&hl =fr&biw=1252&bih=531&site=webhp&source=lnms&tbm=isch&sa=X&ved=0CAYQ _AUoAWoVChMIhYG7kpOSxwIVjDk-Ch0qHwS6#imgrc=coGirYNxwKWS-M%3A, accessed July 25, 2017.

22. Imber, *Abortion and the Private Practice of Medicine* (1986), 77–80.

23. According to the testimony of the genetic counselor June Peters, all the women she had followed in the early 1970s had medical abortions, perhaps because D&E for this indication was not considered safe at that time. Stillwell, "Pretty Pioneering-Spirited People" (2015).

24. Grimes and Gates, "Dilatation and Evacuation" (1981).

25. Such obstacles prompted professionals to find ways to shorten the preparation period for D&E. Maurer, Jacobson, and Turok, "Same-Day Cervical Preparation" (2013). In April 2015 in the United States, the state of Kansas passed a ban of D&E. Similar bills proposed in other states stressed D&E's gruesome effect and called it "dismembering abortion." Eckholm and Robles, "Kansas Limits Abortions" (2015). On recent restrictions on abortion in the United States, see Pollitt, *Pro* (2014).

26. Lee and Ng, "Issues in Second Trimester Induced Abortion" (2010), 517–27; Bryant et al., "Second-Trimester Abortion" (2011); Lyus et al., "Second Trimester Abortion" (2013). In China, too, nearly all the second-trimester abortions are medical.

27. Grossman, Blanchard, and Blumenthal, "Complications" (2008); Bryant et al., "Second-Trimester Abortion" (2011); Whitley et al., "Midtrimester Dilation and Evacuation" (2011).

28. Kaltreider, "Psychological Impact" (1981); T. Kelly et al., "Comparing Medical versus Surgical Termination" (2010); Grimes, "The Choice of Second Trimester Abortion Method" (2008).

29. Grimes, Smith, and Witham, "Mifepristone and Misoprostol versus Dilation and Evacuation" (2004). Such a trial was later conducted in the United Kingdom, where the usual method employed in second-trimester abortions is medical induction. It indicated that when given a choice, the majority of women elected to undergo D&E. T. Kelly et al., "Comparing Medical versus Surgical Termination" (2010).

30. Grimes, "The Choice of Second Trimester Abortion Method" (2008); Lyus et al., "Second Trimester Abortion" (2013). The latter article is a plea to give UK women who

undergo second-trimester abortions the possibility of choosing a surgical termination of pregnancy.

31. This claim is not supported by the evidence in the professional literature. Bryant et al., "Second-Trimester Abortion" (2011).

32. Manderbrodt and Girard, "Aspects techniques des interruptions médicales de grossesse" (2008). In certain fetal anomalies, the dissection of an intact fetal body can indeed provide information that cannot be obtained in other ways. In such cases, the woman's physician can explain to her that a medical abortion will result in a more accurate genetic counseling. On the other hand, some physicians may be tempted to use this argument to learn more about "interesting" fetal malformations.

33. Lohr, "Surgical Abortion" (2008), 152; Grossman, Blanchard, and Blumenthal, "Complications" (2008), 180.

34. Lewit, "D&E Midtrimester Abortion" (1982).

35. Leichtentritt, "Silenced Voices" (2011).

36. Ibid., 751–52. Sara aborted at thirty-eight weeks and three days for cystic fibrosis. Because of a long delay in obtaining permission for termination of pregnancy, she was at the point of giving birth when the feticide took place.

37. Ibid., 750.

38. Breeze et al., "Palliative Care" (2007).

39. Guon et al., "Our Children Are Not a Diagnosis" (2013), 310.

40. Weiner, "The Intensive Medical Care" (2009).

41. Gross, "'The Alien Baby'" (2010).

42. De Moura, Guimarães, and Luz, "Tocar" (2013).

43. Berg, Paulsen, and Carter, "Why Were They in Such a Hurry to See Her Die?" (2012). The lack of empathy for the decision of a couple of physicians who, one may assume, knew more about trisomy 18 than an average patient is puzzling, especially in light of the positive attitude in Norway toward children born with Down syndrome. Rates of pregnancy termination for trisomy 21 in Norway are among the lowest in western Europe. Cocchi et al., "International Trends" (2010).

44. The website of Be Not Afraid, accessed July 25, 2017, http://benotafraid.wordpress.com.

45. Kuebelbeck, "A Perinatal Hospice" (2013); Kuebelbeck and Davis, Gift of Time (2011).

46. Americans United for Life, Perinatal Hospice Information Act, accessed July 25, 2017, http://www.aul.org/wp-content/uploads/2010/12/NEW-Perinatal-Hospice-2011-LG.pdf. The document does not mention the possibility that the child may be severely malformed.

47. Janvier, Farlow, and Wilfond, "The Experience of Families" (2012); Guon et al., "Our Children Are Not a Diagnosis" (2013). Parents who testify on websites dedicated to these conditions often affirm their Christian faith.

48. K. Nelson, Kari, and Feudtner, "Inpatient Hospital Care" (2012).

49. Gammeltoft, Haunting Images (2014).

50. Bolton, "Women's Work, Dirty Work" (2005); Chiapetta-Swanson, "Dignity and Dirty Work" (2005).

51. Quoted in Bolton, "Women's Work, Dirty Work" (2005), 177. According to Bolton, gynecological nurses displayed compassion and care when dealing with women who miscarried, but some harshly judged women who decided to have a late-term abortion for what they perceived as a nonjustified reason.

52. Chiapetta-Swanson, "Dignity and Dirty Work" (2005).

53. Ibid., 103–4, 111.

54. Weber et al., "Les soignants et la décision d'interruption de grossesse" (2008).

55. Garel et al., "French Midwives' Practice" (2007). From my fieldwork in France, and my interviews with health professionals, I too can attest that this seems to be the general opinion.

56. Weber et al., "Les soignants et la décision d'interruption de grossesse" (2008), 114.

57. De Wailly-Galambert et al., "Lorsque la naissance et la mort coincident" (2012), 126–27. In French, one distinguishes between the terms *accouchement* (birthing) and *naissance* (birth/childbirth). The first term describes the process undergone by the mother, the second, the birth of the child. The French term *passage* denotes at the same time a physical description of going through something and a transition, as in transition rites.

58. Ibid., 125.

59. Duden, *Disembodying Women* (1993).

60. Woods, *Death before Birth* (2007), 14–34.

61. Morgan, *Icons of Life* (2009).

62. On the history of the pregnancy test, see Olszynko-Gryn, "Pregnancy Testing in Britain" (2014).

63. Han, "The Chemical Pregnancy" (2015). Chemical pregnancies may play an especially important role in the reproductive trajectories of women who undergo infertility treatment and attempt an in vitro fertilization.

64. There is a vast literature on the rise and current uses of in vitro fertilization. A good overview can be found in Franklin, *Biological Relatives* (2013).

65. Almeling, "Selling Genes, Selling Gender" (2007); Harrington, Becker, and Nachtigall, "Nonreproductive Technologies" (2008).

66. E.g., C. Williams, "Framing the Fetus in Clinical Work" (2005); Nuffield Council, *Human Bodies* (2011); Giraud, "L'embryon humain en AMP" (2014).

67. Radkowska-Walkowicz, *Doświadczenie in vitro* (2013). Radkowska-Walkiewicz linked such reactions of Polish women to the the violent opposition of the Polish Catholic Church to in vitro fertilization (IVF), and its description of the elimination of embryos produced in the test tube as "murder." In a 2009 debate about IVF in the Polish parliament (Sejm), right-wing members spoke, for instance, about "screams of assassinated embryos," a direct allusion to Bernard Nathanson's anti-abortion movie of 1980, *The Silent Scream*. See also Radkowska-Walkowicz, "The Creation of 'Monsters'" (2012); Radkowska-Walkowicz, "Frozen Children and Despairing Embryos" (2014). By contrast, in Brazil, conservative Catholics are engaged in a fervent fight against abortion, but do not express a strong opinion about IVF and seldom discuss the homoparental family; French conservative Catholics rarely mention abortion or IVF, but resist gay marriage and parenthood.

68. Radkowska-Walkowicz; *Doświadczenie in vitro* (2013), 153, 159, 163, 172, 161, 162.

69. Ibid., 153, 154,163.

70. Ebling, *Healthcare and Big Data* (2016). Ebling's distressing experience of the "phantom survival" of her lost pregnancy launched her remarkable investigation of the commercial exploitation of private health-related data.

71. Shaw, "Rituals of Baby Death" (2014). Absence of an open expression of grief, Shaw adds, does not necessarily indicate the absence of pain, but it reflects cultural attitudes toward death before life.

72. Morgan, "Imagining the Unborn" (1997).

73. Van der Sijpta, "The Unfortunate Sufferer" (2014).

74. Morgan and Browner, "Why Worry about Embryos" (1995); Duden, *Disembodying Women* (1993).
75. Conklin and Morgan, "Babies, Bodies, and the Production of Personhood" (1996).
76. Oaks, "Smoke Filled Wombs and Fragile Fetuses" (2000).
77. Eckholm, "Case Explores Rights of Fetus versus Mother" (2013).
78. Jülich, "The Making of a Best-Selling Book" (2015).
79. Ibid., 499–504.
80. Petchesky, "Fetal Images" (1987); Franklin, "Fetal Fascinations" (1991); Duden, *Disembodying Women* (1993).
81. Jülich, "The Making of a Best-Selling Book" (2015), 521.
82. Morgan, "Strange Anatomy" (2006), 15. Historians and sociologists who study taxidermy and the production of dioramas examined the process of specimen "purification" through the elimination of disturbing material elements and a problematic social context. Haraway, "Teddy Bear Patriarchy" (1984); Star, "Craft vs. Commodity" (1992).
83. Duden, *Disembodying Women* (1993); Neuman, *Fetal Positions* (1996).
84. Morgan, *Icons of Life* (2009).
85. Wilson, "Ex Utero" (2014). Hooker did not explain why only some fetuses were baptized, and whether these rites were given at the demand of the woman who gave birth.
86. Dubow, *Ourselves Unborn* (2011).
87. Morgan, "'Properly Disposed Of'" (2002); Morgan, "The Social Biography of Carnegie Embryo" (2004); Morgan, *Icons of Life* (2009).
88. W. Silverman, "Incubator-Baby Side Shows" (1979). In the 1930s and early 1940s, when Couney organized his last shows, attitudes toward "loaned" babies were different. In nearly all cases, the surviving babies were taken back by their parents, and Couney organized celebratory reunions for "graduates" of his shows and their families. On the incubator shows, see also Courtright Barr, "Entertaining and Instructing the Public" (1995).
89. Sociologists and anthropologists have studied the fluidity of the definition of *life* in neonatal resuscitation. Kaufman and Morgan, "The Anthropology of the Beginnings and the Ends of Life" (2005); Paillet, *Sauver la vie, donner la mort* (2007); Paillet, "The Ethnography of 'Particularly Sensitive' Activities" (2012); Christoffersen-Deb, "Viability" (2012); Willems, Verhagen, and van Wijlick, "Infants' Best Interests" (2014).
90. Lupton, *The Social Worlds of the Unborn* (2013), 69.
91. Boltanski, *The Foetal Condition* (2013).
92. Kerr, "Reproductive Genetics" (2009).
93. Sandelowski and Barroso, "The Travesty of Choosing" (2005).
94. Duden, *Disembodying Women* (1993).
95. Ward, *Birth, Weight and Economic Growth* (1973); Littlewood, "From the Invisibility of Miscarriage to an Attribution of Life" (1999).
96. Wthycombe, "From Women's Expectations to Scientific Specimens" (2015); Reagan, "From Hazard to Blessing to Tragedy" (2003). It is reasonable to assume that a view of miscarriage as a relief was but one among many possible reactions to this event, and that, then as now, women who strongly desired a child were devastated by a pregnancy loss.
97. Layne, *Motherhood Lost* (2003).
98. Kovitt, "Babies as Social Products" (1978), 347–51. Kovitt reports a case of literal drowning of an anacephalic baby born alive: ibid., 350.

99. Layne, "Unhappy Endings" (2003), 1883–84.

100. R. Rapp, "Constructing Amniocentesis" (1990), 28.

101. Layne, *Motherhood Lost* (2003). Rapp described her—partly unanticipated—pain and grief following a termination of pregnancy for a fetal anomaly. R. Rapp, "XYLO" (1984).

102. R. Rapp, "The Power of 'Positive' Diagnosis" (1994), 209.

103. Kovitt, "Babies as Social Products" (1978). The dividing line of twenty-eight weeks assumed that obstetricians knew the pregnancy's exact age. It is plausible to assume that in 1978, before the more extensive use of obstetrical ultrasound, the evaluation of a pregnancy's age had not always been accurate.

104. For example, French regulations about the treatment of fetal remains, issued in 2010, distinguish between a "declared" and a "nondeclared" fetus (only the first is entitled to burial), but leave it to the parents to decide whether to declare the fetus. A fetus can be declared beginning in the fourteenth week of pregnancy.

105. The discovery of 342 bodies of fetuses in the mortuary chamber of Saint Vincent-de-Paul Hospital in Paris was followed by two administrative inquiries and a radical modification of French regulations for handling bodies of fetuses and stillborn babies. Mirlesse, "Diagnostic prénatal et médecine fœtale" (2014), 88–89. The Alder Hey Hospital scandal had similar consequences in the United Kingdom. Parry, "The Afterlife of a Slide" (2013).

106. A. Myers, Lohr, and Pfeffer, "Disposal of Fetal Tissue" (2015), 86.

107. D. Davidson, "Reflections on Doing Research" (2011).

108. Lupton, *The Social Worlds of the Unborn* (2013), 87–91.

109. Chiapetta-Swanson, "Dignity and Dirty Work" (2005), 111.

110. Lewis, "The Management of Stillbirth" (1976); Lewis, "Mourning by the Family" (1979), 306.

111. Stillwell, "Pretty Pioneering-Spirited People" (2015).

112. Kaltreider, "Psychological Impact" (1981); D. Davidson, "Reflections on Doing Research" (2011).

113. Hughes et al., "Assessment of Guidelines for Good Practice" (2002), 117.

114. Bernadette Modell's testimony. In E. Jones and Tansey, eds., *Clinical and Molecular Genetics in the UK* (2014), 26.

115. Memmi, *La deuxième vie des bébés morts* (2011).

116. Giraud, "Les 'péri-parents' " (2015).

117. Meunier, "Entretien préparatoire à l'interruption de grossesse" (2002); Esplat, "Les sage femmes face à l'interruption médicale de la grossesse" (2012).

118. Mirlesse et al., "Women's Experience of Termination of Pregnancy" (2011); Esplat, "Les sage femmes face à l'interruption médicale de la grossesse" (2012), 65–66.

119. Giraud, "Les 'péri-parents' " (2015).

120. Mitchell, "Time with Babe" (2016). Mitchell wrote her text in 2014, but it is grounded in observations made between 1998 and 2001. In the meantime, Mitchell notes, the trend toward encouraging women to see and hold dead fetuses has increased and is becoming a new norm of appropriate mourning for pregnancy loss. See, e.g., testimonies of US parents collected by the *New York Times*, "Stillbirth: Your Stories" (2016).

121. Advocates of holding and touching the fetus often quote the work of a Swedish group led by Ingela Radestad; this group, which is still in existence, strongly promoted these practices—e.g., Trulsson and Radestad, "The Silent Child" (2004).

122. Leon, "Perinatal Loss" (1992); Statham, "Prenatal Diagnosis of Fetal Abnormality" (2002).

123. Turton et al., "Incidence, Correlates and Predictors of Post Traumatic Stress Disorder" (2001); Hughes, "Post Traumatic Stress Disorder" (2002), 279–84; Sloan, Kirsh, and Mowbray, "Viewing the Fetus" (2008); Robinson, "Pregnancy Loss" (2014).

124. Hughes et al., "Assessment of Guidelines for Good Practice" (2002), 117.

125. Garel et al., "French Midwives' Practice" (2007), 623.

126. Mitchell, "Time with Babe" (2016).

127. After her miscarriage, Athill decided that she was not made to be a mother; she claimed she never regretted this decision. Athill, *Somewhere towards the End* (2008), 160–65.

128. *New York Times*, "Stillbirth: Your Stories" (2016).

129. Fordyce, "When Bad Mothers Lose Good Babies" (2014).

130. Quoted in Freidenfelds, "Enforcing Death Rituals after Miscarriage" (2016).

131. Reagan, "From Hazard to Blessing to Tragedy" (2003), 358.

132. Mitchell, "Time with Babe" (2016).

133. Reagan, "From Hazard to Blessing to Tragedy" (2003).

134. Layne, "He Was a Real Baby" (2000), 321; Giraud, "Les 'péri-parents'" (2015), 92.

135. The use of memory boxes is recommended by support organizations such as the UK-based charity SANDS (Stillbirth and Neonatal Death Society), founded in 1981.

136. Godel, "Images of Stillbirth" (2008), 254. The cheap plastic cameras probably disappeared with the wider ownership of smartphones.

137. Layne, "He Was a Real Baby" (2000), 323.

138. Now I Lay Me Down to Sleep website, accessed July 25, 2017, https://www.nowilaymedowntosleep.org/.

139. Godel, "Images of Stillbirth" (2008).

140. Keane, "Fetal Personhood" (2009).

141. Godel, "Images of Stillbirth" (2008).

142. Giraud, "Les 'péri-parents'" (2015), 96.

143. Layne, "He Was a Real Baby" (2000), 326.

144. Godel, "Images of Stillbirth" (2008).

145. Ibid., 263–65.

146. Ibid., 261.

147. Ibid., 263.

148. Greig, "Mother Shares Heartbreaking Photos of Baby Son" (2014).

149. Alexis Fretz posted these photos on her blog, *Walter Joshua Fretz*. She explained that one of her goals for the site was to help other women who have experienced pregnancy loss and to dissuade those who are contemplating an abortion. *Walter Joshua Fretz* site, accessed July 25, 2017, https://f2photographybylexi.wordpress.com/2013/06/26/walter-joshua-fretz/.

150. Website In Loving Memory of Cate and Cole, accessed July 25, 2017, https://www.youtube.com/watch?v=1MGndijGHg4.

151. Website In Memory of Jeffry "Baby" Drake, accessed July 25, 2017, https://www.youtube.com/watch?v=WEuzKOu_Ck4.

152. Keane, "Fetal Personhood" (2009), 166.

CHAPTER TWO

1. R. Davidson and Rattazi, "Review: Prenatal Diagnosis of Genetic Disorders" (1972).

2. D. Paul, "From Eugenics to Medical Genetics" (1997); Stern, *Telling Genes* (2012); Stillwell, "Pretty Pioneering-Spirited People" (2015); A. Miller, "A Blueprint for Defining Health" (2000), chap. 6.

3. Korenromp et al., "Termination of Pregnancy on Genetic Ground" (1992); Burgin-ion, Briscoe, and Nemzer, "Genetic Abortions" (1999); Chiapetta-Swanson, "Dignity and Dirty Work" (2005).

4. Latimer et al., "Rebirthing the Clinics" (2006); Latimer, *The Gene, the Clinic, and the Family* (2013).

5. Rabinow and Rose, "Biopower Today" (2006); N. Rose, *Politics of Life Itself* (2006). The terms *molar* and *molecular* were coined by the philosopher Gilles Deleuze and the psychoanalyst Felix Guattari in their 1972 book *Anti Oedipus*, to describe interaction between people: *molar* stands, roughly, for "macro" interactions (among groups and formations), and *molecular* for "micro" ones (among isolated individuals). These terms were adopted by Paul Rabinow and Nikolas Rose to describe a shift from a medical science focused on organs and tissues to one that deals mainly with molecular data and laboratory tests. "Death of the clinics" refers in all probability to Michel Foucault's book *The Birth of the Clinics* (1973).

6. Pickstone, *Ways of Knowing* (2001); Edgerton, *The Shock of the Old* (2007).

7. Another important site of the integration of traditional medical knowledge and cutting-edge development in biomedicine is fetopathology, a medical subspecialty discussed in chapters 3 and 4.

8. Since dysmorphology deals with atypical visible traits, children with developmental delays devoid of such traits (e.g., the majority of children on the autism spectrum) are usually not followed by dysmorphologists.

9. David Smith, "Dysmorphology (Teratology)" (1966); Shepard, "Obituary: David W. Smith" (1981).

10. David Smith, *Recognizable Patterns of Human Malformation* (1970).

11. After Smith's death in 1981, the book was edited by his collaborator Kenneth Jones, and became *Smith's Recognizable Patterns of Human Malformation*.

12. Harper, "Interview: Dian Donnai" (2005); Harper, "Interview: Michael Baraitser" (2005); Donnai, "Professor Robin Michael Winter" (2004).

13. Today, the journal *Birth Defects* is published in three parts. Part A is dedicated to clinical and molecular teratology, part B to developmental and reproductive toxicology, and part C to embryo studies.

14. Children diagnosed as having well-studied genetic conditions such as Down syndrome, thalassemia, or achondroplasia are usually followed by physicians who specialize in the management of each condition.

15. Reardon and Donnai, "Dysmorphology Demystified" (2007), 225.

16. Marion, "The Christmas Present" (1996).

17. Lawrence, "Incommunicable Knowledge" (1985).

18. The British pioneer of dysmorphology Robin Winter was described as being able to spot morphological variations invisible to his colleagues. Donnai, "Professor Robin Michael Winter" (2004); Baraitser, "Professor Robin Winter" (2004); Clayton-Smith, "In Memoriam Robin Michael Winter" (2004); Reardon, "Professor Robin M. Winter" (2004).

19. Donnai, "Professor Robin Michael Winter" (2004); Reardon and Donnai, "Dysmorphology Demystified" (2007).

20. Harper, "Interview: George Fraser" (2005); Latimer, "Diagnosis, Dysmorphology and the Family" (2007).

21. Marion, "The Insolvable Puzzles" (1996).

22. Latimer, *The Gene, the Clinics and the Family* (2013), 163–85.

23. Latimer, "Diagnosis, Dysmorphology and the Family" (2007).

24. Ibid.; Latimer, *The Gene, the Clinics and the Family* (2013).

25. For example, the UK group Unique: The Rare Chromosome Disorders Support Group was created by parents of children with specific chromosomal anomalies. Site of Unique, accessed July 25, 2017, http://www.rarechromo.org/html/home.asp.

26. K. Jones, Crandall Jones, and Del Campo, *Smith's Recognizable Patterns of Human Malformation*, 7th ed. (2013).

27. Noble, "Natural History of Down's Syndrome" (1998).

28. Latimer, *The Gene, the Clinics and the Family* (2013), 181–82. A diagnosis of a hereditary condition in a fetus can lead to the diagnosis of the same condition in other family members. People who viewed themselves as healthy can then unexpectedly learn that they are at risk of serious health difficulties. Robyr et al., "Familial Diseases Revealed" (2006).

29. Nowaczyk and Waye, "The Smith-Lemli-Opitz Syndrome" (2001).

30. Dimond, "Multiple Meanings of a Rare Genetic Syndrome" (2011), 62–93.

31. Habel et al., "Towards a Safety Net" (2014).

32. Scambler, "The 22q11 Deletion Syndromes" (2000); Robin and Shprintzen, "Defining the Clinical Spectrum" (2005).

33. Fung et al., "Practical Guidelines for Managing Adults" (2015).

34. Besseau-Ayasse et al., "A French Collaborative Survey of 272 Fetuses" (2014).

35. Lindee, *Moments of Truth in Genetic Medicine* (2005); De Chadarevian, "Mutations in the Nuclear Age" (2010); Santesmases, "Size and the Centromere" (2010).

36. Hogan, "The 'Morbid Anatomy'" (2014); Hogan, *Life Histories of Genetic Disease* (2016).

37. Hogan, "Visualizing Carrier Status" (2012); Hogan, "Locating Genetic Disease" (2013); Hogan, *Life Histories of Genetic Disease* (2016).

38. Harper, "Interview: André et Joelle Boué" (2005).

39. Pinkel et al., "Fluorescence in Situ Hybridization" (1988); Navon, "Genomic Designation" (2011).

40. Multiple ligation probe amplification, or MPLA, is a variation of the polymerase chain reaction (PCR). A rapid amplification of a selected DNA segment allows a rapid check for whether a parent has an anomaly already identified in a child or a fetus, for example.

41. Shaffer et al., "Detection Rates" (2012); Hillman et al., "Microarray Comparative Genomic Hybridization" (2012).

42. Shaffer et al., "Target Genomic Microarray Analysis" (2006).

43. Shaw-Smith et al., "Microarray Based Comparative Genomic Hybridisation" (2004); Mefford, Batshaw, and Hoffman, "Genomics, Intellectual Disability and Autism" (2012); Navon, "Genomic Designation" (2011).

44. Shaffer et al., "American College of Medical Genetics Guideline" (2005). This recommendation was extended in 2012. Leigh and Milunsky, "Updates in the Genetic Evaluation" (2012).

45. K. Jones, Crandall Jones, and Del Campo, *Smith's Recognizable Patterns of Human Malformation* (2013).

46. Latimer, *The Gene, the Clinics and the Family* (2013); Taussig, *Ordinary Genomes* (2009). Taussig interprets such "normalization" of children with genetic syndromes as representing a specifically Dutch point of view. Some scholars question this interpretation. Leonelli, "Review" (2012); M'charek, "Genetics and Its Others" (2010).

47. Witness seminar, "Histoire de diagnostic prenatal en France" (History of PND in France), held in Paris, October 4, 2012. The full transcript of the seminar (in French) is accessible at http://anr-dpn.vjf.cnrs.fr/?q=node/62, accessed July 25, 2017.

48. French geneticists are divided over the evaluation of the severity of impairment produced by Turner syndrome. Some believe that women with this condition are essentially normal; others think that many among them face significant health and cognitive problems. Witness seminar, "History of PND in France" (2013). In Brazil, where abortion for a fetal indication is illegal, fetal medicine experts have explained that severe structural anomalies observed in 45X0 fetuses early in pregnancy may disappear or be attenuated later. On the other hand, many fetuses with a "severe Turner" are miscarried.

49. Despite great advances in the treatment of children with Down syndrome, in 1988 experts evaluated that between 10 and 20% of children with this condition died before the age of five. Kolker and Burke, *Prenatal Diagnosis* (1998), 217–18.

50. Hong and Reiss, "Cognition and Behavior" (2012).

51. Palmer and Mowat, "Agenesis of the Corpus Callosum" (2014).

52. L. Paul et al., "Review: Agenesis of the Corpus Callosum" (2007); Santo et al., "Counseling in Fetal Medicine" (2012), 513–21; Siffredi et al., "Neuropsychological Profile" (2013).

53. This treatment is based on preventing the sensitization of an Rh-negative woman during her first pregnancy with an Rh-positive child or immediately after the birth of this child. Gaidner, "The Rhesus Story" (1979); Zallen, Christie, and Tansey, eds., *The Rhesus Factor* (2004).

54. Bellini et al., "Etiology of Non-immune Hydrops Fetalis" (2009). When the cause of NIH is not known (systemic NIH), the prognosis is considered especially severe.

55. Fukushima et al., "Short-Term and Long-Term Outcomes" (2011); Santo et al., "Prenatal Diagnosis of Non-immune Hydrops Fetalis" (2011).

56. Howarth et al., "Population Based Study of the Outcome following the Prenatal Diagnosis of Cystic Hygroma" (2005); Lajeunesse et al., "First-Trimester Cystic Hygroma" (2014).

57. Noonan syndrome, an autosomal-dominant congenital disorder produced by a mutation, is associated with heart defects, short stature, and learning difficulties. It is often compared with Turner syndrome. However, Noonan syndrome is found in both males and females, and is not linked to anomalies of sexual development.

58. Baldassarre et al., "Prenatal Features of Noonan Syndrome" (2011); Gaudineau et al., "Postnatal Phenotype" (2013).

59. Alamillo, Fiddler, and Pergament, "Increased Nuchal Translucency" (2012), 102.

60. Bakker, Pajkrt, and Bilardo, "Increased Nuchal Translucency" (2014); Senat and Frydman, "Hyperclarté nucale a caryotype normal" (2007); Souka et al., "Increased Nuchal Translucency" (2005).

61. Baer et al., "Risk of Selected Structural Abnormalities" (2014).

62. Bakker, Pajkrt, and Bilardo, "Increased Nuchal Translucency" (2014), 362.

63. Baer et al., "Risk of Selected Structural Abnormalities" (2014).

64. Results obtained with chorionic villus sampling are usually seen as slightly less accurate than those obtained with amniocentesis; this method is also linked to a somewhat higher risk of miscarriage (spontaneous abortion).

65. Hauerberg et al., "Correlation between Prenatal Diagnosis" (2012).

66. Isacksen et al., "A Correlative Study of Prenatal Ultrasound" (2000).

67. In her study of ultrasonographic diagnosis of fetal malformations in Vietnam, the anthropologist Tine Gammeltoft explains that ultrasound experts were often not sure whether their diagnosis was accurate, but when in doubt recommended an abortion, since personal and social costs of care for severely disabled children in Vietnam were high. It is not clear from her book whether Vietnamese physicians made dissections in order to improve their diagnostic skills. Gammeltoft, *Haunting Images* (2014).

68. Clayton-Smith et al., "Examination of Fetuses" (1990).
69. Isaksen et al., "Comparison of Prenatal Ultrasound" (1999); Carroll et al., "Correlation of Prenatal Diagnosis" (2000).
70. Yeo et al., "Value of a Complete Sonographic Survey" (2002).
71. Kaasen et al., "Correlation between Prenatal Ultrasound" (2006).
72. Akgun et al., "Correlation between Prenatal Ultrasound" (2007).
73. Vimercati et al., "Correlation between Ultrasound Diagnosis" (2012).
74. Ramalho et al., "Critical Evaluation of Elective Termination" (2006).
75. Votino et al., "Minimally Invasive Fetal Autopsy" (2014). In the United Kingdom between 2000 and 2007, consent for fetal autopsies declined from 55 to 45% and for neonatal autopsies fell from 28 to 21%, despite an increase in the number of parents asked to give permission for an autopsy. Thayyil et al., "Post-mortem MRI" (2013).
76. Bassat et al., "Development of a Post-mortem Procedure" (2013); Fligner and Dighe, "Post-mortem Diagnosis" (2013); Thayyil et al., "Post-mortem MRI" (2013). The centralization of analyses of pathological samples is not a new idea. For example, in the 1930s the Brazilian Yellow Fever Service collected small samples of liver tissue from all the people who died of a suspicious fever. These samples were then sent to two central laboratories. Löwy, "Epidemiology, Immunology and Yellow Fever" (1997).
77. Potter summarized the principles of fetal pathology, a subspecialty she helped develop, in her best-known book, *Pathology of the Fetus and the Newborn* (1952).
78. Van Dijck, *The Transparent Body* (2005), 116.
79. Fisher, "First-Trimester Screening" (2011); Viaux-Savelon et al., "Prenatal Ultrasound Screening" (2012). On difficulties generated by the growing resolution power of diagnostic tests, see Welch, Schwartz, and Woloshin, *Overdiagnosed* (2011); Aronowitz, *Risky Medicine* (2015).

CHAPTER THREE

My interest in PND started with a collaborative study funded by the French Agence nationale de la recherche (ANR): Les enjeux du diagnostic prénatal dans la prévention des handicaps: L'usage des techniques entre progrès scientifiques et action publique, ANR-09-SSOC-026-01, http://anr-dpn.vjf.cnrs.fr. The collective results of this study are summarized in *Diagnostic prénatal et prévention des handicaps*, final report of the contract ANR 09-SSOC-026, 2013. The report of this collaborative study can be obtained on demand. Main results are accessible on the study's site: http://anr-dpn .vjf.cnrs.fr., accessed July 25, 2017.

One of the elements of this program was a witness seminar on the history of PND in France; its transcript is accessible at http://anr-dpn.vjf.cnrs.fr/?q.node/62. The full transcript of the witness seminar, sponsored by the same ANR grant, is available at http://anr-dpn.vjf.cnrs.fr/?q=node/62, accessed July 25, 2017.

In the preparatory phase of the witness seminar, my colleagues and I interviewed twelve physicians and researchers who played an important role in the development of PND in France; a few among them later participated in the seminar. My understanding of PND in France is grounded in research conducted in the framework of the ANR program, but opinions expressed in this chapter are mine alone. For two and half years, I observed staff in the fetopathology department of a major French hospital. Researchers who work in France usually do not need to obtain formal permission for observations made in a hospital if their study does not include access to patients, but they need to receive the consent of all the professionals they study. I obtained permission to follow staff and consult patients' files, but not to observe dissections. All the fetopathology department's staff were unfailingly helpful, but they bear no responsibility whatsoever for my interpretation of their work. All the patients' data have been anonymized. Names of physicians and technicians are disguised to protect their anonymity; descriptions of selected case studies are partly modified for the same reason.

1. Samerski, "Genetic Counseling and the Fiction of Choice" (2009), 735–61. See also R. Rapp, *Testing Women, Testing the Fetus* (1999).

2. Some studies investigated nevertheless the diagnostic role of ultrasound—e.g., Krakowsky Chazan, "Vérités, attentes, spectacles et consommations" (2010); Guerra et al., "Breaking Bad News" (2011).

3. Lynn Morgan's studies of the fate of fetuses are a notable exception—e.g., Morgan, "'Properly Disposed Of'" (2002); Morgan, "The Social Biography of Carnegie Embryo" (2004); Morgan, "Strange Anatomy" (2006); Morgan, *Icons of Life* (2009).

4. The term "obligatory passage point" is borrowed from Bruno Latour's study of Pasteur. Latour, *Les microbes, guerre et paix* (1984).

5. Professionals tend to focus on fetal anomalies that can be explained, and rarely discuss cases in which the cause of fetal demise is never elucidated. In 2005, 50.2% of the recorded fetal deaths/stillborn deaths in the United Kingdom remained unexplained. Woods, *Death before Birth* (2007), 187.

6. Fagot Largeault, *L'homme bio-éthique* (1985), 51. Fagot Largeault discussed randomized clinical trials.

7. Witness seminar, "Histoire de diagnostic prenatal en France" (History of PND in France) held in Paris, October 4, 2012. The full transcription of the seminar (in French) is accessible at http://anr-dpn.vjf.cnrs.fr/?q=node/62, accessed July 25, 2017.

8. On inequalities in the French health care system, see, e.g., Nay et al., "Achieving Universal Health Coverage" (2016).

9. Carayol et al., "Les femmes de Seine-Saint-Denis" (2015).

10. In France, many pregnant women are supervised by either a gynecologist (in cities) or a general practitioner (in more remote areas); often, they are also supervised, especially for a more advanced pregnancy, in a secondary referral center, typically a maternity ward of a local hospital or a maternity clinic.

11. I observed similar personalized, informal patterns of patient referral to specialized services when I studied cancer genetics services in France. In the United Kingdom, access to specialists for consultation is more strictly regulated, because specialists and health administrators are concerned about futile or non-urgent referrals "clogging" the system. French experts tend to assume that referrals from other physicians are justified. Some also believe that the value of consultation with a specialist should not be judged solely by simplified utilitarian criteria, and that an "unnecessary" consultation which relieves the patient's anxiety is often an effective health intervention. Löwy and Gaudillière, "Localizing the Global" (2008).

12. Code de la santé publique, Law no. L2213-1 of 17 January 1975.

13. Law no. 2014-873 of 4 August 2014.

14. Law no. 82-1172 of 31 December 1982, on cost reimbursement for voluntary interruption of pregnancy.

15. Lotte and Ville, "Histoire de la prénatalité et de la prévention des handicaps" (2013). On similar attitudes of UK physicians in the 1950s and early 1960s, see Coventry, "The Dynamics of Medical Genetics" (2000).

16. Gaille, "On Prenatal Diagnosis" (2016). Some French experts agree that the woman's/couple's decision should be given more weight, at least in the earlier stages of pregnancy.

17. The definition of an inborn impairment as a "serious, incurable condition" may reflect a situational perception of the efficacy of treating this impairment, for example a disorder of sexual development. R. Raz, "Médecins israeliens face au diagnostic prénatal des fetus intersexués" (2015); Kraus, "Diagnostiquer les fétus intersexués" (2015).

18. Agence de la biomédecine, *Rapport Annuel* (Paris: Agence de la biomédecine, 2011), 72. The report can be downloaded online: www.genethique.org/sites/default/files/agence _biomdecine_rapport_annuel_2011, accessed July 25, 2017.

19. In 2009, the British authorities reported 2,006 terminations of pregnancies for fetal indications, among them 775 terminations for chromosomal abnormalities, mainly (468) for Down syndrome (trisomy 21). United Kingdom, *Statistical Bulletin* (2009). A 2016 paper quotes a slightly higher number of abortions for Down syndrome in the United Kingdom. Morris, Grinsted, and Springetti, "Accuracy of Reporting Abortions" (2016).

20. Although a majority of the French fetal medicine specialists seem to be proud of the high efficacy of detecting fetal malformations in France, some are critical of what they see as excessive use of PND. See, e.g., Amalberti, "Risque et dépistage" (2010).

21. On the other hand, between 2013 and 2017 the more extensive use of non-invasive prenatal testing in France occurred more slowly than in the United States.

22. When the law that legalized abortion in France passed in 1975, the legislators' main intention in allowing late-term abortion might have been to allow late interruption of pregnancy for maternal indications such as preeclampsia.

23. In some cases, French physicians in a neonatal intensive care unit decide, often collectively and with the parents' consent, not to resuscitate a child born with serious health problems. Paillet, *Sauver la vie, donner la mort* (2007).

24. Diaphragmatic hernia is, as a rule, diagnosed during an ultrasound. When the hernia is extended, French physicians usually inform the woman/parents that chances for the child's survival are low. In such cases, the experts explain, the pressure of the intestine had already produced irreversible damage to the lungs and possibly to the brain, which may have been harmed by an insufficient circulation of blood.

25. Merge and Schmoll, "Éthique de l'interruption médicale de grossesse" (2005). Similar dilemmas can arise when deciding whether to resuscitate very premature children, who are at high risk of significant impairments. Paillet, *Sauver la vie, donner la mort* (2007).

26. Garel et al., "French Midwives' Practice" (2007), 624.

27. Information leaflets (recently printed, but undated): *Fiche d'information patiente: Interruption médicale de la grossesse*, CHU Caen; *Livret d'information à l'usage des parents: Interruption médicale de la grossesse*, Hopital Femme, Mère, Enfant, Bron; *Livret d'information à l'usage des parents: Interruption médicale de la grossesse*, Hopitaux de la région Languedoc-Roussilon.

28. Meunier, "Entretien préparatoire à l'interruption de grossesse" (2002), reproduced in Esplat, "Les sages femmes face à l'interruption médicale de la grossesse" (2012).

29. Lavigne, Vernerey, and Vienne, "Pratiques hospitalières" (2006).

30. The main French legal documents modifying the status of stillborn children and miscarried/aborted fetuses are decrees (décrets) no. 2008-798 and no. 2008-800 of 20 August 2008, published in *JORF* (*Journal Officiel de la République Française*) no. 0195 of 22 August 2008.

31. French women are entitled (in 2017) to sixteen weeks of fully paid maternity leave (100% of their salary; they are entitled to twenty-six weeks of fully paid maternity leave if they have more than two children. Fathers are entitled to eleven days of paid paternity leave and three days of "birth leave," a total of two weeks.

32. Interministerial circular of 19 June 2009, relative to fetuses and stillborn children: *Circulaire interministérielle DGCL/DACS/DHOS/DGS/DGS/2009/182 du 19 juin 2009 relative à l'enregistrement à l'état civil des enfants décédés avant la déclaration de naissance* . . . (Paris: Direction des affaires juridiques, 2009). Available online at http://affairesjuridiques .aphp.fr/textes/circulaire-interministerielle-dgcldacsdhosdgsdgs2009182-du-19-juin -2009-relative-a-lenregistrement-a-letat-civil-des-enfants-decedes-avant-la-declaration -de-naissance-et-de-ceux-pouvant-donn/, accessed July 25, 2017.

33. Undated circular, *Traitement des corps foetaux*, CHU de Montpellier.

34. Ville and Mirlesse, "Prenatal Diagnosis" (2015); Mirlesse, "Diagnostic prénatal et médecine fœtale" (2014).

35. Dommergues et al., "Termination of Pregnancy" (2010).

36. Memmi, "Archaïsme et modernité" (2003).

37. Dommergues et al., "Termination of Pregnancy" (2010).

38. Pioneers of PND who fought for legalization of abortion in the 1960s and 1970s tend to have a negative view of Down syndrome; conservative, Catholic pediatricians tend to stress the positive aspects of educating a Down syndrome child. Séminaire à témoins (witness seminar), "Histoire du diagnostique prénatal en France" (2013). See note 7 above for details.

39. Dommergues et al., "Termination of Pregnancy" (2010).

40. Agence de la biomédecine, *Rapport Annuel* (Paris: Agence de la biomédecine, 2012), "Tableau CPDPN1: Résumé des activités des CPDPN de 2008 à 2010," available online at https://www.google.com.br/search?q=Agence+de+la+Biom%C3%A9decine ,+Rapport+Annuel+(2012),&spell=1&sa=X&ved=0ahUKEwjy6ba33anVAhWFf5AKHY _lDxMQBQgoKAA&biw=1158&bih=660, accessed July 25, 2017.

41. Witness seminar, "Histoire du diagnostique prénatal en France" (2013), http://anr-dpn .vjf.cnrs.fr/?q1/4node/62. Testimony of M. Planchu. See note 7 above for details.

42. Ibid. Testimony of M. Voyer.

43. Memmi, "Archaïsme et modernité" (2003).

44. Merge and Schmoll, "Éthique de l'interruption médicale de grossesse" (2005).

45. In the United Kingdom, fetal pathology is but one of the many tasks of pediatric pathologists. N. McIntosh, *Report: Fetal, Perinatal and Paediatric Pathology* (2002).

46. Carles, "Letter of the SOFFŒT's President" (2009).

47. Razavi and Carles, coords., *Pathologie fœtale et placentaire pratique* (2008). SOFFŒT has a very modest budget, and functions mainly thanks to its members' volunteer work.

48. US geneticists similarly argue that clinical and pathological studies are necessary to understand the meaning of microdeletions found in 1% of fetuses. Statements of Mary Norton and Ronald Wapner, quoted in Ratner, "Roche Swallows Ariosa" (2015).

49. When I negotiated access to the fetal pathology department at AH, I received permission to be present at all staff meetings and examine all the department's written records, but, because of legal restrictions, I was not allowed to view autopsies. My knowledge of the practical aspects of dissections comes from interviews with pathologists and technicians, and information gathered during staff meetings.

50. On the history of the standardization of medical records, including those for autopsies, see Hess and Mendelsohn, "Case and Series" (2010).

51. At fourteen weeks of pregnancy, counted from the date of the last period—that is, twelve weeks of fetal development—the fetus is, on average, seven to eight centimeters long. One of the reasons for this notice might have been that the woman had given birth vaginally, after treatment with mifepristone and misoprostol (antiprostaglandins employed to induce an abortion); the fetus, although very small, was probably well formed.

52. Although French obstetricians, geneticists, and fetopathologists invariably speak about a fetus, not a baby in discussing a dissected body, they usually speak about the pregnant woman as "mommy" (*la maman*), even in cases in which they implicitly recommend a termination of pregnancy. In these cases, the term *mommy* may refer to the woman's hope to be a mother.

53. On the embodied knowledge of the medical expert, see Lawrence, "Incommunicable Knowledge" (1985).

54. Until the 1960s, obstetricians and pediatricians automatically assumed that all the babies born small were premature. In 1960, the US pioneer of studies of fetal anomalies, Joseph Warkany, demonstrated that this was not the case. The small size of a newborn is often the consequence of a specific pathology, intrauterine growth delay. Warkany, "The Medical Profession and Congenital Malformations" (1979).

55. In about 29–30% of the cases, identical twins do not share a placenta, because they were separated early in their embryonic development (dichorionic diamniotic twins).

56. Akkermans et al., "A Worldwide Survey of Laser Surgery" (2015); J. McIntosh et al., "Long Term Developmental Outcomes" (2014).

57. Vassy, "From a Genetic Innovation to Mass Health Programmes" (2006).

58. Training programs for non-MD medical counselors were introduced in France in the early 2000s, but AH's medical genetics department did not employ such counselors.

59. Bosk, *All God's Mistakes* (1992).

60. Paillet, *Sauver la vie, donner la mort* (2007).

61. Some psychologists have argued that a woman's decision to undergo a late-term abortion (for her, to kill her child) can produce a severe, long-lasting trauma ("Medea complex"). For some women, waiting until the child dies in the womb or giving birth to a child who will live only for a short time may be less traumatic options. On the other hand, psychologists also have discussed cases of women who gave birth to a severely malformed fetus/child, and were haunted by the image of the "monster" they had produced. Leuzinger-Bohleber, "Introduction: Ethical Dilemmas" (2011). The poem "Hospital Deliveries" describes a C-section delivery of a dead, severely malformed child. The parents refused abortion on religious grounds. The father got over his initial shock of seeing the severely deformed body, and recognized the child (of unknown sex) as his beloved daughter. The poem does not tell whether the mother, too, reached some kind of closure. Chiosi, "Hospital Deliveries" (2013).

62. Long periods of uncertainty followed by termination of pregnancy have been linked to higher rates of a severe post-abortion distress. Mirlesse et al., "Women's Experience of Termination of Pregnancy" (2011).

63. For example, in 1996 the journal *Clinical Dysmorphology* announced the creation of a web-based "WWW Dysmorphology Discussion Board," which invites specialists in this domain to submit their puzzling cases for evaluation by their colleagues. Blisma and Donnai, "WWW Dysmorphology Discussion Board" (1996).

64. Shaw et al., "Surveying 'Slides' " (2003).

65. On "preparations," see Rheinberger, "Preparations" (2010); Löwy, "Microscope Slides in the Life Sciences" (2013).

66. The high percentage of cases that Soraya saw as "solved" may reflect the type of fetuses studied by AH fetopathologists. Since AH is a tertiary referral center, its pathologists more frequently dissect fetuses aborted after a diagnosis of an anomaly than do their colleagues working in nonspecialized hospitals. The colleagues deal more often with unexpected and unexplained pregnancy loss. On the elevated proportion of unexplained fetal and neonatal deaths, see Woods, *Death before Birth* (2007), 187.

67. The introduction to the 2013 edition of *Smith's Recognizable Patterns of Human Malformation* similarly points to a rapid increase in the number of inborn malformations associated with specific mutations. K. Jones, Crandall Jones, and Del Campo, *Smith's Recognizable Patterns of Human Malformation* (2013).

68. Some inborn defects, such as Turner syndrome (45X0), are rare because most of the fetuses with this condition are miscarried. Hook and Warburton, "Turner Syndrome Revisited" (2014).

69. The acronym CHARGE stands for Coloboma of the eye (a hole in one of the eye's structures), Heart defects, Atresia (narrowing) of the inner nostrils, Retardation of growth and/or development, Genital and/or urinary abnormalities, and Ear abnormalities and deafness. Most (but not all) cases of CHARGE syndrome are associated with mutations in the CHD7 gene on chromosome 8.

70. On the role of family history in the construction of genetic knowledge, see Atkinson, Parsons, and Featherstone, "Professional Constructions of Family and Kinship" (2001).

71. A woman's difficult reproductive trajectory may also lead to a categorical refusal of an "imperfect" pregnancy. Gaille, "On Prenatal Diagnosis" (2016).

72. Wapner et al., "Chromosomal Microarray versus Karyotyping" (2012), 2182–83.

73. Equilibrated Robertsonian translocation is a reciprocal exchange of genetic material between two chromosomes. Its frequency in humans is estimated at 1:1000. When no genetic material is lost, people with an equilibrated translocation do not suffer ill effects. Their offspring, however, are at an increased risk of genetic anomalies, because during the production of egg and sperm cells, some cells may receive an abnormal set of chromosomes.

74. In France, only public laboratories are qualified to perform PGD; a majority of the laboratories that perform in vitro fertilization in that country are public, too.

75. Such purification of scientists' narratives was first described in Fleck's pioneering study. Fleck, *Genesis and Development of a Scientific Fact* (1979), first published in 1935.

76. Although French gynecologists and fetal medicine experts often speak about their fear of being sued for a professional mistake, this is a rare occurrence. Fillon, "Quelles stratégies de soins?" (2012). French physicians who work in public hospitals are usually shielded from patients' accusations of professional misconduct by mediation structures established by the hospital; the only exceptions to this protection are obvious professional mistakes. Underestimation of the severity of fetal malformations is probably not perceived as being among such mistakes. Nevertheless, no fear of juridical consequences for a diagnostic error and the understanding of colleagues do not shield professionals from being badly shaken when they make an error.

77. Giving the child with a nonlethal impairment up for adoption seemed to me a good alternative for a very late abortion with feticide until I assisted at a meeting, organized by Ann Fessler, of women who gave their (illegitimate) children up for adoption and never overcame the trauma produced by their decision. Fessler, *The Girls Who Went Away* (2007).

78. Witness seminar, "Histoire du DPN en France" (2013). "Schizophrenic" (the actors' term) refers to the popular—and inaccurate—perception of schizophrenia as representing a split personality.

CHAPTER FOUR

My research in Brazil is grounded in three periods, each two months long, of observing the staff of the anatomic pathology department at the Maternal Health Center (MHC). I received permission to conduct this research from the MHC's ethics committee. At the anatomic pathology department, I had access to all the data and patients' files, and was able to observe all the dissections. I received very generous help from the department's staff, technicians, and trainees, but they do not bear any responsibility for my interpretation of their activity. During my research in Brazil, I conducted interviews with members of the anatomic pathology department, as well as geneticists and physicians at the MHC—a total of twenty-four interviews; some individuals were interviewed more than once. All the personal data about patients were anonymized, names of sites and participants have been disguised to prevent recognition, and some of the case studies were partly modified for the same reason.

1. On "estrangement," see Ginzburg, *Wooden Eyes* (2001).
2. *Synecdoche* may be more accurate than *metonymy*, because the trash bin was a part of the anatomic pathology service, not an external device that could symbolize it. Seeing the trash bin as an object that might stand for fetal pathology studies is obviously an idiosyncratic personal view.
3. Poor women/couples are entitled to a free funeral for the fetus or stillborn child, but many are not aware of this policy and are not adequately informed of their rights.
4. In the 1960s and 1970s, Canadian midwives used containers such as a peanut butter jar, a urine specimen bottle, or a Tupperware box to transport aborted fetuses to the pathology department. Chiapetta-Swanson, "Dignity and Dirty Work" (2005).
5. Morgan, " 'Properly Disposed Of' " (2002).
6. Morgan explains that in the early twentieth century, embryologists usually did not discuss the social conditions that might have affected pregnancy loss. She quotes, however, a 1917 expert who explained that upper-class women miscarried because of venereal disease; they had early miscarriages, and the fetus was often malformed. Poor women frequently miscarried because of unfavorable social conditions. They had late miscarriages, and the fetus often looked normal. Ibid, 265.
7. The MHC was a midsized hospital. Large teaching and research hospitals are usually modern high-rise buildings.
8. Paim et al., "The Brazilian Health System" (2011).
9. The Brazilian health system has more similarities with the United States than with western Europe. Some affluent Brazilians purchase expensive foreign (usually North American) health insurance, which provides wider coverage than Brazilian health insurance plans.
10. Rasella, Aquino, and Barreto, "Reducing Childhood Mortality" (2010).
11. Pereira dos Santos, Malta, and Mehry, "A regulação na saúde suplementar" (2008).
12. The documentary film *Uma historia Severina*, made by the anthropologist Debora Diniz and the filmmaker Eliane Brum, narrates the history of a poor Brazilian woman who was given permission to abort an anencephalic fetus. While she was already hospitalized for termination of pregnancy, she learned that following the decision of the supreme court of 20 October 2004, the hospital could not go

ahead with the procedure. Diniz's film can be seen at https://www.youtube.com /watch?v=65Ab38kWFhE, accessed July 26, 2017.

13. Diniz and Medeiros, "Aborto no Brasil" (2010).

14. Diniz and Corrêa, *Aborto e Saúde Publica* (2008); Diniz and Medeiros, "Itinerários e métodos do aborto ilegal" (2012).

15. This opinion was expressed in 2012; in 2016, there were signs of tightening the repression of illegal abortions.

16. Costa, Goulart, and Nasciento, "Operacoe Herodes" (2014).

17. Brazilian media rarely mention the suffering of women who abort in suboptimal and unsafe conditions, and the high levels of morbidity induced by clandestine abortion. Nunez, Madeiro, and Diniz, "Histórias de aborto provocado" (2013); Le et al., "The Burden of Unintended Pregnancies" (2014).

18. Clavery, "Médico de hospital público no Rio" (2016). It is possible that the gynecologist was denounced by a member of the clinic's staff. I was told in 2016 that higher-end clinics in São Paulo that performed abortions were seen as "safer."

19. Diniz and Medeiros, "Aborto no Brasil" (2010).

20. Stepan, *The Hour of Eugenics* (1991); Fry, "Politics, Nationality, and the Meanings of 'Race'" (2000); Rajao and Duque, "Between Purity and Hybridity" (2014).

21. Horovitz et al., "Atenção aos defeitos congênitos" (2006); Horovitz et al., "Genetic Services and Testing" (2013).

22. Emilia Sanabria has studied the intersection between private and public health services in Brazil—Sanabria, "From Sub- to Super-Citizenship" (2010); Sanabria, *Plastic Bodies* (2016).

23. Horovitz et al., "Atenção aos defeitos congênitos" (2006).

24. Guerra et al., "Confiabilidade das informações" (2004); Guerra, "Avaliaça das informaçoes sobre defeitos congentios" (2006). The two databases are SINASC datasus (Sistema de Informações de Nascidos Vivos), http://datasus.saude.gov.br/sistemas -e-aplicativos/eventos-v/sinasc-sistema-de-informacoes-de-nascidos-vivos, and SIM datasus (Sistema de Informações de Mortalidade), http://datasus.saude.gov.br/sistemas -e-aplicativos/eventos-v/sim-sistema-de-informacoes-de-mortalidade, accessed July 26, 2017.

25. Castilla and Orioli, "ECLAMC" (2004). The accuracy of reporting newborn malformations varies in different Brazilian states.

26. I found one article that explicitly mentions termination of pregnancy for a nonlethal fetal malformation (increased nuchal translucency and/or trisomy 21) in Brazil. Saldanha et al., "Anomalias e prognóstico fetal" (2009).

27. Horovitz et al., "Genetic Services and Testing" (2013).

28. Biehl and Petryna, "Bodies of Rights" (2011).

29. Livingston, *Improvising Medicine* (2012).

30. Brazil, 2014, Ministério da Saúde, portaria GM no. 199, 30 de janeiro de 2014. Institui a Política Nacional de Atenção Integral às Pessoas com Doenças Raras, aprova as Diretrizes para Atenção Integral às Pessoas com Doenças Raras no âmbito do Sistema Único de Saúde (SUS) e institui incentivos financeiros de custeio. Diário Oficial da União 31/01/2014. (Decree of January 30, 2014, establishing a national policy of integrated [or complete] care of people with rare disease within the Unified Health System, and providing financial incentives for such care).

31. Brazil, 2009, Ministério da Saúde, portaria GM no. 81, 20 de janeiro de 2009—Institui, no ambito do Sistema Único de Saúde (SUS), a Politica Nacional de Atenção Integral em Genetica Clinica. Diário Oficial da União 21/01/2009 (Decree of January 20,

2014, establishing national policy of integrated [complete] care in clinical genetics). On the failure to implement this law, see Horovitz et al., "Genetic Services and Testing" (2013).

32. In 2016 and 2017, implementation of genetic services within SUS was hampered by the consequences of a severe economic and political crisis.

33. In principle, Brazilian women have the right to undergo in vitro fertilization (IVF) in the national health service clinics, but in practice such access is very limited. IVF is reserved nearly exclusively for affluent women. Alfano, *Reproducao assistida* (2014). On preimplantation diagnosis in Brazil, see Damian, Bonetti, and Horovitz, "Practices and Ethical Concerns" (2015).

34. Guilam, "O discurso do risco" (2003); Guilam and Correa, "Risk, Medicine and Women" (2002). An alternative explanation for the "disappearance" of women after a diagnosis of a fetal impairment may be that some radically reject this diagnosis, do not want to learn more about their future child's problems, and hope for a miracle.

35. Zlot, *Anomalias congênitas* (2008).

36. Ibid. On the proliferation of C-sections in Brazil, see, e.g., Béhague, Victora, and Barros, "Consumer Demand for Caesarean Sections" (2002); Sass and Hwang, "Dados epidemiológicos, evidências e reflexões" (2009); De Vasconcellos et al., "Nascer no Brasil" (2014).

37. Novaes Machado, "A necropsia prenatal" (2012).

38. The term *genetic counseling* in this dissertation may resonate with the term *genetic abortion* as shorthand for an abortion for a fetal indication. Burginion, Briscoe, and Nemzer, "Genetic Abortions" (1999).

39. The organizers of the RADIUS trial (Routine Antenatal Diagnosis Imaging with Ultrasound), a large-scale randomized trial of routine ultrasonography in the United States—and probably one of the rare randomized trials of efficacy of this technique worldwide—concluded that the routine use of ultrasound in pregnancy does not provide a measurable beneficial effect on pregnancy outcomes. Ewigman et al., "Effect of Prenatal Ultrasound" (1993). Ulterior investigations mainly confirmed this observation, although they found that obstetrical ultrasound has some positive effects on maternal health. Bricker, Medley, and Pratt, "Routine Ultrasound" (2015).

40. *Manual tecnico pre-natal* [Technical directory of prenatal tests], published by the Brazilian Ministry of Health in 2006, quoted in Baião, "A decisao informada no rastramento prenatal" (2009).

41. Guerra, "Avaliaça das informaçoes sobre defeitos congentios" (2006).

42. Sampaio Rodrigues, "Sentidos, limites e potentialidades da medecina fetal" (2010).

43. Bomfim, "A antecipação ultra-sonográfica de malformação fetal" (2009); Bomfim, Coser, and Lopes Moreira, "Unexpected Diagnosis of Fetal Malformations" (2014).

44. Krakowsky Chazan, "O aparelho é como um automovel" (2011).

45. Medical services for the elites are seldom studied by Brazilian sociologists and anthropologists of medicine. Sanabria, "From Sub- to Super-Citizenship" (2010), 396.

46. Mirlesse, "Diagnostic prénatal et médecine fœtale" (2014).

47. Bomfim, Coser, and Lopes Moreira, "Unexpected Diagnosis of Fetal Malformations" (2014). Lilian Krakowsky Chazan describes the key role of learning fetal sex in including the future children in their families. Krakowsky Chazan, "É . . . tá grávida mesmo!" (2008).

48. Bomfim, "A antecipação ultra-sonográfica de malformação fetal" (2009), 33–37.

49. Bomfim, Coser, and Lopes Moreira, "Unexpected Diagnosis of Fetal Malformations" (2014). The authors studied the reactions of lower-class women treated in SUS clinics

and hospitals to the announcement of a severe fetal malformation; these women's only option was to continue the pregnancy. Lilian Krakowsky Chazan observed the ultrasonographic diagnosis of a fetal malformation in private ultrasound facilities, but did not investigate what happened next. Krakowsky Chazan, "Vérités, attentes, spectacles et consommations" (2010).

50. Ávila da Costa, "O que os olhos veem" (2015).

51. Ville and Mirlesse, "Prenatal Diagnosis" (2015). 25.

52. Mirlesse, "Diagnostic prénatal et médecine fœtale" (2014), 130.

53. Veronique Mirlesse and Patrica Ávila da Costa record similar statements. Mirlesse, "Diagnostic prénatal et médecine fœtale" (2014), 127–29. Ávila da Costa, "O que os olhos veem" (2015).

54. Germain, "Brazil Log Notes, 1966–1970." Ford Foundation Archives, New York. Germain was sent to Brazil by the Ford Foundation to investigate the possibility of introducing family planning there.

55. BEMFAM (1986); Sobrinho, *Estado e população* (1993); Brasil, Ministério da Saúde, *Pesquisa nacional de demografia e saúde* (2008).

56. Mirlesse, "Diagnostic prénatal et médecine fœtale" (2014), 126–28.

57. E.g., Faúndes et al., "The Closer You Are" (2004); Diniz and Corrêa, *Aborto e Saúde Publica* (2008); Diniz and Medeiros, "Itinerários e métodos do aborto ilegal" (2012); Dias et al., "Association between Educational Level" (2015).

58. For example, in the maternity hospital of Ribeirão Preto University (São Paulo State), only one out of fourteen attempted postnatal separations of conjoined twins did not end with the death of both twins. Berezoski et al., "Gêmeos conjugados" (2010).

59. Brizot et al., "Conjoined Twins Pregnancies" (2011). Professional evaluation of a woman's capacity to understand that the twins she was carrying will not survive was perhaps made to deflect the judge's possible objection.

60. Nomura et al., "Conjoined Twins" (2011). Because of their size and their position in the womb, conjoined twins are usually born through a C-section which involves a long incision in the uterine wall. Such an incision takes a long time to heal and increases risks for future pregnancies. A second-trimester abortion of conjoined twins is less risky, because the fetuses are much smaller.

61. Mirlesse, "Diagnostic prénatal et médecine fœtale" (2014), 141–43.

62. Sampaio Rodriques, "Sentidos, limites e potencialidades da medecina fetal" (2010), 22–24.

63. On the complexity of racial divisions in Brazil, and the difficulty in correlating "race" and genetic data, see Parra et al., "Color and Genomic Ancestry" (2003), 177–82; Santos and Chor, "Race, Genomics, Identities and Politics" (2004).

64. Romero, "Brazil Enacts Affirmative Action Law" (2012). The Brazilian "positive discrimination" law takes into consideration both the self-declared skin color ("black" or "brown") and parents' income. The law is binding only in state universities, but in Brazil these are usually the best universities.

65. Sanabria, "From Sub- to Super-Citizenship" (2010).

66. For example, Jessica Gregg concludes that all the women with cervical malignancies treated in an SUS facility in Recife received the same high dose of radiation (and many suffered debilitating side effects), because individualized therapies were considered too expensive. J. Gregg, *Virtually Virgins* (2002).

67. Biehl, "The Activist State" (2004). *Citizenship* is often an ambivalent rather than a positively connoted term in Brazil. One's capacity to stand out in the mass of undifferentiated "individuals" submitted to impersonal rules and laws and to be a unique

person is often a significant mark of higher status. DaMatta, " 'Do You Know Who You're Talking To?!' " (1979).

68. Sanabria, "From Sub- to Super-Citizenship" (2010), 382–83; Gonçalves et al., "Contraceptive Medicalisation" (2011); 201–15; Sanabria, *Plastic Bodies* (2016).

69. The program has four divisions: prenatal care, childbirth, postpartum care, and a division of logistics which deals with transportation and regulatory issues. Additional goals of this program are the training of nurses and an effort to limit the number of C-sections performed in public hospitals. For more information, go to http://dab .saude.gov.br/portaldab/ape_redecegonha.php, accessed July 26, 2017.

70. Brazilian data on maternal mortality can be found on the government's system of health-related data analysis (DATASUS): http://www2.datasus.gov.br, accessed July 26, 2017. For the French data, see the INSERM report on maternal mortality: http:// presse-inserm.fr/mortalite-maternelle-diminution-de-la-mortalite-par-hemorragies /10335/, accessed July 26, 2017.

71. Fordyce, "When Bad Mothers Lose Good Babies" (2014).

72. Zlot, *Anomalias congênitas* (2008); Baião, "A decisao informada no rastramento prenatal" (2008); Mirlesse, "Diagnostic prénatal et médecine fœtale" (2014).

73. Beserra, "Alloimunizacao RhD" (2015). An additional problem has been the difficulty to assess the number of earlier pregnancies that did not end with a live birth. Since abortion is illegal in Brazil, women are often reluctant to admit that they had induced abortions.

74. Bustos, Sierre, and Diaz del Castillo, "Working with Care" (2014).

75. Rubin, Greene, and Baden, "Zika Virus and Microcephaly" (2016); Araujo, Silva, and Araujo, "Zika Virus–Associated Neurological Disorders" (2016).

76. The 2015–16 Zika epidemics in Brazil favored the emergence of much closer collaborations between MHC fetopathologists and other experts both within and outside the country. Such collaborations are, however, mainly limited to fundamental scientific investigations of anomalies induced by the Zika virus, and do not (in 2017) modify the routine monitoring of pregnant women.

77. Some patients in private practice can also have access to fetal surgery, developed in a few cutting-edge Brazilian clinics. The Zika epidemics reopened the debate about abortion for fetal indications. E.g., Editorial, "Microcephalia e aborto" (2016); Diniz, "The Zika Virus" (2016); Galli and Deslandes, "Threats of Retrocession" (2016).

78. Bernard, *Introduction à la médecine expérimentale* (1865). Claude Bernard (1813– 1878) is one of the founders of modern physiology.

79. The reconstruction of early operating rooms in the popular television series *The Knick*, directed by Steven Soderbergh and first shown in 2014, may recall the MHC's dissection room.

80. The "adhesive tape" management of samples at the MHC, in which the same piece of marked white adhesive tape is transferred between boxes and test tubes, contrasts with a full computerization of tubes, samples, and patients' files in Brazil's high-end private gynecological clinics. On the other hand, French public hospital experts, who often conduct more advanced tests than those performed by Brazilian doctors in private clinics, use a "mixed economy" of inscriptions, in which computerized patients' files, data, photographs, and labels coexist with handwritten or typed notes and letters, handwritten case summaries, circulating kraft envelopes, and handwritten labels on tubes and slides.

81. For example, the pathology department of London Hospital shifted from handwritten to machine-typed notes in the early 1950s. The intervention between direct

observation and its transcription favored the standardization of pathology reports. Löwy, *Preventive Strikes* (2009), 98–99.

82. Lawrence, "Incommunicable Knowledge" (1985).

83. It may become less theoretical in the future, with the predicted spread of sequencing platforms in Brazil.

84. With the exception of Doctora Theresa, all the pathologists called each other by their first name, but were called Doctora Luisa, Doctora Raquel, Doctora Simone by the technicians. I was Ilana for the pathologists and Doctora Ilana for the technicians. All the people who worked at the anatomic pathology department—technicians, administrative staff, trainees, and pathologists—chatted informally all the time, exchanged professional and personal information, and engaged in banter.

85. I was told in 2017 that this internal rule had been modified recently. Thus, women who miscarried or gave birth to a dead child (*neomorto*) may sign an autopsy permit.

86. Lynn Morgan described such exclusion of women from church ceremonies in Ecuador. Morgan, "Imagining the Unborn" (1997).

87. The dissection reports rarely mention maternal disease as a cause of a fetal anomaly. The reason may be the MHC's specialization in fetal pathologies. Another teaching and research hospital specializes in the follow-up of women with pathologies such as pre-eclampsia, heart disease, diabetes, and kidney disease, which put the fetus at risk.

88. The MHC's geneticists apply advanced genetic technologies mainly to research of rare hereditary conditions; they seldom use them to refine a postmortem diagnosis.

89. Guilam and Correa, "Risk, Medicine and Women" (2002).

90. Mirlesse and Ville, "The Uses of Ultrasonography" (2013), 168–75; Vassy, Rosman, and Rousseau, "From Policy Making to Service Use" (2014), 67–74; Mirlesse, "Diagnostic prénatal et médecine fœtale" (2014); Ville and Mirlesse, "Prenatal Diagnosis" (2015).

91. Löwy, *Preventive Strikes* (2009). On overdiagnosis, see, e.g., Welch, Schwartz, and Wolochin, *Overdiagnosed* (2011).

92. Brazilian public health care experts rarely mention in this context deaths from complications of illegal abortion, for some specialists responsible for up to 20% of maternal deaths. Guilhem and Azevedo, "Brazilian Public Policies for Reproductive Health" (2007).

93. The only exceptions are women who receive a diagnosis of a lethal anomaly of the fetus and who successfully apply for juridical permission to terminate the pregnancy. These cases are, however, very rare, even in the MHC, where geneticists strongly support such applications.

94. Brazilian fetal pathologists claim that they fear being sued for the failure to detect a fetal anomaly, but it is not clear how often this happens, if ever. Sampaio Rodrigues, "Sentidos, limites e potencialidades da medecina fetal" (2010), 41, 52.

CHAPTER FIVE

1. Stern, *Telling Genes* (2012).

2. Quoted in Hayden, "Prenatal Screening Companies" (2014).

3. On the "self-evidence" of such a view in the 1970s, see Stern, *Telling Genes* (2012).

4. Discussion, Henry Nadler's text "Risks in Amniocentesis" (1970), 144. Later, an explicit endorsement of abortion for Down syndrome became rare; one exception is a statement made in 2014 by the biologist Richard Dawkins. Dawkins, "'Immoral' Not to Abort" (2014).

5. Theodore Cooper's statement, in a press conference on the implication of the Amniocentesis Registry findings. Cooper read this text on 20 October 1975, during the presentation of conclusions of the United States Department of Health, Education, and Welfare study on the safety of amniocentesis sponsored by the National Institutes of Health and the National Institute of Child Health and Human Development. Cooper, "Implication of the Amniocentesis Registry Findings" (1975), quoted in Coulliton, "Amniocentesis" (1975), 537.

6. Stein, Susser, and Guterman, "Screening Programme for Down" (1973), 308.

7. Herzog, "Abortion, Christianity, Disability" (2014), 257.

8. Ibid., 255–56. Such positions were firmly rejected by advocates of the abortion law. Accessed July 28, 2017, http://www2.assemblee-nationale.fr/14/evenements/2015/anniversaire-loi-veil#node_9805.

9. D. Williams, "The Partisan Trajectory" (2015).

10. Luker, *Abortion and the Politics of Motherhood* (1984), 236. "Genocide against the disabled" might have migrated from pro-life to disability activism.

11. E.g., Hubbard, "Prenatal Diagnosis and Eugenic Ideology" (1985); Hubbard, "Eugenics: New Tools, Old Ideas" (1987).

12. Some women who terminated a pregnancy following diagnosis of a fetal malformation shared with pro-life activists a disdain for women who aborted for "social" reasons. These women presented themselves as "good mothers" who made the heartbreaking choice to prevent the suffering of their future child, and abhorred an association with "bad mothers" who ended a pregnancy for "selfish" reasons. McCoyd, "What Do Women Want?" (2009).

13. E.g., Hubbard, "Prenatal Diagnosis and Eugenic Ideology" (1985); Hubbard, "Eugenics: New Tools, Old Ideas" (1987); Wertz and Fletcher, "A Critique of Some Feminist Challenges" (1993).

14. On the (presumed) aspiration to have "flawless" children, see Rothschild, *The Dream of the Perfect Child* (2005), and on the striving to perfect human beings, Camporesi, *From Bench to Bedside* (2014); Bateman et al., eds., *Inquiring into Human Enhancement* (2015).

15. There is an abundant literature on the history of eugenics and clinical genetics. In the context of PND, especially useful studies are D. Paul, *Controlling Human Heredity* (1995); D. Paul, *The Politics of Heredity* (1998); Stern, *Telling Genes* (2012); Comfort, *The Science of Human Perfection* (2012).

16. Koch, "The Meanings of Eugenics" (2004), 318, 329.

17. Paul, *The Politics of Heredity* (1998), 97.

18. Rosenberg, "The Tyranny of Diagnosis" (2002).

19. Linder, "On the Borderland of Medical and Disability History" (2013); Cooter, "The Disabled Body" (2000).

20. Whyte and Ingstad, "Disability and Culture" (1995), 7–8.

21. Weisz, *Chronic Disease* (2014); Linder, "On the Borderland of Medical and Disability History" (2013).

22. The US Army continues to promote a broad definition of *disability*. In 2014, 50% of US soldiers returning from tours of duty in Iraq or Afghanistan claimed invalidity pensions; one-third of these claims were related to mental health issues. Panzzanese, "Disarray at the VA" (2014).

23. This is still the case for selected infectious diseases such as poliomyelitis or encephalitis, and for infections (rubella, toxoplasmosis, cytomegalovirus) that induce fetal malformations.

24. For a more detailed discussion, see Gaudillière and Löwy, eds., *Heredity and Infection* (2001).

25. Linder, "On the Borderland of Medical and Disability History" (2013), 515. In French, the term *handicap* also covers *disability*. In Brazil, the usual term for "disability" is *deficienia* (deficiency), despite its potentially negative connotations.

26. Linder, "On the Borderland of Medical and Disability History" (2013), 519, 525.

27. Wendell, "Unhealthy Disabled" (2001).

28. Carpenter, "Is Health Politics Different?" (2012), 291.

29. Rosenberg, "The Tyranny of Diagnosis" (2002).

30. Scully, "What Is a Disease?" (2004); Whitmarsh et al., "A Place for Genetic Uncertainty" (2007).

31. Rabinow, "Artificiality and Enlightenment" (1996).

32. Chadwick, "Can Genetic Counseling Avoid the Charge of Eugenics?" (1998); Chadwick, "Genetic Choice and Responsibility" (1999).

33. Geneticists who developed prenatal testing in the 1970s testified about the contribution of women at risk of giving birth to a child with a given hereditary disease to the development of prenatal testing for this disease. Christie and Tansey, eds., *Genetic Testing* (2001). Leslie Reagan describes the promotion of anti-rubella vaccine by mothers of children impaired by infection with the rubella virus. Reagan, *Dangerous Pregnancies* (2010).

34. Reagan's *Dangerous Pregnancies* has been criticized by the disability activist Claudia Malacrida, because Reagan, following the perception of some of the mothers she studied occasionally presents birth defects induced by rubella and thalidomide as a "tragedy." Such presentation, Malacrida argues, strengthens negative perceptions of disabled people. Malacrida, "Review of *Dangerous Pregnancies*" (2014).

35. Saxton, "Disability Rights and Selective Abortion" (1998). See also Saxton, "Prenatal Screening and Discriminatory Attitudes" (1987). Saxton, a professor in the Interdisciplinary Studies Department at the University of California–Berkeley, is a wheelchair user.

36. Degener, "Female Self Determination" (1990).

37. Ibid., 89. Degener presents gynecologists and geneticists as adversaries of a pro-woman standpoint. For a different view of the geneticists'/genetic counselors' standpoint in the 1970s and 1980s, see Stillwell, "Pretty Pioneering-Spirited People" (2015).

38. Asch, "Prenatal Diagnosis and Selective Abortion" (1999); Parens and Asch, "The Disability Rights Critique" (2000).

39. Edwards, "Disability, Identity and the 'Expressivist Objection'" (2004).

40. Press, "Assessing the Expressive Character" (2000).

41. Piepmeier, "The Inadequacy of 'Choice'" (2013).

42. Malacrida, "Review of *Dangerous Pregnancies*" (2014), 336. Ann Furedi, director of the British Pregnancy Advisory Service (BPAS), the main provider of abortions in the United Kingdom, commented that promoters of the expressivist objection to an abortion for a fetal indication are pro-choice, except when they don't like the choice. Quoted by Wilkinson, "Prenatal Screening" (2015), 30.

43. Steinbock, "Disability, Prenatal Testing and Selective Abortion" (2000).

44. Kristin Luker has provided a lucid overview of arguments of "pro-life" and "pro-choice" groups in the United States. Luker, *Abortion and the Politics of Motherhood* (1984). For a twenty-first-century description of pro-choice arguments, see Pollitt, *Pro* (2014).

45. O'Toole, "What Hope Has Pope Francis Offered" (2016).

46. Teixiera et al., "The Epidemic of Zika Virus–Related Microcephaly" (2016), 603.

47. Diniz, "The Protection to Women's Fundamental Rights" (2016), e9. Psychological torture in this context refers to the severe psychological suffering of women obliged to continue their pregnancy while facing extreme uncertainty about the pregnancy's outcome. Teixiera and her colleagues agree with Diniz's argument, but add that in countries where terminations are legal (e.g., French Polynesia, the United States, and Slovenia), many affected women who learned about an already existing brain anomaly of the fetus decided to terminate the pregnancy. Teixiera and Rodriguez, "Response" (2016).

48. Mol, *The Logic of Care* (2008), xi. Netherlands has a low uptake of screening for Down and a low level of women's employment, a possible source of Mol's insistence that she has work that fascinates her and is reluctant to give it up to become a full-time caregiver for a special-needs child.

49. R. Rapp and Ginsburg, "Enabling Disability" (2001), 542.

50. Gaille, "On Prenatal Diagnosis" (2016). In the sample studied by Gaille, two women who became pregnant following an in vitro fertilization explained that after all the difficult reproductive procedures they underwent, they cannot bear the idea of raising a severely impaired child. As one of them put it, "I have suffered enough to do without having anyone preaching me about my duty."

51. Ivry, "The Predicaments of Koshering Prenatal Diagnosis" (2015), 288.

52. McCoyd, "I'm Not a Saint" (2008).

53. Mirlesse, "Diagnostic prénatal et médecine fœtale" (2014), 57.

54. Piepmeier, "Choosing to Have a Child" (2012); Piepmeier, "Outlawing Abortion" (2013).

55. Piepmeier, "Would It Be Better for Her Not to Be Born?" (2015).

56. Parens and Asch, eds., *Prenatal Testing and Disability Rights* (2000).

57. The utilitarian argument—high costs of treating severely disabled people and keeping them alive limits resources available to "socially useful" individuals—may be more frequent in periods of economic hardship. Eugenicists' discourse on "useless lives" became more popular following the economic crisis of 1929. D. Paul, *Controlling Human Heredity* (1995), 134–35.

58. Shakespeare, "Losing the Plot?" (1999); C. Williams, Alderson, and Farsides, "What Constitutes Balanced Information" (2002).

59. Ford, "Prenatal Testing and Disability Rights" (2001).

60. Shakespeare, "The Content of Individual Choice" (2005). For Brock's views on the parental duty to avoid impairment in offspring, see Brock, "Preventing Genetically Transmitted Disabilities" (2005). Shakespeare's focus here is the social determinants of dis/ability. He does not discuss the possibility that a severe physical or mental impairment may be a source of suffering for disabled people and their families, even when society provides adequate support to its impaired members. Andrew Solomon discusses such suffering of parents of schizophrenic children. Solomon, *Far from the Tree* (2012), chapter 6.

61. Gedge, "Reproductive Choice" (2011); McKinney, "Selective Abortion" (2016).

62. Solomon, *Far from the Tree* (2012), 692.

63. Ibid., 693–97.

64. Caplan, Foreword to *Prenatal Testing* (1988), xiii–xviii.

65. Ibid., xiv–xv.

66. Joan Rothschild presents a similar view of PND. Rothschild, *The Dream of the Perfect Child* (2005).

67. On the role of fear in the expansion of the scope of diagnostic tests, see, e.g., Welch, *Should I Be Tested for Cancer?* (2006); Löwy, *Preventive Strikes* (2009); Rosenberg, "Managed Fear" (2009); Welch, Schwartz, and Woloshin, *Overdiagnosed* (2011); Aronowitz, *Risky Medicine* (2015).

68. Weir, *Pregnancy, Risk and Biopolitics* (2006); Ruhl, "Liberal Governance" (2009); Lupton, " 'Precious Cargo' " (2012).

69. Rosenberg, "The Tyranny of Diagnosis" (2002).

70. Ville and Lotte, "Évolution des politiques publiques" (2013); Hashiloni Dolev,. *Life (Un)worthy of Living* (2007).

71. Wahlberg, "Serious Disease" (2009), 106.

72. Jesudason, "Editorial: The Paradox of Disability" (2011).

73. E.g., Benson, ed., *Cellular Organelles and Membranes* (1971); Simpson et al., *Genetics in Obstetrics and Gynecology* (1982), 95–120.

74. Homsy et al., "De Novo Mutations" (2015).

75. Carey, "Parents and Professionals" (2014), 71.

76. Osteen, introduction to *Autism and Representation* (2008), 3.

77. Eyal et al., *The Autism Matrix* (2010), 261–63.

78. On conflicts between the caregivers and the parents of intellectually disabled people, see, e.g., Jingree and Finlay, "It's Got So Politically Correct Now" (2012).

79. Fishely, "I Am John" (1992).

80. Ibid., 157.

81. An article published in the *New York Times* in May 2016 makes a similar argument about society's duty to prevent the mistreatment and exploitation of people with mild intellectual disabilities. However, focusing on people able to lead autonomous and semiautonomous lives, like focusing on the "healthy disabled," avoids the thorny question of the cost of care of more severely impaired people. Barry, "Giving a Name, and Dignity, to a Disability" (2016).

82. Solomon, *Far from the Tree* (2012), 23.

83. Featherstone, *A Difference in the Family* (1981), 223–29.

84. Ibid., 219, 226.

85. Menzel et al., "The Role of Adaptation" (2002); Dylan Smith et al., "Happily Hopeless" (2009).

86. Culver et al., "Informed Decisions" (2000), 3201.

87. Harrison, "Making Lemonade," (2001), 239–40. Helen Harrison was an activist on behalf of children with severe inborn impairments, especially those resulting from extreme prematurity. Roberts, "Helen Harrison" (2015).

88. Skotko, Levine, and Goldstein, "Having a Son or Daughter with Down Syndrome" (2011); Skotko, Levine, and Goldstein, "Having a Brother or Sister with Down Syndrome" (2011).

89. Landsman, *Reconstructing Motherhood and Disability* (2009); Kittay, "When Caring Is Just and Justice Is Caring" (2001); Kittay, "The Personal Is Philosophical Is Political" (2010); Bérubé, *Life As We Know It* (1996). These important and moving studies focus on life with disabled children. In most of the described cases, the child's disability was not detected before birth, sometimes because the parents elected to eschew PND.

90. Olsson and Hwang, "Depression in Mothers and Fathers" (2001); Leiter et al., "The Consequences of Caring" (2004); Eisenhower, Baker, and Blacher, "Preschool Children with Intellectual Disability" (2005).

91. Hacking, *The Taming of Chance* (1990).

92. My description of DS is focused mainly on the difficulties involved in educating

a child with this condition, for two reasons. First, this is a study of PND as a risk management technology. Second, while in the 1960s and 1970s Down syndrome was often presented by PND promoters in an excessively negative light and many physicians viewed a woman's decision to continue a pregnancy with a DS fetus as "irrational," in the twenty-first century, women who learn that they carry a DS fetus and decide to terminate the pregnancy may be harshly criticized for their "eugenic" choice. Löwy, "Review: Down's" (2014).

93. C. Miller, "Discovering the Upside of Down" (2012); Becker, "Has Down Syndrome Hurt Us?" (2011).

94. Helen Featherstone described a babysitter's refusal to take care of her severely brain-damaged child. She was mortified by this refusal, then realized that she sees her son as beautiful and lively, but the babysitter who faced her son for the first time saw only a drooling child who makes strange noises. Featherstone, *A Difference in the Family* (1981), 41.

95. E.g., Brighter Tomorrows website, http://www.brightertomorrows.org/, accessed July 28, 2017. By contrast, stories of women who decided to abort DS fetuses found on the site A Heartbreaking Choice (http://www.aheartbreakingchoice.com/, accessed July 28, 2017) focus on the material and emotional difficulties of caring for DS children.

96. Baily, "Why I Had Amniocentesis" (2000); Press, "Assessing the Expressive Character" (2000).

97. Choi, Van Riper, and Thoyre, "Decision Making" (2012), 161.

98. McCoyd, "I'm Not a Saint" (2008), 1494. Women who testified about their decision to interrupt pregnancy on the website A Heartbreaking Choice often legitimated their decision by the "good mother" stance: their willingness to suffer themselves to prevent the suffering of their future child. "Given his diagnosis, he would have known only suffering. As his mother, I couldn't allow that to happen." "I would rather suffer every day for the rest of my life than to allow her to suffer one moment in life." "A mother will stop at nothing, including her own physical and mental hurt, to protect her child." Vietnamese women interviewed by the anthropologist Tine Gammeltoft similarly explained that they elected to suffer themselves rather than condemn their child to the harsh life of a severely disabled person in a resource-poor environment. Gammeltoft, *Haunting Images* (2014).

99. McCoyd, "Pregnancy Interrupted" (2007), 39.

100. McCoyd, "I'm Not a Saint" (2008), 1493. This testimony may reflect an idiosyncratic problem: this woman's difficulty in coping with an uncontrolled expression of teenagers' sexuality. It shows nevertheless that a firsthand familiarity with a given condition can lead to both optimistic and pessimistic views of this condition.

101. Rosman, "Down Syndrome Screening Information" (2016); Schwennesen, Svendsen, and Koch, "Beyond Informed Choice" (2010); Gross, " 'The Alien Baby' " (2010); Vassy, Rosman, and Rousseau, "From Policy Making to Service Use" (2014).

102. US activists' attitude may be connected with the greater value in the United States placed on the "wish to know," expressed for example in an enthusiastic support of screening for cancer. Schwartz et al., "Enthusiasm for Cancer Screening" (2004).

103. Tredgold, *Mental Deficiency* (1908), 189.

104. Lionel Penrose's conference at the meeting on birth defects at the Pediatric Section of the Royal Medical Society, 27 May 1960. Lionel Penrose's papers, UCL Archives, file 62/5; Lionel Penrose to Klaus Patau, letters of 29 November 1961 and 6 December 1961. Klaus Patau's papers, University of Wisconsin, Madison.

105. Skotko, Levine, and Goldstein, "Having a Son or Daughter with Down Syndrome" (2011).

106. Reactions to David Perry's editorial. Perry, "Down Syndrome Isn't Just Cute" (2014), accessed July 28, 2017, http://america.aljazeera.com/opinions/2014/10/down-s-behind -thesmiles.html.

107. Letters to the British Down Syndrome Association from the 1960s, quoted in Zihni, "The History of the Relationship" (1989), 18–19.

108. Monckton, " 'When Will I Be Normal?' " (2012).

109. Skotko, Levine, and Goldstein, "Having a Son or Daughter with Down Syndrome" (2011); Skotko, Levine, and Goldstein, "Having a Brother or Sister with Down Syndrome" (2011). On adaptation to difficult situations through "response shift," see Menzel et al., "The Role of Adaptation" (2002).

110. Piepmeier, "Would It Be Better for Her Not to Be Born?" (2015).

111. https://www.govtrack.us/congress/bills/110/s1810/text, accessed July 28, 2017. Senator Edward Kennedy and his sister, Eunice Kennedy Shriver, who had an intellectually impaired sister, became very active in promoting the rights of people with intellectual disabilities. I'm grateful for Diane Paul for attracting my attention to the US legislation on DS information.

112. Information on the National Down Syndrome Society website: http://www.ndss .org/Advocacy/Advocacy-Programs/NDSS-Government-Affairs-Committee-GAC -Program/NDSS-Prenatal-Information-State-Law-Toolkit/, accessed July 28, 2017. The law does not differentiate between information given to expectant mothers and information given to parents who have learned that their newborn child has DS.

113. The Prenatally and Postnatally Diagnosed Conditions Awareness Act and similar legislation systematically equate "positive prenatal diagnosis" with the diagnosis of trisomy 21, reinforcing the concept that the only/main goal of PND is the detection of this inborn anomaly.

114. Perry, "Down Syndrome Isn't Just Cute" (2014). Perry strongly advocates for honesty about DS and opposes "sugarcoated" and "cute" images of this condition; he also affirms that families of DS children are lucky and blessed. For a positive view of DS people, see, e.g., the video How Do You See Me, made in 2016 for World Down Syndrome Day: https://www.youtube.com/watch?v=YhCEoL1pics, accessed July 28, 2017.

115. North, "New Down Syndrome Law" (2014). Some parents of children with DS explain that they oppose termination of pregnancy for this condition because they fear that a significant decrease in the number of children with Down will limit resources available for their children. Not all the DS activists adopt this view, however. On debates among DS activists, see, e.g., Mark Leach's blog, NDSC, GDSF, Prenatal Testing, and Abortion, July 22, 2014, accessed July 28, 2017, http://www.downsyndromeprenatal testing.com/ndsc-gdsf-prenatal-testing-and-abortion/.

116. Lewin, "Ohio Bill Would Ban Abortion" (2015).

117. Schrad, "Does Down Syndrome Justify Abortion?" (2015).

118. Testimonies of Annie, Rick Love, Gregory Fairchild, and Diane. In readers' comments to ibid.

119. Testimonies of 'Virginia, Jules, and G. B. In readers' comments to ibid.

120. Testimony of K, from Tennessee. In readers' comments to ibid.

121. Many parents of DS children strongly resist having their child perceived exclusively through the lens of health and cognitive problems. E.g., Jericho, "I No Longer See My Daughter's Down Syndrome" (2016).

122. Caplan, "Chloe's Law" (2015).

123. Choi, Van Riper, and Thoyre, "Decision Making" (2012), 163.
124. Fitzgerald et al., "Hospital Admissions" (2013); Kucik et al., "Trends in Survival among Children" (2013). Differences in the mortality of children with DS among countries and sites may reflect the availability of PND of this condition, including diagnostic ultrasound, coupled with higher rates of abortion of trisomic fetuses diagnosed as having structural malformations such as a severe heart defect.
125. For example, the "Fact Sheet about Down Syndrome for New and Expectant Parents," reviewed by the National Society of Genetic Counselors Down Syndrome Information Act Working Group (nsgc.org/d/do/4640), provides quantitative data on the frequency of health problems experienced by DS children (e.g., 40–60% have a heart condition and 12% a gastrointestinal condition), but does not provide similar information about the percentage of these children/people with severe intellectual and psychiatric difficulties. The Fact Sheet explains that thanks to changes in attitudes toward this condition, children with DS *can* go to school, participate in sports, and socialize, and *many* people with DS are thriving as active and valued members of the community. But it does not attach any number to the "can" and "many" statements.
126. One of the rare exceptions is Alison Piepmeier's study. She interviewed two women with disabilities who experienced sexual assault, and as a consequence were acutely aware of the higher risk of sexual abuse for impaired people when deciding whether to continue a pregnancy with a DS fetus. One of these women elected to have an abortion; the other decided to continue the pregnancy. Piepmeier, "Would It Be Better for Her Not to Be Born?" (2015), 11–12.
127. Noble, "Natural History of Down's Syndrome" (1998), 175.
128. Kolker and Burke, afterword to *Prenatal Testing* (1988), 217–18.
129. The Italy and Netherlands fable was written by the DS activist Emilie Pearl Knight, a mother of a DS child, in 1987. It is reproduced in Solomon, *Far from the Tree* (2012), 169–70. Solomon adds that Knight's son, Jason King, a "high-performing" person with DS, studied in a regular school, learned foreign languages, and wrote a book about DS. Nevertheless, he needs a significant level of support—partly provided by the state—because he cannot live on his own, is not able to hold a job, and had suffered from several depressive episodes.
130. Standard medical textbooks provide an overview of the range of symptoms linked with trisomy 21—e.g., K. Jones, Crandall Jones, and Del Campo, *Smith's Recognizable Patterns of Human Malformation* (2013), 7–10.
131. Noble, "Natural History of Down's Syndrome" (1998).
132. Bertoli et al., "Needs and Challenges of Daily Life" (2011). The authors of this survey contacted general practitioners in Rome and asked them to identify all their DS patients. In other studies, e.g., Skotko, Levine, and Goldstein, "Having a Son or Daughter with Down Syndrome" (2011); Skotko, Levine, and Goldstein, "Having a Brother or Sister with Down Syndrome'" (2011), participating families were recruited through networks of DS activists.
133. Kolker and Burke, afterword to *Prenatal Testing* (1988), 218.
134. Lindean, "Down Syndrome Screening" (2015); O. Gordon, "Living with Down" (2015).
135. Quoted by O. Gordon, "Living with Down" (2015).
136. M. Gordon, "Review of 'Rosemary'" (2015). Frontal lobotomy is no longer used to treat agitated people with psychiatric illness/intellectual disability, but drug treatments of these people can in some cases leave them calmer but mentally incapacitated. The documentary film made in 2008 by the French actress and director Sandrine Bonnaire, *Elle s'appelle Sabine*, follows the dramatic fate of Bonnaire's sister,

Sabine, a young woman on the autism spectrum who was interned in a psychiatric hospital in her late twenties because of her increasingly disruptive and erratic behavior. Sabine was heavily medicated and lost nearly all her previous skills, including an ability to conduct a conversation, read a book, and play music.

137. Olsson and Hwang, "Depression in Mothers and Fathers" (2001); Emerson, "Prevalence of Psychiatric Disorders" (2003); Saint Louis, "When Caregivers Need Healing" (2014).

138. Blum, "Mother Blame in the Prozac Nation" (2007), 207.

139. Jane Berenstein's daughter was such a case. An easy child, she became very difficult to manage as a teenager. Finally, her parents found her a residential home, where she became much happier and calmer. She lost, however, her ability to keep herself clean and speak proper English, and became severely overweight. Berenstein, *Rachel in the World* (2007).

140. Wynn and Wynn, *Prevention of Handicap* (1979), 1–23.

141. Tredgold, *Mental Deficiency* (1908), provides many examples of the conflation of mental impairment and mental disease.

142. Eisenhower, Baker, and Blacher, "Preschool Children with Intellectual Disability" (2005). The authors analyzed data on more than ten thousand children in Great Britain aged five to fifteen years. They found that 8% of controls and 39% of children with intellectual disabilities were diagnosed as having at least one psychiatric disorder.

143. Ibid.; Solomon, *Far from the Tree* (2012), 195–96.

144. One study found that 22% of people with DS have a psychiatric disorder; another reported that 25.3% have psychiatric or behavioral issues; still another described occasional psychiatric difficulties in 33% and severe psychiatric problems in 22% of people with DS. B. Myers and Pueschel, "Psychiatric Disorders" (1991); Määttä et al., "Mental Health, Behaviour and Intellectual Abilities" (2006); Bertoli et al., "Needs and Challenges of Daily Life" (2011). As far as I know, information provided to pregnant women rarely mentions the risk of psychiatric/behavioral problems in people with DS. The only example I found was the testimony of a UK woman who, when diagnosed as having a trisomic fetus, was told by the midwife that DS "is a spectrum, from a child who can grow up, function and live alone, to a child who is possibly violent, incontinent and unable to communicate." O. Gordon, "Living with Down" (2015). It is possible that this midwife painted a bleak picture because she wanted to encourage an abortion for DS, but if the information is accurate, it may be difficult to argue that it should remain concealed. The claim that it is legitimate to provide an optimistic view of DS because when the parents discover that their child is more severely affected than they believed s/he will be, they are already deeply attached to this child and love her/him unconditionally, assumes that all the parents/families successfully adapt to the presence of a seriously impaired child.

145. Eyal et al., *The Autism Matrix* (2010). France, Eyal and his colleagues suggest, has a higher level of institutionalization of intellectually impaired children and, possibly for this reason, lower rates of diagnosis of autism.

146. Autism, like DS, is a spectrum, with "high-functioning" and "low-functioning" individuals. Some high-functioning people with autism reject the classification of their condition as a disability and argue that their behavior should be recognized as a legitimate way of being in the world. E.g., Ginsburg and Rapp, "No Judgment" (2015).

147. Donvan and Zuker, *In a Different Key* (2016); Silberman, *Neurotribes* (2015).

148. Kanne and Mazurek, "Aggression in Children and Adolescents" (2011).

149. The meaning of autism has changed over time. A review of *In a Different Key* notes that in 1966, epidemiologists evaluated the frequency of autism in the United Kingdom at less than 0.05% (4.5 out of 10,000 children), while in their 2016 book, Donovan and Zuker estimate that the frequency of autism is close to 1.5% (one out of 68 children), "raising the question what exactly are we calling autism in 2016?" Baron, "Review of 'In a Different Key'" (2016). This statement resonates with Robert Aronowitz's argument that the rapid extension of definitions of cancer led to the inclusion of many individuals with manageable conditions in the group "cancer patients." Ironically, the new, positive image of "cancer survivor" might have contributed to the stigmatization and invisibility of people dying of cancer and increased their isolation and exclusion. Aronowitz, *Risky Medicine* (2015), 154.

150. Testimony of What3231, Illinois, in comments to Schrad, "Does Down Syndrome Justify Abortion?" (2015).

151. Testimony of JMF, Hartford, CT, in comments to ibid.

152. Osteen, "Urinatown" (2008).

153. Belkind, "The Unvarnished Reality of Autism" (2009).

154. Lutz, *Each Day I Like It Better* (2014). Lutz collected harrowing (and possibly exceptional) testimonies of parents who were unable to cope with their autistic child's extreme anger until that child was treated with electroconvulsive therapy.

155. Simplican, "Care, Disability and Violence" (2015), 230. On the risk of violent behavior in children with autism, and the parallel risk that these children will become victims of physical abuse from caregivers, including parents, see Kanne and Mazurek, "Aggression in Children and Adolescents" (2011).

156. Marion, "Autism Spectrum Disorders" (2013), 33–36; Betancur, "Etiological Heterogeneity" (2011); Mefford, Batshaw, and Hoffman, "Genomics, Intellectual Disability and Autism" (2012). Sex chromosome anomalies such as Klinefelter syndrome (47, XXY) and 47, XYY are also associated with an increased risk of autism spectrum disorder, although such risk is much lower for individuals with an abnormal number of sex chromosomes than in those with fragile X and DiGeorge syndromes.

157. Hercher and Bruenner, "Living with a Child at Risk for Psychotic Illness" (2008).

158. Karas et al., "Perceived Burden and Neuropsychiatric Morbidities" (2014), 203.

159. Quoted in McCoyd, "I'm Not a Saint" (2008), 1495.

160. Quoted in Waldman, "Rocketship" (2010), 128. Disability rights advocates may argue that believing that a child with severe physical and intellectual disabilities will be a "burden" for his parents and siblings reflects prejudice against disabled people. This is a valid argument. On the other hand, it is difficult to deny the existence of a risk (rather than a certainty) that the birth of a severely disabled child will have a negative impact on the child's family.

161. Gammeltoft and Wahlberg, "Selective Reproductive Technologies" (2014), 208.

CHAPTER SIX

1. Asch, "Prenatal Diagnosis and Selective Abortion" (1999); Parens and Asch, "The Disability Rights Critique" (2000); Landsman, *Reconstructing Motherhood and Disability* (2009).

2. Gillis, *A World of Their Own Making* (1997).

3. Badinter, *L'amour en plus* (1980).

4. Present-day debates about disability are often dominated by idealized images of disabled persons and their families. This is often true for recent literature, drama, and cinema, too. A truly "bad" fictional hero is rarely disabled, with the obvious

exception of those who became disabled as a result of their criminal activities. One not-so-recent exception is P. D. James's novel *Unnatural Causes*, in which a disabled woman reacts to the mixture of pity and contempt she inspires in others with cruelty and violence. James, *Unnatural Causes* (1967).

5. Kerr, "Reproductive Genetics" (2009).
6. Luker, *Abortion and Politics of Motherhood* (1984).
7. Wertz and Fletcher, "A Critique of Some Feminist Challenges" (1993).
8. Harper, Reynolds, and Tansey, eds., *Clinical Genetics in Britain* (2010).
9. Mara Buchbinder and Stefan Timmermans have analyzed the use of emotions in shaping health care policies. Buchbinder and Timmermans, "Affective Economies" (2013).
10. France et al., "How Personal Experiences Feature in Women's Accounts" (2011); Place, "Motherhood and Genetic Screening" (2008).
11. Atkinson, Parsons, and Featherstone, "Professional Constructions of Family and Kinship" (2001).
12. D. Paul and Brosco, *The PKU Paradox* (2013). Peter Coventry reports that in the 1960s and 1970s, hemophilia was perceived in the United Kingdom not as a "genetic disease" but as a blood disorder that can be treated successfully, decreasing the demand for PND of this condition. Coventry, "The Dynamics of Medical Genetics" (2000), chapter 2.
13. The term *geneticization* has been coined by the the feminist geneticist Abby Lippman to describe the phenomenon through which many aspects of health and illness, but also human traits and abilities, are framed with reference to genetic profiles. Lippman, "Prenatal Genetic Testing and Screening" (1991). The concept of "biosociality" has been proposed by the anthropologist Paul Rabinow to depict a new type of socialization around a "molecularized" understanding of health and disease. Such understanding, Rabinow has argued, is replacing older definitions of disease. Rabinow, "Artificiality and the Enlightenment" (1996).
14. Cox and Starzomski, "Genes and Geneticization?" (2004); Bahradwaj, Atkinson, and Clarke, "Medical Classifications" (2006).
15. Zlotogora and Lewenthal, "Screening for Genetic Disorders" (2000); Cowan, *Heredity and Hope* (2008); Beck and Niewöhner, "Localising Genetic Testing" (2009); Cousens et al., "Carrier Screening" (2010).
16. Zlotogora et al., "A Targeted Population Carrier Screening Program" (2009). Preconceptional diagnosis alerts potential parents, or potential spouses, to risks of having children with hereditary disease. Preimplantation diagnosis allows a couple who elects in vitro fertilization a selective implantation of embryos devoid of a targeted mutation.
17. Alderson, Scotte, and Thapar, "Living with a Congenital Condition" (2001), 166. The interviewed people with Down syndrome were probably on the high-functioning end of this condition's spectrum, because they were able to answer a complex questionnaire, while some among the people with spina bifida had a moderate impairment, invisible to people around them.
18. A best-selling history of cancer proposes such a view of malignant tumors. Mukherjee, *The Emperor of All Maladies* (2010).
19. Dekeuwer and Bateman, "Much More Than a Gene" (2011), 239.
20. Lloyd, "Duchenne Muscular Dystrophy" (2009), 199–200. All the mothers interviewed by Lloyd were familiar with the expressivist objection to PND.
21. Ibid., 203–7.

22. Boardman, "The Role of Experiential Knowledge" (2010); Boardman, "Knowledge Is Power?" (2013); Boardman, "The Expressivist Objection" (2014).
23. Boardman, "The Role of Experiential Knowledge" (2010), 38.
24. The story of a British mother of three small children with SMA who killed them in November 2014, then tried to commit suicide, opened a debate about the plight of some parents/mothers of children with this condition. Clarence, "Case against Mother" (2014).
25. Boardman, "The Role of Experiential Knowledge" (2010), 248.
26. S. Kelly, "Choosing Not to Choose" (2008); Boardman, "The Expressivist Objection" (2014). Some mothers of children with a hereditary condition elected preimplantation genetic diagnosis (PGD) rather than PND in their next pregnancy to avoid the emotional hardship of aborting a fetus similar to their affected child. PDG has, however, a high failure rate. Lloyd, "Duchenne Muscular Dystrophy" (2009); Boardman, "The Role of Experiential Knowledge" (2010).
27. Discourse on "disability pride," like the one about "gay pride," may mask the persistence of shame about a (still) stigmatized condition. Solomon, *The Noonday Demon* (2002), 205.
28. A. Raz, "Important to Test" (2004); Gammeltoft, "Childhood Disability" (2008).
29. Chen and Schiffman, "Attitudes toward Genetic Counseling" (2000), 148.
30. Boesky, introduction to *The Story Within* (2013), 3.
31. Marion, "The Skeleton in Mr. Anderson's Closet" (1995), 2.
32. Leuzinger-Bohleber, "Introduction: Ethical Dilemmas" (2011), 18–21. In that case, PND revealed that the fetus was female, and therefore not at risk of hemophilia, an X-linked disorder.
33. Scholars criticizing treatment of pregnant women at risk of giving birth to CAH girls with dexamethasone described CAH as "difference" rather than disability. Dreger, Feder, and Tamar-Mattis, "Prenatal Dexamethasone" (2012).
34. Collste, "Moral Decision-Making" (2011), 171–73.
35. Gaille, "On Prenatal Diagnosis" (2016).
36. Dimond, "Patient and Family Trajectories" (2013); Weintraub, "Three Biological Parents" (2013); Watts, coord., *Novel Techniques for Mitochondrial DNA* (2015).
37. UK Parliament, Commons debate statutory instrument on mitochondrial donation, February 3, 2015, , accessed July 28, 2017, http://www.parliament.uk/business/news/2015/february/commons-debate-statutory-instrument-on-mitochondrial-donation/.
38. Data on reproductive patterns in families with a disabled child are contradictory. A 2011 study has found that families in which the firstborn child is disabled had on average more children than families in which the firstborn child is not disabled. The study followed a hereditary condition, cystic fibrosis, and a genetic but usually nonhereditary impairment, Down syndrome. Burke, Urbano, and Hodapp, "Subsequent Births" (2011). An earlier study arrived at an opposite conclusion. D'Amico et al., "Reproductive Choices" (1992).
39. S. Kelly, "Choosing Not to Choose" (2008), 87.
40. Ibid., 107.
41. Ibid., 108.
42. E. Rapp, "Dear Dr. Frankenstein" (2013), 226–28.
43. Ibid.
44. Raspberry and Skinner, "Enacting Genetic Responsibility" (2010), 6–8. Raspberry and Skinner do not mention an additional problem: some carriers of the fragile-X trait

may suffer later in life from a debilitating neurological condition, fragile-X tremor ataxia syndrome. Dunsford, "The Long Arm" (2013).

45. Raspberry and Skinner, "Enacting Genetic Responsibility" (2010), 8.

46. Ibid., 9–11. Sarah Franklin and Celia Roberts have studied the use of preimplantation diagnosis by couples at risk of having a child with a hereditary disease. Franklin and Roberts, *Born and Made* (2006).

47. Clarke, "Stigma, Self-Esteem and Reproduction" (2013).

48. Rough, "Three Phone Calls" (2013), 2119.

49. Powell, "On the Precarious Cusp" (2012).

50. Ibid., 799.

51. Ibid., 800. Some feminist critics of PND rejected the possibility that a decision to terminate a pregnancy for a fetal indication can be perceived as empowering. Rowland, *Living Laboratories* (1992), 112.

52. Powell, "On the Precarious Cusp" (2012), 801.

53. Ibid., 802.

54. Valenti, "My 28-Week Pregnancy" (2014).

55. Raspberry and Skinner, "Experiencing the Genetic Body" (2007), 374.

56. Clarke, "Stigma, Self-Esteem and Reproduction" (2013), 901.

57. Raspberry and Skinner, "Enacting Genetic Responsibility" (2010), 8.

58. Serres and Latour, *Conversations on Science, Culture, and Time* (1995).

59. Daum, "I Nearly Died" (2014).

60. This was, for example, the argument of the bioethicist Dan Brock. Brock, "Preventing Genetically Transmitted Disabilities" (2005).

61. Shakespeare, "The Content of Individual Choice" (2005), 218.

62. Atkins, "The Choice of Two Mothers" (2008).

63. Ibid., 122, 115.

64. Bérubé, *Life as We Know It* (1998), 88.

65. I discuss this point in my book *L'emprise du genre* (2005).

66. Rousseau, *Émile, or Treatise on Education* ([1762] London: J. M. Dent and Sons, 1914), bk. 5, 321.

67. Ginsburg, *Contested Lives* (1998), 210.

68. Blum, *At the Breast* (1999); Bryder, "From Breast to Bottle" (2009); Wolf, "Is Breast Really Best?" (2007); Sacker et al., "Breast Feeding and Intergenerational Social Mobility" (2013).

69. Readers' comments on Young, "Overselling Breast-Feeding" (2015).

70. Featherstone, *A Difference in the Family* (1981), 71.

71. Roberts, "Helen Harrison" (2015).

72. "Why Isn't My Child as Clever as Me?" (2013).

73. Blum, *Raising Generation Rx* (2015), 8–9.

74. Hopcroft and McLaughlin, "Why Is the Sex Gap in Feelings of Depression Wider" (2012).

75. Wahlberg, "Serious Disease" (2009), 104–6.

76. N. Gregg, "Further Observations of Congenital Defects" (1944), 129. Gregg supported the right of women infected with rubella early in pregnancy to have a "therapeutic abortion." Reagan, Dangerous Pregnancies (2010), 36–48.

77. Bell, "Corrrespondence" (1959); italics mine.

78. For example, the Harvard physician Adam Wolfberg lavishly praised his wife, who left her work as a clinical psychologist and dedicated herself entirely to an intensive physical reeducation of their impaired, prematurely born daughter. He did not,

however, seem to consider the possibility that he could have taken a similar step. Wolfberg, *Fragile Beginnings* (2012).

79. Home, "Challenging Hidden Oppression" (2002).
80. Ashtana, "Take Care " (2017).
81. Dykens et al., "Reducing Distress in Mothers" (2014). For a comment, see Saint Louis, "When Caregivers Need Healing" (2014).
82. R. Rapp, "The Power of 'Positive' Diagnosis" (1994).
83. Some lower-class women become highly effective advocates for their disabled children, and through activism create strong links with women/parents of disabled children from more privileged social strata. The complexity of services for disabled children, and the frequent need to fight with the administrators to obtain help for one's child, are nevertheless obstacles to lower-class families' equal access to these services.
84. Featherstone, *A Difference in the Family* (1981), 83.
85. Hollander, *When the Bough Breaks* (2008); Solomon, *Far from the Tree* (2012), 402–4.
86. Shur, "The Real Tiger Mothers" (2011), 2088. The article's title alludes to Amy Chua's best-selling book *Battle Hymn of a Tiger Mother*, which praises the "extreme parenting" provided by Chinese mothers, who push their children to be high achievers. Chua, *Battle Hymn of a Tiger Mother* (2011).
87. Hanisch, "Politics of Love" (2013), 11.
88. Ibid., 4.
89. Ibid., 8, 9.
90. Landsman, *Reconstructing Motherhood and Disability* (2009); Kittay, "When Caring Is Just and Justice Is Caring" (2001); Kittay, "The Personal Is Philosophical Is Political" (2010); Bérubé, *Life As We Know It* (1996); E. Rapp, *The Still Point of the Turning World* (2013), 13–14.
91. Hanisch, "Politics of Love" (2013), 12.
92. Ibid., 10.
93. Ibid, 13.
94. E.g., Rowbotham, *Woman's Consciousness, Man's World* (1973); Jonasdottir, *Why Women Are Oppressed* (1994); Ingraham, *White Weddings* (1999).
95. Shakespeare, *Disabilities Rights and Wrongs* (2006), 197. Shakespeare evokes the feminist critique of the patriarchal family, infrequently mentioned in disability rights activists' writing.
96. Featherstone, *A Difference in the Family* (1981), 118.
97. "A Letter to . . . My Beloved Older Sister" (2014).
98. Ibid.
99. Hanisch, "Politics of Love" (2013), 13.
100. R. Rapp, "The Power of 'Positive' Diagnosis" (1994), 217. See also R. Rapp, "Moral Pioneers" (1987).
101. J. Rose, "Mothers" (2014).
102. Quindlen, "Public and Private" (1992).

CONCLUSION

1. Testimony of "Diane," quoted by Piepmeier, "Would It Be Better for Her Not to Be Born?" (2015), 1.
2. McCoyd, "Women in No Man's Land" (2010), 135.
3. Ahmed, "Affective Economies" (2004); Buchbinder and Timmermans, "Affective Economies" (2013).

4. Bosk, "Bioethics, Raw and Cooked" (2010). See also Hilgartner and Bosk, "The Rise and Fall of Social Problems" (1988).

5. Bosk, "Bioethics, Raw and Cooked" (2010), S135–S136.

6. A third group, feminists opposed to the perception of women as producers of "good-quality children," were active in the early stages of the PND debate. Hubbard, "Prenatal Diagnosis and Eugenic Ideology" (1985); Wertz and Fletcher, "A Critique of Some Feminist Challenges" (1993). Feminists were less present in the PND debate from the 1990s onward.

7. Kolker and Burke, afterword to *Prenatal Testing* (1988), 218–19. On the uneasy relationships between public health goals and PND, see Wilkinson, "Prenatal Screening" (2015).

8. In the original, "il fallait arriver à enrober ça." The French term *enrober* can be translated as "to coat," "to wrap up," or "to disguise." One can speak about coating nuts with chocolate (*enrober les noix*) or wrapping up/disguising/masking a lie (*enrober un mensonge*). Quoted in Mirlesse, "Diagnostic prénatal et médecine fœtale" (2014), 140.

9. Snochowska-Gonzales, ed., *A jak hipokryzja* (2011).

10. Jørgensen et al., "Including Ethical Considerations in Models" (2014); Dondorp, Page-Christiaens, and de Wert, "Genomic Futures of Prenatal Screening" (2015); Ville, "Politiques du handicap et périnatalité" (2011).

11. In practice, the conditions under which a woman can terminate a pregnancy for a fetal indication vary in different countries. Wertz and Fletcher, "A Critique of Some Feminist Challenges" (1993).

12. Statham, "Prenatal Diagnosis of Fetal Abnormality" (2002); Gaille, "On Prenatal Diagnosis" (2016).

13. Lippman, "Prenatal Genetic Testing and Screening" (1991); Samerski, "Genetic Counseling and the Fiction of Choice" (2009).

14. Poster children for the March of Dimes represented such a "positive" image of disability. The disability activist Emily Rapp was one of those poster children. E. Rapp, *Poster Child* (2007).

15. Kerr and Cunningham-Burley, "On Ambivalence and Risk" (2000).

16. Kerr, "Reproductive Genetics" (2009), 72.

17. Mirlesse, "Diagnostic prénatal et médecine fœtale" (2014); Ville and Mirlesse, "Prenatal Diagnosis" (2015); Ávila da Costa, "O que os olhos veem" (2015).

18. Gammeltoft, *Haunting Images* (2014).

19. R. Rapp, *Testing Women, Testing the Fetus* (1999), 225.

20. Bosk, "Bioethics, Raw and Cooked" (2010).

21. For a critique of schematic representations of women's decision-making process, see Piepmeier, "Would It Be Better for Her Not to Be Born?" (2015).

22. Kerr, Cunningham-Burley, and Amos, "Eugenics and the New Genetics" (1998).

23. Mol, *The Logic of Care* (2008), 12.

24. Gammeltoft and Wahlberg. "Selective Reproductive Technologies" (2014), 208. According to Rozsika Parker, maternity, too, is infused with ambivalence. Such ambivalence, she argued, can be highly productive, but our society tends to stigmatize and not embrace it. Parker, *Torn in Two* (1995).

25. Kerr, "Reproductive Genetics" (2009).

26. Dommergues et al., "Termination of Pregnancy" (2010).

27. Shakespeare, *Disabilities Rights and Wrongs* (2006), 197. Shakespeare is one of the rare disability activists who explicitly evoked the abuse of disabled people by their families.

28. Z. Williams, "Mary Beard" (2016).
29. Brandt, "From Analysis to Advocacy" (2004); D. Paul, "Reflections" (2016).
30. Murphy, "Unsettling Care" (2015).
31. Amsterdamska, "Editorial" (1994).
32. Liana Woskie, reflections at the memorial service for her grandmother, Barbara Gutman Rosenkrantz, Radcliffe Institute, Harvard University, August 10, 2014. Text was sent to the me by Ms. Woskie as a personal comunication, August 14, 2014.
33. The quotation is from the conclusion of the opinion of the French National Consultative Ethics Committee for Health and Life Sciences on fetal genetic testing. French National Consultative Ethics Committee for Health and Life Sciences, "Opinion no. 120" (2013).
34. Haraway, "When Species Meet" (2010).

CODA

In Greco-Roman mythology, the owl is the symbol for Minerva, the goddess of wisdom and learning, while the lyre is the symbol for Apollo, the god of "sunlight, prophecy, music, and poetry," according to *Webster's*.

1. The cultural historian Carlo Ginzburg explains thus that literature can greatly enrich the understanding of historical and present-day developments. Illuz and Vidal, "Carlo Ginzburg" (2003).
2. Simon, "Interview avec Eugène Green" (2015), 26.
3. Bargielska, *Obsoletki* (2010).
4. Szostak, "Justyna Bargielska" (2011).
5. Bargielska, *Obsoletki* (2010), 18–19.
6. Ibid., 27–28. Alas, in this short story Bargielska makes some unfortunate remarks about the supposedly "tribal" Black physician who, she believes, curses her in his "shamanic language." Bargielska probably employed these expressions not because she is racist, but because she wanted to convey her extreme confusion, distress, and fear. Her language nonetheless reflects insensitivity toward racial questions frequently found in Poland.
7. Eight weeks after fertilization (ten weeks of absence of menses), a shapeless embryo is transformed into a formed fetus. Judging from Bargielska's description of "Hanutka," the fetus was probably older than eight weeks.
8. Bargielska, *Obsoletki* (2010), 39–40.
9. The organization is Now I Lay Me Down to Sleep (NILMDTS), https://www.nowilayme downtosleep.org/, accessed July 28, 2017.
10. Bargielska, *Obsoletki* (2010), 31–42.
11. Ibid., 57–58. Bargielska's description can be juxtaposed to Linda Layne's study of miscarriage. Layne, *Motherhood Lost* (2003).
12. Sandelowski and Barroso, "The Travesty of Choosing" (2005).
13. Bargielska, *Obsoletki* (2010), 82.
14. Szostak, "Justyna Bargielska" (2011).
15. Bargielska's description of life before, during, and after pregnancy loss resonates with the last lines of Elizabeth Bishop's poem "The Bight": "All the untidy activity continues, / awful but cheerful." Bishop, *The Complete Poems* (1983). Bishop requested that these lines be her epitaph.
16. Today the painting is in the Dolores Olmedo Museum, Xochimilco, Mexico.
17. Rosenthal, "Diego and Frida" (2015), 60.
18. Herrera, *Frida* (1983), 138.

19. Quoted in ibid., 141.
20. Ibid., 143.
21. Begun's testimony was reproduced in Raquel Tibol's biography of Kahlo. Tibol, *Frida Kahlo* (1977).
22. Zamora, *Frida: El pincel de la angustia* (1987), 131–32. On the contradictions in reports about Kahlo's reproductive history, see also Zetterman, "Frida Kahlo's Abortions" (2006). Early miscarriage may be perceived as a delayed menstruation and vice versa. Laboratory methods of early diagnosis of pregnancy were not available in the 1920s and early 1930s. Kahlo might have believed that she was pregnant and miscarried when that was not the case.
23. Quoted in Grimberg, "The Lost Desire of Frida Kahlo" (2015), 148.
24. Quoted in ibid., 149–51.
25. E.g., Herrera, *Frida* (1983); Rosenthal, "Diego and Frida" (2015).
26. Grimberg, "The Lost Desire of Frida Kahlo" (2015), 152–53. Herrera, *Frida* (1983), 143. Dead fetuses may indeed be expulsed several weeks after their death in the womb, and when it happens, the fetus can disintegrate. However, Kahlo's fetus might have died of natural causes, not because of her abortion attempt.
27. E.g., Dannenberg and Johnson, "Use of Quinine" (1983); J. P. Smith, "Risky Choices" (1998).
28. Kahlo was later reported to have kept in her studio a preserved human fetus (a present from her Mexican physician) and medical textbooks. Lomas and Howell, "Medical Imagery" (1989).
29. The drawing is now in a private collection. On its genesis, see Rosenthal, "Diego and Frida" (2015), 60–62. Kahlo's drawing that depicts her abortion, and then the painting *Henry Ford Hospital*, show Kahlo surrounded with floating symbolic images. Kahlo's drawing recalls the painting by the Dadaist artist Hannah Hoch, *The Bride (Pandora)* of 1927, in which the bride is surrounded by floating symbolic objects. Hoch's painting is usually interpreted as depicting woman's ambivalent status in society.
30. Writings about miscarriage and induced abortion seldom mention the physical changes that mark the woman's body long after the pregnancy has ended, and which may increase her difficulty in coping with pregnancy loss.
31. Herrera, *Frida* (1983), 144–45.
32. Geissler, "Public Secrets in Public Health" (2013); Marris, Jefferson, and Lentzos, "Negotiating the Dynamics of Uncomfortable Knowledge" (2014).
33. Alice Dreger similarly argues that historians nearly always come too late to the scene. Dreger, *Galileo's Middle Finger* (2015), 276.
34. The expression is from Hegel's preface to his book *Elements of Philosophy of Right* (1820), 23. The full quotation: "Philosophy, at any rate, always comes too late to perform this function. As the thought to the world, it appears only at a time when actuality has gone through its formative process and attained its completed state. . . . When philosophy paints its grey in grey, a shape of life has grown old, and it cannot be rejuvenated, but only recognized, by the grey in grey of philosophy; the owl of Minerva begins its flight only with the onset of dusk."

BIBLIOGRAPHY

Ahmed, Sara. "Affective Economies." *Social Text* 22 (2004): 117–39.

Akgun, Huyla; Mustafa Basbug; Mahmut Tuncay Ozgun; Ozlem Canoz; Fatma Tokat; Nurcan Murat; and Figen Ozturk. "Correlation between Prenatal Ultrasound and Fetal Autopsy Findings in Fetal Anomalies Terminated in the Second Trimester." *Prenatal Diagnosis* 27 (2007): 457–62.

Akkermans, J.; S. H. P. Peeters; J. M. Middeldorp; F. J. Klumper; E. Lopriore; G. Ryan; and D. Oepkes. "A Worldwide Survey of Laser Surgery for Twin-Twin Transfusion Syndrome." *Obstetrics and Gynecology* 45 (2015): 168–74.

Al-Gailani, Salim. "The Making of Antenatal Life: Monsters, Obstetrics and Maternity Care in Edinburgh, c. 1900." PhD diss., University of Cambridge, 2010.

Alamillo, Christina; Morris Fiddler; and Eugene Pergament. "Increased Nuchal Translucency in the Presence of Normal Chromosomes: What's Next?" *Current Opinion in Obstetrics and Gynecology* 24 (2012): 102–8.

Alderson, Priscilla; Penny Scotte; and Neelam Thapar. "Living with a Congenital Condition: The Views of Adults Who Have Cystic Fibrosis, Sickle Cell Anemia, Down's Syndrome, Spina Bifida or Thalassemia." In *Before Birth: Understanding Prenatal Screening*, edited by Elisabeth Ettore, 156–71. Aldershot: Ashgate, 2001.

"A Letter to . . . My Beloved Older Sister." *Guardian Saturday* (Manchester), February 22, 2014.

Alfano, Bianca. "Reproducao assistida: A organização da atenção às infertilidades e o acesso às técnicas reprodutivas em dois serviços público-universitários no Estado do Rio de Janeiro." PhD diss., Instituto de Medecina Social, UERJ, 2014.

Almeling, Rene. "Selling Genes, Selling Gender: Egg Agencies, Sperm Banks, and the Medical Market in Genetic Material." *American Sociologial Review* 72 (2007): 319–40.

Amalberti, René. "Risque et dépistage: Les paradoxes d'une ambition excessive." *Médecine de la Reproduction, Gynécologie Endocrinologie* 12 (2010): 77–81.

Americans United for Life. *Perinatal Hospice Information Act: Model Legislation and Policy Guide for the 2011 Legislative Year.* PDF format. Accessed July 17, 2017. http://www.aul .org/wp-content/uploads/2010/12/NEW-Perinatal-Hospice-2011-LG.pdf.

Amsterdamska, Olga. "Editorial." *Science, Technology and Human Values* 19 (1994): 3–4.

Araujo, Abelardo Q. C.; Marcus Tulius T. Silva; and Alexandra P. Q. C. Araujo. "Zika Virus–Associated Neurological Disorders: A Review." *Brain* 139 (2016): 2122–30.

Ariss, Rachel. "Theorizing Waste in Abortion and Foetal Ovarian Use." *Canadian Journal of Women and the Law* 15 (2003): 255–81.

Aronowitz, Robert. *Risky Medicine: Our Quest to Cure Fear and Uncertainty*. Chicago: University of Chicago Press, 2015.

Asch, Adrienne. "Prenatal Diagnosis and Selective Abortion: A Challenge to Practice and Policy." *American Journal of Public Health* 89 (1999): 1649–57.

Ashtana, Anushka. " 'Take Care of Your Elderly Mothers and Fathers,' Says Tory Minister." *Guardian* (Manchester), January 31, 2017.

Athill, Diana. *Somewhere towards the End*. London: Granta, 2008.

Atkins, Chloe. "The Choice of Two Mothers: Disability, Gender, Sexuality and Prenatal Testing." *Cultural Studies* 8 (2008): 106–29.

Atkinson, Paul; Evelyn Parsons; and Katie Featherstone. "Professional Constructions of Family and Kinship in Medical Genetics." *New Genetics and Society* 20 (2001): 5–24.

Ávila da Costa, Patricia. "O que os olhos veem, o coração sente: Estudo sobre a detecção da malformação fetal na experiência das gestantes na Bahia." PhD diss., Instituto de Medicina Social, Universidade do Estado do Rio de Janeiro (UERJ), 2015.

Badinter, Elisabeth. *L'Amour en plus: Histoire de l'amour maternel (XVII–XXe siècle)*. Paris: Flammarion, 1980. Translated by Barbara Wright as *Mother Love, Myth and Reality: Motherhood in Modern History* (New York: Macmillan 1981).

Baer, Rebecca; Mary Norton; Gary Shaw; Monica Flessel; Sara Goldman; Robert Currier; and Laura Jelliffe-Pawlowski. "Risk of Selected Structural Abnormalities in Infants after Increased Nuchal Translucency Measurement." *American Journal of Obstetrics and Gynecology* 211 (2014): 675.e1–19.

Bahradwaj, Aditya; Paul Atkinson; and Agnus Clarke. "Medical Classifications and the Experience of Genetic Hematochromosis." In *New Genetics, New Identities*, edited by Paul Atkinson, Peter Glasner, and Helen Greenslade, 120–38. London: Routledge, 2006.

Baião, Ana Elisa Rodrigues. "A decisao informada no rastramento prenatal das aneuploidias." Master's thesis, Instituto Fernandes Figueira-Fiocruz, 2009.

Baily, Anne Marie. "Why I Had Amniocentesis." In Parens and Asch, *Prenatal Testing and Disability Rights*, 64–71.

Bakker, Merel; Eva Pajkrt; and Caterina Bilardo. "Increased Nuchal Translucency with Normal Karyotype and Anomaly Scan: What Next?" *Best Practice and Research Clinical Obstetrics and Gynaecology*, 28 (2014): 355–66.

Baldassarre, G.; A. Mussa; A. Dotta; E. Banaudi; S. Forzano; A. Marinosci; C. Rossi; et al. "Prenatal Features of Noonan Syndrome: Prevalence and Prognostic Value." *Prenatal Diagnosis* 31 (2011): 349–54.

Bang, J., and A Northeved. "A New Ultrasonic Method for Transabdominal Amniocentesis." *American Journal of Obstetrics and Gynecology* 114 (1972): 599–601.

Baraitser, Michael. "Professor Robin Winter, Internationally Renowned Clinical Geneticist and Leading Expert on Human Malformations." *Guardian* (London), January 21, 2004.

Bargielska, Justyna. *Obsoletki*. Wolowiec: Wydawnictwo Czarne, 2010.

Baron, Saska. "Review of 'In a Different Key.' " *The Observer* (London), January 17, 2016.

Barry, Dan. "Giving a Name, and Dignity, to a Disability." *New York Times*, May 7, 2016.

Bassat, Quique; Jaume Ordi; Jordi Vila; Mamudo Ismail; Carla Carrilho; Marcus Lacerda; Khátia Munguambe; et al. "Development of a Post-mortem Procedure to Reduce the Uncertainty regarding Causes of Death in Developing Countries." *Lancet Global Health* 3 (2013): 3e125–e126.

Bateman, Simone; Sylvie Allouche; Jean Gayon; Michaela Marzano; and Jerome Goffrette, eds. *Inquiring into Human Enhancement: Interdisciplinary and International Perspectives*. Basingstoke: Palgrave Macmillan, 2015.

Beck, Stefan, and Jörg Niewöhner. "Localising Genetic Testing and Screening in Cyprus and Germany: Contingencies, Continuities, Ordering Effects and Bio-cultural Intimacy." In *Handbook of Genetics and Society*, edited by Paul Atkinson, Peter Glaser, and Margaret Lock, 76–93. London: Routledge, 2009.

Becker, Amy Julia. "Has Down Syndrome Hurt Us?" *New York Times*, October 3, 2011.

Béhague, Dominique; Cesar Victora; and Fernando Barros. "Consumer Demand for Caesarean Sections in Brazil: Informed Decision-Making, Patient Choice, or Social Inequality? A Population-Based Birth Cohort Study Linking Ethnographic and Epidemiological Methods." *British Medical Journal* 324 (2002): 942–45.

Bekker, Hilary; Michael Modell; Gill Dennis; Anne Silver; Christopher Mathew; Martin Bobrow; and Theresa Marteau. "Uptake of Cystic Fibrosis Testing in Primary Care: Supply Push or Demand Pull?" *British Medical Journal* 306 (1993): 1584–86.

Belkind, Lisa. "The Unvarnished Reality of Autism." *New York Times*, July 22, 2009.

Bell, Julia. "Corrrespondence." *British Medical Journal* 1, no. 5132 (1959): 1302.

———. "On Rubella in Pregnancy." *British Medical Journal* 1, no. 5123 (1959): 686–88.

Bellini, Carlo; Raoul Hennekam; Ezio Fulcheri; Mariangela Rutigliani; Guido Morcaldi; Francesco Boccardo; and Eugenio Bonioli. "Etiology of Non-immune Hydrops Fetalis: A Systematic Review." *American Journal of Medical Genetics* 149A (2009): 844–51.

BEMFAM (Sociedade Civil Bem-estar Familiar no Brasil). *Pesquisa Nacional sobre Saúde Materno-Infantil e Planejamento Familiar.* São Paulo: BEMFAM, 1986.

Benson, P. F., ed. *Cellular Organelles and Membranes in Mental Retardation.* Edinburgh: Churchill Livingstone, 1971.

Berenstein, Jane. *Rachel in the World.* Urbana: University of Illinois Press, 2007.

Berezoski, Anderson Tadeu; Geraldo Duarte; Reinaldo Rodriguez; Ricardo de Carvalho Cavalli; Roberto de Oliveira Cardoso dos Santos; Yvone Avalloni de Moraes Villela de Andrade Vicente; and Maria de Fátima Galli Sorita Tazim. "Gêmeos conjugados: Experiência de um hospital terciário do sudeste do Brasil." *Revista Brasileira de Ginecologia e Obstetrica* 32 (2010): 61–65.

Berg, Siri; Odd Paulsen; and Brian Carter. "Why Were They in Such a Hurry to See Her Die?" *American Journal of Hospice and Palliative Medicine* 30 (2012): 406–8.

Bermel, Joyce. "Update On Genetic Screening: Views on Early Diagnosis." *Hastings Center Report* 13 (1983): 4–5.

Bernard, Claude. *Introduction à la médecine expérimentale.* Paris, 1865.

Bertoli, M.; G. Biasini; M. T. Calignano; G. Celani; G. De Grossi; M. C. Digilio; C. C. Fermariello; et al. "Needs and Challenges of Daily Life for People with Down Syndrome Residing in the City of Rome, Italy." *Journal of Intellectual Disability Research* 55 (2011): 801–20.

Bérubé, Michael. *Life as We Know It: A Father, a Family and an Exceptional Child.* New York: Pantheon Books, 1996.

Beserra, Ana Heloisa Nascimento. "Alloimunizacao RhD em gestantes: Um problema evitave." Master's thesis, Instituto Fernandes Figueira-Fiocruz, 2015.

Besseau-Ayasse, J.; C. Violle-Poirsier; A. Bazin; N. Gruchy; A. Moncla; F. Girard; M. Till; et al. "A French Collaborative Survey of 272 Fetuses with 22q11.2 Deletion: Ultrasound Findings, Fetal Autopsies and Pregnancy Outcomes." *Prenatal Diagnosis* 34 (2014): 424–30.

Betancur, Catalina. "Etiological Heterogeneity in Autism Spectrum Disorders: More Than 100 Genetic and Genomic Disorders and Still Counting." *Brain Research* 1380 (2011): 42–77.

Bianchi, Diana. "From Prenatal Genomic Diagnosis to Fetal Personalized Medicine: Progress and Challenges." *Nature Medicine* 18 (2012): 1041–51.

Biehl, João. "The Activist State: Global Pharmaceuticals, AIDS, and Citizenship in Brazil." *Social Text* 22 (2004): 105–32.

Biehl, João, and Adriana Petryna. "Bodies of Rights and Therapeutical Markets." *Social Research* 78 (2011): 359–94.

Bishop, Elizabeth. "The Bight." In *The Complete Poems, 1927–1979*, 84. New York: Farrar, Straus and Giroux, 1983.

Blisma, B., and D. Donnai. "WWW Dysmorphology Discussion Board." *Clinical Dysmorphology* 5 (1996): 89–91.

Blizzard, Deborah. *Looking Within: A Sociocultural Examination of Fetoscopy*. Cambridge, MA: MIT Press, 2007.

Blum, Linda. *At the Breast: Ideologies of Breastfeeding and Motherhood in the Contemporary United States*. Boston: Beacon Press, 1999.

———. "Mother Blame in the Prozac Nation: Raising Children with Invisible Disabilities." *Gender and Society* 21 (2007): 202–26.

———. *Raising Generation Rx: Mothering Kids with Invisible Disabilities in the Age of Inequality*. New York: New York University Press, 2015.

Blume, Stuart. *Insight and Industry: On the Dynamics of Technological Change in Medicine*. Cambridge, MA: MIT Press, 1992.

Boardman, Felicity Kate. "The Expressivist Objection to Prenatal Testing: The Experiences of Families Living with Genetic Disease." *Social Science and Medicine* 197 (2014): 18–25.

———. "Knowledge Is Power? The Role of Experiential Knowledge in Genetically 'Risky' Reproductive Decisions." *Sociology of Health and Illness* 36 (2013): 137–50.

———. "The Role of Experiential Knowledge in the Reproductive Decision Making in Families Genetically at Risk: The Case of Spinal Muscular Atrophy." PhD diss., University of Warwick, 2010.

Boccardo, Francesco, and Eugenio Bonioli. "Etiology of Non-immune Hydrops Fetalis: A Systematic Review." *American Journal of Medical Genetics* 149A (2009): 844–51.

Boesky, Amy. Introduction to *The Story Within: Personal Essays on Genetics and Identity*, edited by Amy Boesky, 1–15. Baltimore: Johns Hopkins University Press, 2013.

Boltanski, Luc. *The Foetal Condition: A Sociology of Engendering and Abortion*. Translated by Catherine Porter. Cambridge: Polity Press, 2013 [2004].

Bolton, Sharon. "Women's Work, Dirty Work: The Gynaecology Nurse as 'Other.'" *Gender, Work and Organization* 12 (2005): 169–86.

Bomfim, Olga Luiza. "A antecipação ultra-sonográfica de malformação fetal sob a ótica da mulher." Master's thesis, Instituto Fernandes Figueira-Fiocruz, 2009.

Bomfim, Olga Luiza; Orlando Coser; and Maria Elisabeth Lopes Moreira. "Unexpected Diagnosis of Fetal Malformations: Therapeutic Itineraries." *Physis-Revista de Saúde Coletiva* 24 (2014): 607–22.

Bosk, Charles. *All God's Mistakes: Genetic Counseling in a Pediatric Hospital*. Chicago: University of Chicago Press, 1992.

———. "Bioethics, Raw and Cooked: Extraordinary Conflict and Everyday Practice." *Journal of Health and Social Behavior* 51 (2010): S133–S144.

Brandt, Allan. "From Analysis to Advocacy: Crossing Boundaries as a Historian of Health Policy." In *Locating Medical History: The Stories and Their Meanings*, edited by Frank Huisman and John Harley Warner, 460–84. Baltimore: Johns Hopkins University Press, 2004.

Brazil. Ministério da Saúde. Data on maternal mortality, in the government's system of health-related data analysis (DATASUS). Accessed July 26, 2017. http://www2.datasus.gov.br.

———. *Pesquisa nacional de demografia e saúde da criança e da mulher*. PNDS-2006. Brasília, DF: Ministério da Saúde, 2008.

Breeze, A. C.; C. C. Less; A. Kumar; H. H. Missfelder-Lobos; and E. M. Murdoch. "Palliative Care for Prenatally Diagnosed Lethal Fetal Anomaly." *Archives of Diseases in Childhood: Fœtal and Neonatal Edition* 92 (2007): F56–F58.

Bricker, L.; N. Medley; and J. J. Pratt. "Routine Ultrasound in Late Pregnancy (after 24 Weeks' Gestation)." *Cochrane Database of Systematic Reviews*, no. 6 (June 29, 2015), article no. CD001451. doi: 10.1002/14651858.CD001451.pub4.

British Pregnancy Advisory Service (BPAS). Information on second-trimester surgical abortions. Accessed July 20, 2017. https://www.bpas.org/abortion-care/abortion-treat ments/surgical-abortion/dilatation-and-evacuation/.

Brizot, M. L.; A. W. Liao; L. M. Lopes; M. Okumura; M. S. Marques; V. Krebs; R. Schultz; et al. "Conjoined Twins Pregnancies: Experience with 36 Cases from a Single Center." *Prenatal Diagnosis* 31 (2011): 1120–25.

Brock, D. J., and R. G. Sutcliffe. "Alpha-Fetoprotein in the Antenatal Diagnosis of Anen-cephaly and Spina Bifida." *Lancet* 300 (1972): 197–99.

Brock, Dan. "Preventing Genetically Transmitted Disabilities while Respecting People with Disabilities." In *Quality of Life and Human Difference: Genetic Testing and Disability*, edited by David Wasserman, Jerome Bickenbach, and Robert Wachbroit, 67–100. Cambridge: Cambridge University Press, 2005.

Bryant, Amy; David Grimes; Joanne Garrett; and Gretchen Stuart. "Second-Trimester Abor-tion for Fetal Anomalies or Fetal Death: Labor Induction Compared with Dilation and Evacuation." *Obstetrics and Gynecology* 117 (2011): 788–92.

Bryder, Linda. "From Breast to Bottle: A History of Modern Infant Feeding." *Endeavour* 33 (2009): 54–59.

Buchbinder, Mara, and Stefan Timmermans. "Affective Economies and the Politics of Sav-ing Babies' Lives." *Public Culture* 26 (2013): 101–26.

Burginion, A.; B. Briscoe; and L. Nemzer. "Genetic Abortions: Consideration for Patient's Care." *Journal of Perinatal and Neonatal Nursing* 13 (1999): 47–58.

Burke, Meghan; Richard Urbano; and Robert Hodapp. "Subsequent Births in Families of Chil-dren with Disabilities: Using Demographic Data to Examine Parents' Reproductive Pat-terns." *American Journal on Intellectual and Developmental Disabilities* 116 (2011): 233–45.

Bustos, Tania Pérez; Maria Fernanda Olarte Sierre; and Adriana Diaz del Castillo. "Work-ing with Care: Narratives of Invisible Women Scientists Practicing Forensic Genetics in Colombia." In *Beyond Imported Magic: Essays on Science, Technology and Society in Latin America*, edited by Eden Medina, Ivan de Costa Marquez, and Christina Holmes, 67–83. Cambridge, MA: MIT Press, 2014.

Camporesi, Silvia. *From Bench to Bedside, to Track and Field: The Context of Enhancement and Its Ethical Relevance.* Berkeley: University of California Press, 2014.

Caplan, Arthur. "Chloe's Law: A Powerful Legislative Movement Challenging a Core Ethi-cal Norm of Genetic Testing." *PLoS Biology* 13 (2015): e1002219. doi:10.1371/ journal .pbio.1002219.

———. Foreword to *Prenatal Testing: A Sociological Perspective*, Aliza Kolker and B. Meredith Burke, xvii–xx. Westport, CT: Bergin and Garvey, 1988.

Carayol, M.; M. Bucourt; J. Cuesta; B. Blondel; and J. Zeitlin. "Les femmes de Seine-Saint-Denis ont-elles un suivi prénatal différent de celui des autres femmes d'Île-de-France?" *Journal de Gynécologie Obstétrique et Biologie de la Reproduction* 44 (2015): 258–68.

Carey, Allison. "Parents and Professionals: Parents' Reflections on Professionals, the Sup-port System, and the Family in the Twentieth-Century United States." In *Disability Histories*, edited by Susan Burch and Michael Rembis, 58–76. Champaign: University of Illinois Press, 2014.

Carles, Dominique. "Letter of the SOFFŒT's President." *Bulletin de la SOFFŒT* (Spring 2009): 1–2.

Carpenter, Daniel. "Is Health Politics Different?" *Annual Review of Political Science* 15 (2012): 287–311.

Carroll, S. G. M.; H. Porter; S. Abdel-Fattah; P. M. Kyle; and P. W. Soothill. "Correlation of Prenatal Diagnosis and Pathological Findings in Fetal Brain Anomalies." *Ultrasound in Obsterics and Gynecology* 16 (2000): 149–53.

Casper, Monica. *The Making of the Unborn Patient: A Social Anatomy of Fetal Surgery.* New Brunswick, NJ: Rutgers University Press, 1998.

Castilla, Eduardo, and Ieda Orioli. "ECLAMC: The Latin-American Collaborative Study of Congenital Malformations." *Community Genetics* 7 (2004): 76–94.

Chadwick, Ruth. "Can Genetic Counseling Avoid the Charge of Eugenics?" *Science in Context* 11 (1998): 471–80.

———. "Genetic Choice and Responsibility." *Health, Risk and Society* 1 (1999): 293–300.

Chen, Elizabeth, and Judith Schiffman. "Attitudes toward Genetic Counseling and Prenatal Diagnosis among a Group of Individuals with Physical Disabilities." *Journal of Genetic Counseling* 9 (2000): 137–52.

Chiapetta-Swanson, Catherine. "Dignity and Dirty Work: Nurses' Experiences in Managing Genetic Termination for Fetal Anomaly." *Quantitative Sociology* 26 (2005): 93–116.

Chiosi, Christine. "Hospital Deliveries." *American Journal of Medical Genetics*, part A, 161A (2013): 2122–33.

Chitty, Lyn. "Prenatal Screening for Chromosome Abnormalities." *British Medical Bulletin* 54 (1998): 839–56.

Choi, Hyunkyung; Marcia Van Riper; and Suzanne Thoyre. "Decision Making following a Prenatal Diagnosis of Down Syndrome: An Integrative Review." *Journal of Midwifery and Women's Health* 21 (2012): 156–64.

Christie, Daphne, and E. M. Tansey, editors. *Genetic Testing.* Transcript, witness seminar held by Wellcome Trust Centre for the History of Medicine, UCL, London, July 13, 2001.

———, editors. *Wellcome Witnesses to Twentieth Century Medicine*, vol. 17. London: Wellcome Trust Centre for the History of Medicine at UCL, 2003.

Christoffersen-Deb, Astrid. "Viability: A Cultural Calculus of Personhood at the Beginnings of Life." *Medical Anthropology Quarterly* 26 (2012): 575–93.

Chua, Amy. *Battle Hymn of a Tiger Mother.* New York: Penguin Press, 2011.

Clarence, Tania. "Case against Mother Who Killed Her Three Disabled Children." *Telegraph* (London), November 14, 2014.

Clarke, Agnus. "Stigma, Self-Esteem and Reproduction: Talking with Men about Life with Hypohidrotic Ectodermal Dysplasia." *Sociology* 47 (2013): 887–905.

Clavery, Elisa. "Médico de hospital público no Rio e assessor parlamentar são presos em clínica de aborto." *Extra, O Globo* (Rio de Janeiro), April 6, 2016.

Clayton-Smith, Jill. "In Memoriam Robin Michael Winter 1950–2004." *Clinical Dysmorphology* 13 (2004): 61.

Clayton-Smith, Jill; P. A. Farndon; Carole McKeown; and Dian Donnai. "Examination of Fetuses after Induced Abortion for Fetal Abnormality." *British Medical Journal* 300 (1990): 295–97.

Cocchi, G.; S. Gualdi; C. Bower; J. Halliday; B. Jonsson; A. Myerlid; B. Doray; et al. "International Trends of Down Syndrome 1993–2004: Births in Relation to Maternal Age and Terminations of Pregnancies." *Birth Defects*, part A 88 (2010): 474–79.

Colgrove, James. "The McKeown Thesis: A Historical Controversy and Its Enduring Influence." *American Journal of Public Health* 92 (2002): 725–29.

Collste, Göran. "Moral Decision-Making, Narratives and Genetic Diagnosis." In *Ethical Dilemmas in Prenatal Diagnosis*, edited by Tamara Fischmann and Elisabeth Hildt, 167–76. Dordrecht: Springer, 2011.

Comfort, Nathaniel. *The Science of Human Perfection: How Genes Became the Heart of American Medicine.* New Haven, CT: Yale University Press, 2012.

Conklin, Beth, and Lynn Morgan. "Babies, Bodies, and the Production of Personhood in North America and a Native Amazonian Society." *Ethos* 24 (1996): 657–94.

Cooper, Theodore. "Implication of the Amniocentesis Registry Findings." Paper read at the National Institute of Health, Bethesda, MD, October 20, 1975. Quoted in Barbara Coulliton, "Amniocentesis: HEW Backs Tests for Prenatal Diagnosis of Disease." *Science* 190 (1975): 537–39.

Cooter, Roger. "The Disabled Body." In *Body in the Twentieth Century*, edited by Roger Cooter and John Pickstone, 367–83. Amsterdam: Harwood, 2000.

Costa, Ana Claudia; Goustavo Goulart; and Rafael Nasciento. "Operacoe Herodes: Os doutores do aborto." *O Globo* (Rio de Janeiro), October 15, 2014.

Courtright Barr, Bernadine. "Entertaining and Instructing the Public: John Zahorsky's 1904 'Incubator Institute.'" *Social History of Medicine* 8 (1995): 17–36.

Cousens, N. E.; C. L. Gaff; S. A. Metcalfe; and M. B. Delatycki. "Carrier Screening for Beta Thalassemia: A Review of International Practice." *European Journal of Human Genetics* 18 (2010): 1077–83.

Coventry, Peter. "The Dynamics of Medical Genetics: The Development and Articulation of Clinical and Technical Services under the NHS, Especially at Manchester c. 1945–1979." PhD diss., Manchester University, 2000.

Cowan, Ruth Schwartz. *Heredity and Hope: The Case for Genetic Screening.* Cambridge, MA: Harvard University Press, 2008.

Cox, Susan, and Rosalie Starzomski. "Genes and Geneticization? The Social Construction of Autosomal Dominant Polycystic Kidney Disease." *New Genetics and Society* 23 (2004): 137–66.

Crombag, Neeltje; Ynke Vellinga; Sandra Kluijfhout; Louise Bryant; Pat Ward; Rita Ledema Kuiper; Peter Schliem; et al. "Explaining Variation in Down's Syndrome Screening Uptake: Comparing the Netherlands with England and Denmark Using Documentary Analysis and Expert Stakeholder Interviews." *BMC Health Services Research* 14 (2014): 437. http://www.biomedcentral.com/1472-6963/14/437.

Cuckle, H. S.; N. J.Wald; and R. H. Lindenbaum. "Maternal Serum Alpha-Fetoprotein Mesurements: A Screening for Down Syndrome." *Lancet* 323 (1984): 926–29.

Cuckle, Howard. "Rational Down Syndrome Screening Policy." *American Journal of Public Health* 88 (1998): 558–59.

Culver, Gloria; Kristina Fallon; Ronnie Londer; Nancy Montalvo; Brian Vila; Brenda James Ramsey; Carl Scott Ramsey; ct al. "Informed Decisions for Extremely Low-Birth-Weight Infants." *Journal of the American Medical Association* 283 (2000): 3201–2.

DaMatta, Roberto. "'Do You Know Who You're Talking To?!': The Distinction between Individual and Person in Brazil." In *Carnivals, Rogues and Heroes: An Interpretation of the Brazilian Dilemma*, translated by John Druty, 137–97. Notre Dame, IN: University of Notre Dame Press, 1991 [1979].

Damian, B. B.; T. C. S. Bonetti; and D. D. G. Horovitz. "Practices and Ethical Concerns regarding Preimplantation Diagnosis: Who Regulates Preimplantation Genetic Diagnosis in Brazil." *Brazilian Journal of Medical and Biological Research* 48 (2015): 25–33.

D'Amico, R.; G. Jacopini; G.Vivona; and M. Frontali. "Reproductive Choices in Couples at Risk for Genetic Disease: A Qualitative and Quantitative Analysis." *Birth Defects: Original Article Series* 28 (1992): 41–46.

Dannenberg, A. L., and S. F. Johnson. "Use of Quinine for Self-Induced Abortion." *Southern Medicine Journal* 76 (1983): 846–49.

Daum, Meghan. "I Nearly Died. So What?" *New York Times*, November 15, 2014.

Davidson, Deborah. "Reflections on Doing Research Grounded in My Experience of Perinatal Loss: From Autobiography to Autoethnography." *Sociological Research Online* 16 (February 2011). http://www.socresonline.org.uk/16/1/6.html.

Davidson, Ronald, and Mario Rattazi. "Review: Prenatal Diagnosis of Genetic Disorders." *Clinical Chemistry* 18 (1972): 179–87.

Dawkins, Richard. " 'Immoral' Not to Abort If Foetus Has Down's Syndrome." *Guardian* (Manchester), August 21, 2014.

De Chadarevian, Soraya. "Mutations in the Nuclear Age." In *Making Mutations: Objects, Practices, Contexts*, edited by Luis Campos and Alexander von Schwerin, 179–88. Preprint 393. Berlin: Max Plank Institute for the History of Science, 2010.

Degener, Theresia. "Female Self Determination between 'Voluntary' Eugenics, 'Rights' and 'Ethics.' " *Reproductive and Genetic Engineering: Journal of International Feminist Analysis* 3 (1990): 87–92.

Dekeuwer, Catherine, and Simone Bateman. "Much More Than a Gene: Hereditary Breast and Ovarian Cancer, Reproductive Choices and Family Life." *Health Care and Philosophy* 16 (2011): 231–44.

Deleuze, Gilles, and Felix Guattari. *Capitalisme et schizophrénie*. Vol.1, *L'anti-Œdipe*. Paris: Editions de minuit, 1972. Translated by Robert Hurley, Mark Seem, and Helen R. Lane as *Anti-Oedipus: Capitalism and Schizophrenia*. New York: Viking Press, 1977.

De Moura, Maria Martha Duque; Guimarães, Maria Beatriz Lisbôa; and Madel Luz. "Tocar: Atenção ao vínculo no ambiente hospitalar" [Touch: Attention to Boundaries in a Hospital Setting]. *Interface (Botucatu)* 17 (2013): 393–404.

De Vasconcellos, Mauricio Teixeira Leite; Pedro Luis do Nascimento Silva; Ana Paula Esteves Pereira; Arthur Orlando Correa Schilithz; Paulo Roberto Borges de Souza Junior; and Celia Landmann Szwarcwald. "Nascer no Brasil: Pesquisa nacional sobre parto e nascimento." *Cadernos do Saúde Pública* 30 suppl. (2014): S49–S58.

De Wailly-Galambert, Diane; Dominique Vernier; Pascale Rossigneux-Delage; and Sylvain Missionier. "Lorsque la naissance et la mort coincident: Quel vécu pour les sages femmes?" *Médicine et Hygiene* 24 (2012): 117–39.

Dias, Tabata; Renato Passini Jr.; Graciana Duarte; Maria Sousa; and Aníbal Faúndes. "Association between Educational Level and Access to Safe Abortion in a Brazilian Population." *International Journal of Gynecology and Obstetrics* 128 (2015): 224–27.

Dimond, Rebecca. "Multiple Meanings of a Rare Genetic Syndrome: 22q11 Deletion Syndrome." PhD diss., Cardiff University, 2011.

———. "Patient and Family Trajectories of Mitochondrial Disease: Diversity, Uncertainty and Genetic Risk." *Life Sciences, Society and Policy* 9, no. 2 (2013). doi:10.1186 /2195-7819-9-2.

Diniz, Debora. "The Protection to Women's Fundamental Rights Violated by the Zika Epidemics." *American Journal of Public Health* 106, no. 8 (2016): e9.

———. "The Zika Virus and Brazilian Women's Right to Choose." *New York Times*, February 8, 2016.

Diniz, Debora, and Marilena Corrêa. *Aborto e Saúde Publica: 20 anos de pesquisas no Brasil.* Brasilia DF: Editora do Ministério da Saúde, 2008.

Diniz, Debora, and Marcelo Medeiros. "Aborto no Brasil: Uma pesquisa domiciliar com técnica de urna." *Ciencia e Saúde Publica* 15, no. 1 suppl. (2010): S2105–S2112.

———. "Itinerários e métodos do aborto ilegal em cinco capitais brasileiras." *Ciência e Saúde Coletiva* 17 (2012): 1671–81.

Discussion. Henry Nadler's text "Risks in Amniocentesis." In *Early Diagnosis of Human Genetic Defects: Scientific and Ethical Considerations; A Symposium Sponsored by the John E. Fogarty International Center for Advanced Study in the Health Sciences, National Institutes of Health, Bethesda, Maryland, May 18-19, 1970*, edited by Maureen Harris, 139–48. Washington, DC: US Government Printing Office, [1971].

Dommergues, Marc; Laurent Mandelbrot; Dominique Mahieu-Caputo; Noel Boudjema; Isabelle Durand-Zaleski; and the ICI Group-Club de médecine fœtale. "Termination of Pregnancy following Prenatal Diagnosis in France: How Severe Are the Foetal Anomalies?" *Prenatal Diagnosis* 30 (2010): 531–39.

Donald, Ian. "Ultrasonic in Diagnosis (Sonar)." *Proceedings of the Royal Society of Medicine* 62 (1969): 442–43.

Dondorp, Wybo; Guido De Wert; Yvonne Bombard; Diana Bianchi; Carsten Bergmann; Pascal Borry; Lyn S. Chitty; et al., on behalf of the European Society of Human Genetics (ESHG) and the American Society of Human Genetics (ASHG). "Non-invasive Prenatal Testing for Aneuploidy and Beyond: Challenges of Responsible Innovation in Prenatal Screening." *European Journal of Human Genetics* 23 (2015): 1438–50.

Dondorp, Wybo; C. G. M. Page-Christiaens; and Guido de Wert. "Genomic Futures of Prenatal Screening: Ethical Reflection." *Clinical Genetics* 89 (2016): 531–38.

Donnai, Dian. "Professor Robin Michael Winter 1950–2004: An Appreciation." *American Journal of Medical Genetics* 128A, no. 2 (2004): 107–9.

Donvan, John, and Caren Zuker. *In a Different Key: The Story of Autism.* New York: Crown, 2016.

Dreger, Alice. *Galileo's Middle Finger: Heretics, Activists and the Search for Justice in Science.* New York: Penguin Press, 2015.

Dreger, Alice; Ellen Feder; and Anne Tamar-Mattis. "Prenatal Dexamethasone for Congenital Adrenal Hyperplasia: An Ethics Canary in the Modern Medical Mine." *Bioethical Inquiry* 9 (2012): 277–94.

Dubow, Sara. *Ourselves Unborn: A History of the Fetus in Modern America.* Oxford: Oxford University Press, 2011.

Duden, Barbara. *Disembodying Women: Perspectives on Pregnancy and the Unborn.* Translated by Lee Hoinacki. Cambridge, MA: Harvard University Press, 1993 [1991].

Dunsford, Clare. "The Long Arm." In Boesky, *The Story Within*, 201–13.

Dykens, Elisabeth; Marisa Fisher; Julie Lounds Taylor; Warren Lambert; and Nancy Miodrag. "Reducing Distress in Mothers of Children with Autism and Other Disabilities: A Randomized Trial." *Pediatrics* 134 (2014): e454–e463.

Ebling, Mary. *Healthcare and Big Data: Digital Spectres and Phantom Objects.* New York: Palgrave Macmillan, 2016.

Eckholm, Erik. "Case Explores Rights of Fetus versus Mother." *New York Times*, October 23, 2013.

Eckholm, Erik, and Frances Robles. "Kansas Limits Abortions, Opening a New Line of Attack." *New York Times*, April, 7, 2015.

Edgerton, David. *The Shock of the Old: Technology and Global History since 1900.* Oxford: Oxford University Press, 2007.

Editorial. "Microcephalia e aborto." *Folha de São Paulo*, January 16, 2016. http://www1 .folha.uol.com.br/opiniao/2016/01/1730182-microcefalia-e-aborto.shtml.

Edwards, S. D. "Disability, Identity and the 'Expressivist Objection.'" *Journal of Medical Ethics* 30 (2004): 418–20.

Eisenhower, A. S.; B. Baker; and J. Blacher. "Preschool Children with Intellectual Disability: Syndrome Specificity, Behaviour Problems, and Maternal Well-Being." *Journal of Intellectual Disability Research* 49 (2005): 657–71.

Emerson, E. "Prevalence of Psychiatric Disorders in Children and Adolescents with and without Intellectual Disability." *Journal of Intellectual Disability* 47 (2003): 51–58.

Esplat, Delphine. "Les sages femmes face à l'interruption médicale de la grossesse." Master's thesis, Ecole des Sages Femmes de Clermont Ferrand, 2012.

Ewigman, Bernard; James Crane; Frederic Frigoletto; Michael Lefevre; Raymond Bain; Donald McNellis; and the RADIUS study group. "Effect of Prenatal Ultrasound on Perinatal Outcome." *New England Journal of Medicine* 329 (1993): 821–26.

Eyal, Gil; Brenden Hart; Emine Onculer; Neta Oren; and Natasha Rossi. *The Autism Matrix: The Social Origins of the Autism Epidemics.* London: Polity Press, 2010.

Fagot Largeault, Anne. *L'homme bio-éthique: Pour une déontologie de la recherche sur le vivant.* Paris: Maloine, 1985.

Fassin, Didier. "A Case for Critical Ethnography: Rethinking the Early Years of the AIDS Epidemic in South Africa." *Social Science and Medicine* 99 (2013): 119–26.

Faúndes, Aníbal; Graciana Alves Duarte; Jorge Andalaft Neto; and Maria Helena de Sousa. "The Closer You Are, the Better You Understand: The Reaction of Brazilian Obstetricians-Gynaecologists to Unwanted Pregnancy." *Reproductive Health Matters* 12, no. 24 suppl. (2004): S47–S56.

Featherstone, Helen. *A Difference in the Family: Life with a Disabled Child.* New York: Penguin Books, 1981.

Fessler, Ann. *The Girls Who Went Away: The Hidden History of Women Who Surrendered Children for Adoption in the Decades before Roe v. Wade.* New York: Penguin Books, 2007.

Fillon, Emmanuelle. "Quelles stratégies de soins face à des risques concurrents? Les professionnels de la grossesse et de la naissance aux prises avec des conflits de légitimité." *Science Sociales et Santé* 30 (2012): 5–27.

Finley, Sara; Wayne Finley; and Charles Flowers, eds. *Birth Defects: Clinical and Ethical Considerations.* New York: Allan Liss, 1983.

Fishely, Pat. "I Am John." *Health and Social Work* 17 (1992): 151–57.

Fisher, Jane. "First-Trimester Screening: Dealing with the Fall-Out." *Prenatal Diagnosis* 31 (2011): 46–49.

Fitzgerald, P.; H. Leonard; T. J. Pikora; J. Bourke; and G. Hammond. "Hospital Admissions in Children with Down Syndrome: Experience of a Population-Based Cohort Followed from Birth." *PloS One* 8, no. 8 (August 13, 2013): e70401. doi:10.1371/journal.pone.0070401.

Fleck, Ludwik. *Genesis and Development of a Scientific Fact.* Translated by Fred Bradley and Thaddeus J. Trenn. Chicago: University of Chicago Press, 1979 [1935].

Fletcher, John. "Ethics and Trends in Applied Human Genetics." In *Birth Defects: Clinical and Ethical Considerations,* edited by Sara Finely, Wayne Finley, and Charles Flowers, 143–58. New York: Allan Liss, 1983.

Fligner, Corinne, and Manjiri Dighe. "Post-mortem Diagnosis: Evolving a Team Approach." *Lancet* 382 (2013): 186–88.

Ford, Paul. "Prenatal Testing and Disability Rights." *Medical Anthropology Quarterly* 15 (2001): 560–61.

Fordyce, Lauren. "When Bad Mothers Lose Good Babies: Understanding Fetal and Infant Mortality Case Reviews." *Medical Anthropology* 33 (2014): 379–94.

Foucault, Michel. *The Birth of the Clinics: An Archaeology of Medical Perception*. Translated by A. M. Sheridan Smith. New York: Pantheon Books, 1973 [1963].

———. "Le jeu de Michel Foucault" (1977). In *Dits et Ecrits (1954–1988)*, edited by Daniel Defert and François Ewald, 3:298–329. Paris: Gallimard, 1994. Translated by Colin Gordon as "The Confession of the Flesh" in *Power/Knowledge: Selected Interviews and Other Writings, 1972–1977*, edited by Colin Gordon, 194–228 (New York: Pantheon, 1980).

Fox, Maurice, and John Littlefield. "Editorial: Reservation concerning Gene Therapy." Science 173 (1971): 195.

France. Agence de la biomédecine. *Rapport Annuel*. Paris: Agence de la biomédecine, 2011. Available online at www.genethique.org/sites/default/files/agence_biomedecine_rapport_annuel_2011. Accessed July 25, 2017.

———. "Tableau CPDPN1: Résumé des activités des CPDPN de 2008 à 2010." In *Rapport Annuel*. Paris: Agence de la biomédecine, 2012. Available online at https://www.google.com.br/search?q=Agence+de+la+Biom%C3%A9decine,+Rapport+Annuel+(2012),&spell=1&sa=X&ved=0ahUKEwjy6ba33anVAhWFf5AKHY_lDxMQBQgoKAA&biw=1158&bih=660. Accessed July 25, 2017.

France. Agence nationale de la recherche. Witness seminar, "Histoire de diagnostic prenatal en France" (History of PND in France), Paris, October 4, 2012. The full transcript of the seminar (in French) is accessible at http://anr-dpn.vjf.cnrs.fr/?q=node/62. Accessed July 25, 2017.

France. Direction des affaires juridiques. Interministerial circular of 19 June 2009, relative to fetuses and stillborn children: *Circulaire interministérielle DGCL/DACS/DHOS/DGS/DGS/2009/182 du 19 juin 2009 relative à l'enregistrement à l'état civil des enfants décédés avant la déclaration de naissance* . . . (Paris: Direction des affaires juridiqes, 2009). Available online at http://affairesjuridiques.aphp.fr/textes/circulaire-interministerielle-dgcldacsdhosdgsdgs2009182-du-19-juin-2009-relative-a-lenregistrement-a-letat-civil-des-enfants-decedes-avant-la-declaration-de-naissance-et-de-ceux-pouvant-donn/. Accessed July 25, 2017.

France. French National Consultative Ethics Committee for Health and Life Sciences. "Opinion no. 120: Ethical Issues in Connection with the Development of Foetal Genetic Testing on Maternal Blood," April 25, 2013. http://www.comite-ethique.fr.

France, Emma; Sally Wykea; Sue Ziebland; Vikki Entwistle; and Kate Hunt. "How Personal Experiences Feature in Women's Accounts of Use of Information for Decisions about Antenatal Diagnostic Testing for Foetal Abnormality." *Social Science and Medicine* 72 (2011): 755–62.

Franklin, Sarah. *Biological Relatives: IVF, Stem Cells, and the Future of Kinship*. Durham, NC: Duke University Presss, 2013.

———. "Fetal Fascinations: New Dimensions to the Medical-Scientific Construction of Fetal Personhood." In *Off-Centre: Feminism and Cultural Studies*, edited by Sarah Franklin, Celia Lury, and Jackie Stacey, 190–205. London: HarperCollins, 1991.

Franklin, Sarah, and Celia Roberts. *Born and Made: An Ethnography of Preimplantation Genetic Diagnosis*. Princeton, NJ: Princeton University Press, 2006.

Freidenfelds, Lara. "Enforcing Death Rituals after Miscarriage Is Just Plain Cruel." *Clio Nursing*, April 13, 2016. Accessed July 21, 2017. https://nursingclio.org/2016/04/13/enforcing-death-rituals-after-miscarriage-is-just-plain-cruel/.

Fry, Peter. "Politics, Nationality, and the Meanings of 'Race' in Brazil." *Daedalus* 129 (2000): 83–118.

Fukushima, K.; S. Morokuma; Y. Fujita; K. Tsukimori; S. Satoh; M. Ochiai; T. Hara; et al. "Short-Term and Long-Term Outcomes of 214 Cases of Non-Immune Hydrops Fetalis." *Early Human Development* 87 (2011): 571–75.

Fung, Wai Lun Alan; Nancy Butcher; Gregory Costain; Daniellen Andrade; Erik Boot; Eva Chow; Brian Chung; et al. "Practical Guidelines for Managing Adults with 22q11.2 Deletion Syndrome." *Genetics in Medicine* 17 (2015): 599–609.

Gaidner, Douglas. "The Rhesus Story." *British Medical Journal* 2, no. 6192 (1979): 709–11.

Gaille, Marie. "On Prenatal Diagnosis and the Decision to Continue or Terminate a Pregnacy in France: A Clinical Ethics Study of Unknown Moral Territories." *Medicine, Health Care and Philosophy* 19 (2016): 381–91. doi:10.1007/s11019-016-9689-2.

Galli, Beatriz, and Suely Deslandes. "Threats of Retrocession in Sexual and Reproductive Health Policies in Brazil during the Zika Epidemic." *Cadernos de Saúde Pública* 32, no. 4 (2016), e00031116, abr, 2016.

Gammeltoft, Tine. "Childhood Disability and Parental Moral Responsibility in Northern Vietnam." *Journal of the Royal Anthropology Institute* 14 (2008): 825–42.

———. *Haunting Images: A Cultural Account of Selective Reproduction in Vietnam*. Berkeley: University of California Press, 2014.

Gammeltoft, Tine, and Ayo Wahlberg. "Selective Reproductive Technologies." *Annual Review of Anthropology* 43 (2014): 201–16.

Garel, M.; E. Etienne; B. Blondel; and M. Dommergues. "French Midwives' Practice of Termination of Pregnancy for Fetal Abnormality: At What Psychological and Ethical Cost?" *Prenatal Diagnosis* 27 (2007): 622–28.

Gaudillière, Jean-Paul, and Ilana Löwy, eds. *Heredity and Infection: The History of Disease Transmission*. London: Routledge, 2001.

Gaudineau, A.; B. Doray; E. Schaefer; N. Sananès; G. Fritz1; M. Kohler; Y. Alembik; et al. "Postnatal Phenotype According to Prenatal Ultrasound Features of Noonan Syndrome: A Retrospective Study of 28 Cases." *Prenatal Diagnosis* 33 (2013): 238–41.

Gedge, Elisabeth. "Reproductive Choice and the Ideal of Parenting." *International Journal of Feminist Approaches to Bioethics* 4 (2011): 32–47.

Geissler, Paul Wenzel. "Public Secrets in Public Health: Knowing Not to Know while Making Scientific Knowledge." *American Ethnologist* 40 (2013): 13–34.

Gillis, John. *A World of Their Own Making: Myth, Ritual and Quest for Family Values*. Cambridge MA: Harvard University Press, 1997.

Ginsburg, Faye. *Contested Lives: The Abortion Debate in an American Community*. 2nd ed. Berkeley: University of California Press, 1998 [1989].

Ginsburg, Faye, and Rayna Rapp. "No Judgment: Fieldwork on the Spectrum." Somatosphere, July 20, 2015. http://somatosphere.net/2015/07/no-judgments-fieldwork-on-the-spectrum.html.

Ginzburg, Carlo. *Wooden Eyes: Nine Reflections on Distance*. Translated by Martin Ryle and Kate Soper. New York: Columbia University Press, 2001 [1998].

Giraud, Anne-Sophie. "L'embryon humain en AMP, éléments pour une approche relationnelle." *Enfances, Familles, Géneration* 21 (2014): 48–69.

———. "Les 'péri-parents': À la recherche d'un statut spécifique après une mort périnatale." *Recherches Familiales* 5 (2015): 85–97.

Godel, Margaret. "Images of Stillbirth: Memory, Mourning and Memorial." *Visual Studies* 22 (2008): 253–69.

Gonçalves, Helen; Ana D. Souza; Patrícia Tavares; Suélen Cruz; and Dominique Béhague. "Contraceptive Medicalisation, Fear of Infertility and Teenage Pregnancy in Brazil's

Culture." *Health and Sexuality: An International Journal for Research, Intervention and Care* 13 (2011): 201–15.

Gordon, Meryl. "Review of 'Rosemary: The Hidden Kennedy Daughter,' by Kate Clifford Larson." *New York Times*, October 6, 2015.

Gordon, Olivia. "Living with Down." *Guardian* (London), October 17, 2015.

Greely, Henry. "Get Ready for the Flood of Fetal Gene Screening." *Nature* 469 (2011): 289–91.

Gregg, Jessica. *Virtually Virgins: Sexual Strategies and Cervical Cancer in Recife, Brazil*. Stanford, CA: Stanford University Press, 2002.

Gregg, Norman McAlister. "Further Observations of Congenital Defects in Infants following German Measles in the Mother." *Transactions of the Ophthalmological Society of Australia* 4 (1944): 119–31.

Greig, Alex. "Mother Shares Heartbreaking Photos of Baby Son Born at Just 19 Weeks." *Daily Mail On Line* (London), January19, 2014. http://www.dailymail.co.uk/news/article-2542212/Mother-shares-heartbreaking-photos-baby-miscarried-19-weeks.html.

Grimberg, Salomon. "The Lost Desire of Frida Kahlo in Detroit." In *Diego Rivera and Frida Kahlo in Detroit*, edited by Mark Rosenthal, 144–63. New Haven, CT: Yale University Press, 2015.

Grimes, David. "The Choice of Second Trimester Abortion Method: Evolution, Evidence and Ethics." *Reproductive Health Matters* 16, suppl. (2008): S183–S188.

Grimes, David, and Willard Gates. "Dilatation and Evacuation." In *Second Trimester Abortions: Perspectives from a Decade of Experience*, edited by Gary Berger, William Brener, and Louis Keith, 119–35. Boston: John Wright, 1981.

Grimes, David; Susan Smith; and Angela Witham. "Mifepristone and Misoprostol versus Dilation and Evacuation for Midtrimester Abortion: A Pilot Randomised Controlled Trial." *BJOG: An International Journal of Obstetrics and Gynaecology* 111 (2004): 148–53.

Gross, Sky. "'The Alien Baby': Risk, Blame and Prenatal Indeterminacy." *Health, Risk and Society* 12 (2010): 21–31.

Grossman, Daniel; Kelly Blanchard; and Paul Blumenthal. "Complications after Second Trimester Surgical and Medical Abortion." *Reproductive Health Matters* 16 suppl. (2008): S173–S182.

Guerra, Fernando. "Avaliaça das informaçoes sobre defeitos congentios no municipio do Rio de Janeiro atraves do SINASC." PhD diss., Instituto Fernandes Figueira-Fiocruz, 2006.

Guerra, Fernando; Antônio Ramos; Juan Clinton Llerena Jr.; Silvana Granado Nogueira da Gama; Cynthia Braga da Cunha; and Mariza Miranda Theme Filha. "Confiabilidade das informações das declarações de nascido vivo com registro de defeitos congênitos no Município do Rio de Janeiro, Brasil." *Cadernos de Saúde Pública* 24 (2004): 438–46.

Guerra, Fernando; Antônio Ramos; Véronique Mirlesse; and Ana Elisa Rodrigues Baião. "Breaking Bad News during Prenatal Care: A Challenge to Be Tackled." *Ciência e Saúde Coletiva* 16 (2011): 2361–67.

Guilam, Maria Cristina. "O discurso do risco na prática do aconselhamento genético pré-natal." PhD diss., Insitiuto de Medicina Social, UERJ, 2003.

Guilam, Maria Cristina, and Marilena Correa. "Risk Medicine and Women: A Case Study on Prenatal Diagnosis in Brazil." *Developing World Bioethics* 7 (2002): 78–85.

Guilhem, Dirce, and Anamaria Fereira Azevedo. "Brazilian Public Policies for Reproductive Health: Family Planning, Abortion and Prenatal Care." *Developing Worlds Bioethics* 7 (2007): 68–77.

Guon, Jennifer; Benjamin Wilfond; Barbara Farlow; Tracy Brazg; and Annie Janvier. "Our Children Are Not a Diagnosis: The Experience of Parents Who Continue Their Pregnancy after a Prenatal Diagnosis of Trisomy 13 or 18." *American Journal of Medical Genetics*, part A, 164A (2013): 308–18.

Habel, Alex; Richard Herriot; Dinakantha Kumararatne; Jeremy Allgrove; Kate Baker; Helen Baxendale; Frances Bu'Lock; et al. "Towards a Safety Net for Management of 22q11.2 Deletion Syndrome: Guidelines for Our Times." *European Journal of Pediatrics* 173 (2014): 757–65.

Hacking, Ian. *The Taming of Chance*. Cambridge: Cambridge University Press, 1990.

Han, Sallie. "The Chemical Pregnancy: Technology, Mothering and the Making of a Reproductive Experience." *Journal of the Motherhood Initiative* 5 (2015): 42–53.

Hanisch, Halvor. "Politics of Love: Narrative Structures, Intertextuality and Social Agency in the Narratives of Parents with Disabled Children." *Sociology of Health and Illness* 20 (2013): 1–15.

Haraway, Donna. "Teddy Bear Patriarchy: Taxidermy in the Garden of Eden, New York City, 1908–1936." *Social Text* 11 (1984): 20–64.

———. "When Species Meet: Staying with the Trouble." *Environment and Planning D: Society and Space* 28 (2010): 53–55.

Harper, Peter. *First Years of Human Chromosomes: The Beginnings of Human Cytogenetics*. Oxford: Scion, 2006.

———. "Interview: Michael Baraitser." January 3, 2005. Genetics and Medicine Historical Network. Accessed July 24, 2017. https://genmedhist.eshg.org/interviews/recorded -interviews/.

———. "Interview: André et Joelle Boué." April 22, 2005. Genetics and Medicine Historical Network. Accessed July 24, 2017. https://genmedhist.eshg.org/interviews/recorded -interviews/.

———. "Interview: Dian Donnai." February 6, 2005. Genetics and Medicine Historical Network. Accessed July 24, 2017. https://genmedhist.eshg.org/interviews/recorded -interviews/.

———. "Interview: George Fraser." February 3, 2005. Genetics and Medicine Historical Network. Accessed July 24, 2017. https://genmedhist.eshg.org/interviews/recorded -interviews/.

Harper, Peter; Lois Reynolds; and Elisabeth Tansey, eds. *Clinical Genetics in Britain: Origins and Development; the Transcript of a Witness Seminar Held by the Wellcome Trust Centre for the History of Medicine at UCL, London, on 23 September 2008*. Wellcome Witnesses to Twentieth Century Medicine, vol. 39. London: Wellcome Trust Centre for the History of Medicine at UCL, 2010.

Harrington, Jennifer; Gay Becker; and Robert Nachtigall. "Nonreproductive Technologies: Remediating Kin Structure with Donor Gametes." *Science, Technology and Human Values* 33 (2008): 393–418.

Harris, R., and T. Andrews. "Prenatal Screening for Down's Syndrome." *Archives of Disease in Childhood* 63 (1988): 705–6.

Harrison, Helen. "Making Lemonade: A Parent's View of 'Quality of Life' Studies." *Journal of Clinical Ethics* 12 (Fall 2001): 239–50.

Hashiloni Dolev, Yael. *A Life (Un)worthy of Living: Reproductive Genetics in Israel and Germany*. Dordrecht: Springer, 2007.

Hauerberg, Lara; Lillian Skibsted; Niels Graem; and Lisa Leth Maroun. "Correlation between Prenatal Diagnosis by Ultrasound and Fetal Autopsy Findings in Second-Trimester Abortions." *Acta Obstetricia et Gynecologica Scandinavica* 91 (2012): 386–90.

Hayden, Erika Check. "Prenatal Screening Companies Expand the Scope of DNA Tests." *Nature* 507 (2014): 19.

Hegel, Georg Wilhelm Friedrich. Preface to *Elements of the Philosophy of Right*, 9–23. Translated by H. B. Nisbet. Cambridge: Cambridge University Press, 1991 [1820].

Hercher, Laura, and Georgette Bruenner. "Living with a Child at Risk for Psychotic Illness: The Experience of Parents Coping with 22q11 Deletion Syndrome; An Exploratory Study." *American Journal of Medical Genetics*, part A, 146A (2008): 2355–60.

Herrera, Hayden. *Frida: A Biography of Frida Kahlo*. New York: Harper and Row, 1983.

Herzog, Dagmar. "Abortion, Christianity, Disability: Western Europe, 1960s–1970s." In *Gender and Sexualities in History*, edited by Gert Hekma and Alain Giami, 249–63. Basingstoke: Palgrave Macmillan, 2014.

Hess, Volker, and Andrew Mendelsohn. "Case and Series: Medical Knowledge and Paper Technology, 1600–1900." *History of Science* 48 (2010): 287–314.

Hilgartner, Stephan, and Charles Bosk. "The Rise and Fall of Social Problems: A Public Arenas Model." *American Journal of Sociology* 94 (1988): 53–76.

Hillman, S.; J. McMulan; D. Williams; R. Maher; and M. D. Kilby. "Microarray Comparative Genomic Hybridization in Prenatal Diagnosis: A Review." *Ultrasound Obstetrics and Gynecology* 40 (2012): 385–91.

Hogan, Andrew. *Life Histories of Genetic Disease: Patterns and Prevention in Postwar Medical Genetics*. Baltimore: Johns Hopkins University Press, 2016.

———. "Locating Genetic Disease: The Impact of Clinical Nosology on Biomedical Conceptions of the Human Genome, 1966–1990." *New Genetics and Society* 32 (2013): 78–96. Related content in Hogan, *Life Histories of Genetic Disease* (2016), 56–86.

———. "The 'Morbid Anatomy' of the Human Genome: Tracing the Observational and Representational Approaches of Postwar Genetics and Biomedicine." *Medical History* 58 (2014): 315–36.

———. "Visualizing Carrier Status: Fragile X Syndrome and Genetic Diagnosis since the 1940s." *Endeavour* 35 (2012): 77–84.

Hollander, Julia. *When the Bough Breaks: A Mother's Story*. London: John Murray, 2008.

Home, Alice. "Challenging Hidden Oppression: Mothers Caring for Children with Disabilities." *Critical Social Work* 3 (2002): 88–103.

Homsy, Jason; Samir Zaidi; Yufeng Shen; James Ware; Kaitlin Samocha; Konrad Karczewski; Steven DePalma; et al. "De Novo Mutations in Congenital Heart Disease with Neurodevelopmental and Other Congenital Anomalies." *Science* 350 (2015): 1262–66.

Hon, Giora; Jutta Schickore; and Frederich Steinle. "Introduction: Mapping 'Going Amiss.'" In *Going Amiss in Experimental Research*, edited by G. Hon, J. Schickore, and F. Steinle, 1–7. Dordrecht: Springer, 2009.

Hong, D. S., and A. L. Reiss. "Cognition and Behavior in Turner Syndrome: A Brief Review." *Pediatric Endocrinological Review* 9 (2012): 710–12.

Hook, Ernest, and Geraldine Chambers. "Estimated Rates of Down Syndrome in Live Births by One Year Maternal Age Intervals for Mothers Aged 20–49 in a New York State Study." In *Birth Defects, Original Article Series* 13 [3A] (1977): 123–41.

Hook, Ernest, and Dorothy Warburton. "Turner Syndrome Revisited: Review of New Data Supports the Hypothesis That All Viable 45,X Cases Are Cryptic Mosaics with a Rescue Cell Line, Implying an Origin by Mitotic Loss." *Human Genetics* 133 (2014): 417–24.

Hopcroft, Rosemary, and Julie McLaughlin. "Why Is the Sex Gap in Feelings of Depression Wider in High Gender Equity Countries? The Effect of Children on the Psychological Well-Being of Men and Women." *Social Science Research* 41 (2012): 501–13.

Horovitz, Dafne Dain Gandelman; Maria Helena Cabral de Almeida Cardoso; Juan Clinton Llerena Jr.; and Ruben Araújo de Mattos. "Atenção aos defeitos congênitos no Brasil: Características do atendimento e propostas para formulação de políticas públicas em genética clínica." *Cadernos de Saúde Pública* 22 (2006): 2599–609.

Horovitz, Dafne Dain Gandelman; Victor Evangelista de Faria Ferraz; Sulamis Dain; and Antonia Paula Marques-de-Faria. "Genetic Services and Testing in Brazil." *Journal of Community Genetics* 4 (2013): 355–75.

Howarth, Edmund; Elisabeth Draper; Judith Budd; Justin Konje; Michael Clarke; and Jennfer Kurinczuk. "Population Based Study of the Outcome following the Prenatal Diagnosis of Cystic Hygroma." *Prenatal Diagnosis* 25 (2005): 286–91.

Hubbard, Ruth. "Eugenics: New Tools, Old Ideas." In *Embryos, Ethics and Women's Rights: Exploring the New Reproductive Technologies*, edited by Elaine Hoffman Baruch, Amadeo F. D'Adamo, and Joni Saeager, 225–35. New York: Haworth Press, 1987.

———. "Prenatal Diagnosis and Eugenic Ideology." *Women's Studies International Forum* 8 (1985): 567–76.

Hughes, P. "Post Traumatic Stress Disorder and Management of Stillbirth." *British Journal of Psychiatry* 180 (2002): 279–84.

Hughes, P.; P. Turton; E. Hopper; and C. D. H. Evans. "Assessment of Guidelines for Good Practice in Psychosocial Care of Mothers after Stillbirth: A Cohort Study." *Lancet* 360 (2002): 114–18.

Illuz, Charles, and Laurent Vidal. "Carlo Ginzburg, l'historien et l'avocat du diable." *Genèses* 53 (2003): 113–38.

Imber, Jonathan. *Abortion and the Private Practice of Medicine*. New Haven, CT: Yale University Press, 1986.

Ingraham, Chrys. *White Weddings: Romancing Heterosexuality in Popular Culture*. New York: Routledge, 1999.

Institut nationale de la santé et de la recherche médicale, France. INSERM 2008 report on maternal mortality. http://presse-inserm.fr/mortalite-maternelle-diminution-de-la-mortalite-par-hemorragies/10335/. Accessed July 26, 2017.

Isacksen, C. V.; S. H. Eik-Nes; H. G. Blaas; and S. H. Trop. "A Correlative Study of Prenatal Ultrasound and Post-mortem Findings in Fetuses and Infants with Abnormal Karyotype." *Ultrasound in Obstetrics and Gynecology* 16 (2000): 37–45.

Isaksen, C. V.; S. H. Eik-Nes; H.-G. Blaas; S. H. Trop; and C. B. van derHagen. "Comparison of Prenatal Ultrasound and Postmortem Findings in Fetuses and Infants with Congenital Heart Defects." *Ultrasound Obstetrics and Gynecology* 13 (1999): 117–26.

Ivry, Tsipy. "The Predicaments of Koshering Prenatal Diagnosis and the Rise of New Rabbinic Leadership." *Ethnologie Française* 45 (2015): 281–92.

James, P. D. *Unnatural Causes*. London: Farber and Farber, 1967.

Janvier, Annie; Barbara Farlow; and Benjamin S. Wilfond. "The Experience of Families with Children with Trisomy 13 and 18 in Social Networks." *Pediatrics* 130 (2012): 293–98.

Jericho, Greg. "I No Longer See My Daughter's Down Syndrome, I Only See a Beautiful Girl Called Emma." *Guardian* (Manchester), March 21, 2016.

Jesudason, Sujatha. "Editorial: The Paradox of Disability in Abortion Debates: Bringing the Pro-choice and Disability Rights Communities Together." *Contraception* 84 (2011): 541–43.

Jingree, Treena, and W. M. L. Finlay. "'It's Got So Politically Correct Now': Parents' Talk about Empowering Individuals with Learning Disabilities." *Sociology of Health and Illness* 34 (2012): 412–28.

Jonasdottir, Anna. *Why Women Are Oppressed*. Philadelphia: Temple University Press, 1994.

Jones, E. M., and E. M. Tansey, eds. *Clinical and Molecular Genetics in the UK, c. 1975–c. 2000, the Transcript of a Witness Seminar Held by the History of Modern Biomedicine Research Group, Queen Mary, University of London, on 5 February 2013*. London: Queen Mary, University of London, 2014.

Jones, Kenneth Lyons; Marilyn Crandall Jones; and Miguel Del Campo. *Smith's Recognizable Patterns of Human Malformation*. 7th ed. Philadelphia: Elsevier Saunders, 2013.

Jørgensen, J. M; P. L. Hedley; M. Gjerris; and M. Christiansen. "Including Ethical Considerations in Models for First-Trimester Screening for Pre-eclampsia." *Reproductive Biomedicine Online* 28 (2014): 638–43.

Jülich, Solveig. "The Making of a Best-Selling Book on Reproduction: 'A Child Is Born.'" *Bulletin of the History of Medicine* 89 (2015): 491–525.

Julsingha, B.; J. M. Tesh; and G. M. Fara, eds. *Advances in the Detection of Congenital Malformations*. Chelmford, Essex: Michael Robin Printers, 1976.

Kaasen, A.; J. Tuveng; A. Heberg; H. Scott; and G. Haugen. "Correlation between Prenatal Ultrasound and Autopsy Findings: A Study of Second-Trimester Abortions." *Ultrasound Obstetrics and Gynecology* 28 (2006): 925–33.

Kaltreider, Nancy. "Psychological Impact on Patients and Staff." In *Second Trimester Abortions: Perspectives from a Decade of Experience*, edited by Gary Berger, William Brener, and Louis Keith, 239–50. Boston: John Wright, 1981.

Kanne, Stephen, and Mical Mazurek. "Aggression in Children and Adolescents with Autistic Spectrum Disorders: Prevalence and Risk Factors." *Journal of Autism Developmental Disorders* 41 (2011): 926–37.

Karas, D. J.; G. Costain; E. W. C. Chow; and A. S. Bassett. "Perceived Burden and Neuropsychiatric Morbidities in Adults with 22q11.2 Deletion Syndrome." *Journal of Intellectual Disability Research* 56 (2014): 198–210.

Kaufman, Sharon, and Lynn Morgan. "The Anthropology of the Beginnings and the Ends of Life." *Annual Review of Anthropology* 34 (2005): 317–41.

Keane, Helen. "Fetal Personhood and Representation of the Absent Child in Pregnancy Loss Memorialization." *Feminist Theory* 10 (2009): 153–71.

Kelly, Susan. "Choosing Not to Choose: Reproductive Responses of Parents of Children with Genetic Conditions or Impairments." *Sociology of Health and Illness* 31 (2008): 81–97.

Kelly, T.; J. Suddes; D. Howel; J Hewison; and S. Robson. "Comparing Medical versus Surgical Termination of Pregnancy at 13–20 Weeks of Gestation: A Randomised Controlled Trial." *BJOG: An International Journal of Obstetrics and Gynaecology* 117 (2010): 1512–22.

Kent, Julie. "The Fetal Tissue Economy: From the Abortion Clinic to the Stem Cell Laboratory." *Social Science and Medicine* 67 (2008): 1747–56.

Kerr, Anne. "Reproductive Genetics: From Choice to Ambivalence and Back Again." In *Handbook of Genetics and Society*, edited by Paul Atkinson, Peter Glaser, and Margaret Lock, 59–75. London: Routledge, 2009.

Kerr, Anne, and Sarah Cunningham-Burley. "On Ambivalence and Risk: Reflexive Modernity and the New Human Genetics.'" *Sociology* 34 (2000): 283–304.

Kerr, Anne; Sarah Cunningham-Burley; and Amanda Amos. "Eugenics and the New Genetics in Britain: Examining Contemporary Professionals' Accounts." *Science, Technology and Human Values* 23 (1998): 175–98.

Kevles, Daniel. *In the Name of Eugenics: Genetics and the Uses of Human Heredity*. New York: Knopf, 1985.

Khoshnood, Babak; Catherine De Vigan; Véronique Vodovar; Gérard Bréart; François Goffinet; and Béatrice Blondel. "Advances in Medical Technology and Creation of

Disparities: The Case of Down Syndrome." *American Journal of Public Health* 96 (2006): 2139–44.

Kittay, Eva Feder. "The Personal Is Philosophical Is Political: A Philosopher and Mother of a Cognitively Disabled Person Sends Notes from the Battlefield." In *Cognitive Disability and Its Challenge to Moral Philosophy*, edited by Eva Feder Kittay and Licia Carson, 393–411. Malden, MA: Wiley-Blackwell, 2010.

——. "When Caring Is Just and Justice Is Caring: Justice and Mental Retardation." *Public Culture* 13 (2001): 557–80.

Koch, Lene. "The Meanings of Eugenics: Reflection on the Government of Genetic Knowledge in the Past and the Present." *Science in Context* 17 (2004): 315–31.

Kolker, Aliza, and B. Meredith Burke. Afterword to *Prenatal Testing: A Sociological Perspective*, 201–23. Westport, CT: Bergin and Garvey, 1988.

——. *Prenatal Diagnosis: A Sociological Perspective*. 2nd ed. Wesport, CT: Bergin and Garvey, 1998.

Korenromp, M. J.; H. J. Idema Kuiper; H. G. van Spijker; G. Christianens; and J. Bergsma. "Termination of Pregnancy on Genetic Ground." *Psychosomatic Obstetrics and Gynecology* 13 (1992): 93–103.

Kovitt, Leonard. "Babies as Social Products: The Social Determinants of Classification." *Social Science and Medicine* 12 (1978): 347–51.

Krakowsky Chazan, Lilian. "'É . . . tá grávida mesmo! E ele é lindo!' A construção de 'verdades' na ultra-sonografia obstétrica." *Manguinhos* 15 (2008): 99–116.

——. "O aparelho é como um automovel; a pista é a paciente." *Physis Revista de Saúde Coletiva*, Rio de Janeiro 21 (2011): 601–27.

——. "Vérités, attentes, spectacles et consommations: A propos de l'échographie obstétriquedans les cliniques de Rio de Janeiro." In *Les technologies de l'espoir: La fabrique d'une historie à accomplir*, edited by Annette Leibing and Virginie Turnay, 163–90. Quebec: Presses Universitaires de Laval, 2010.

Kraus, Cynthia. "Diagnostiquer les fétus intersexués: Quoi de neuf, docteurs?" *Sciences Sociales et Santé* 33 (2015): 35–46.

Kucik, J. E.; M. Shin; L. Marengo; and A. Correa for the Congenital Anomaly Multistate Prevalence and Survival Collaborative. "Trends in Survival among Children with Down Syndrome in 10 Regions of the United States." *Pediatrics* 131 (2013): 27–36.

Kuebelbeck, Amy. "A Perinatal Hospice." *NAPSW Forum* 33 (Spring 2013): 1–4.

Kuebelbeck, Amy, and Deborah Davis. *Gift of Time: Continuing Your Pregnancy When Your Baby's Life Is Expected to Be Brief*. Baltimore: Johns Hopkins University Press, 2011.

Lajeunesse, C.; A. Stadler; B. Trombet; M. N. Varlet; H. Patural; F. Prieur; and G. Chene. "First-Trimester Cystic Hygroma: Prenatal Diagnosis and Fetal Outcome." *Journal de Gyneacologie, Obstetrique et Biologie de la Réproduction* 43 (2014): 455–62.

Landsman, Gail. *Reconstructing Motherhood and Disability in the Age of "Perfect" Babies*. New York: Routledge, 2009.

Latimer, Joanna. "Diagnosis, Dysmorphology and the Family: Knowledge, Motility and Choice." *Medical Anthropology* 26 (2007): 97–138.

——. *The Gene, the Clinic, and the Family: Diagnosing Dysmorphology, Reviving Medical Dominance*. London: Routledge, 2013.

Latimer, Joanna; Kate Featherstone; Paul Atkins; Adele Clarke; and Alison Shaw. "Rebirthing the Clinics: The Interaction of Clinical Judgement and Genetic Technology in the Production of Medical Science." *Science, Technology and Human Values* 31 (2006): 599–630.

Latour, Bruno. *Les microbes, guerre et paix.* Paris: La Découverte, 1984. Translated by Alan Sheridan and John Law as *The Pasteurization of France* (Cambridge, MA: Harvard University Press, 1988).

Lavigne, Claude; Michel Vernerey; and Patrice Vienne. "Pratiques hospitalières concernant les fetus et nouveau nés décédés: CHU de Paris, Lyon et Marseille." Report no. 2006-024. Paris: Inspection générale des affaires sociales, 2006.

Lawrence, Christopher. "Incommunicable Knowledge: Science, Technology and the Clinical Art in Britain 1850–1914." *Journal of Contemporary History* 20 (1985): 503–20.

Layne, Linda. "He Was a Real Baby with Real Baby Things: A Material Culture Analysis of Personhood, Parenthood, and Pregnancy Loss." *Journal of Material Culture* 5 (2000): 321–45.

———. *Motherhood Lost: A Feminist Account of Pregnancy Loss in America.* New York: Routledge, 2003.

———. "Unhappy Endings: A Feminist Reappraisal of the Women's Health Movement from the Vantage of Pregnancy Loss." *Social Science and Medicine* 56 (2003): 1881–91.

Le, Hoa; Mark Connolly; Luis Bahamondes; Jose Cecatti; Jingbo Yu; and Henry Hu. "The Burden of Unintended Pregnancies in Brazil: A Social and Public Health System Cost Analysis." *International Journal of Women's Health* 16 (2014): 663–70.

Lee, Vivian, and Ernest Ng. "Issues in Second Trimester Induced Abortion (Medical/Surgical Methods)." *Best Practice and Research Clinical Obstetrics and Gynaecology* 24 (2010): 517–27.

Leichtentritt, Ronit. "Silenced Voices: Israeli Mothers' Experience of Feticide." *Social Science and Medicine* 72 (2011): 747–54.

Leigh, Anne Flore, and Jeff Milunsky. "Updates in the Genetic Evaluation of the Child with Global Developmental Delay or Intellectual Disability." *Seminars in Pediatric Neurology* 19 (2012): 173–80.

Leiter, Valerie; Marty Wyngaarden Krauss; Betsy Anderson; and Nora Wells. "The Consequences of Caring: Effects of Mothering a Child with Special Needs." *Journal of Family Issues* 25 (2004): 379–403.

Leon, Irving. "Perinatal Loss: A Critique of Current Hospital Practices." *Clinical Pediatrics* 31 (1992): 366–74.

Leonelli, Sabina. "Review of Karen-Sue Taussig, *Ordinary Genomes: Science, Citizenship and Genetic Identities.*" *Acta Biotheoretica* 60 (2012): 319–22.

Leuzinger-Bohleber, Marianne. "Introduction: Ethical Dilemmas Due to Prenatal and Genetic Diagnosis; An Interdisciplinary, European Study (EDIG, 2005–2008)." In *Ethical Dilemmas in Prenatal Diagnosis*, edited by Tamara Fischmann and Elisabeth Hildt, 1–35. Dordrecht: Springer, 2011.

Lewin, Tamar. "Ohio Bill Would Ban Abortion if Down Syndrome Is Reason." *New York Times*, August 22, 2015.

Lewis, Emanuel. "The Management of Stillbirth: Coping with an Unreality." *Lancet* 308 (1976): 618–21.

———. "Mourning by the Family after Stillbirth or Neonatal Death." *Archives of Disease in Childhood* 54 (1979): 303–6.

Lewit, Sarah. "D&E Midtrimester Abortion: A Medical Innovation." *Women and Health* 7 (1982): 49–55.

Lindean, Renate. "Down Syndrome Screening Isn't about Public Health. It's about Eliminating a Group of People." *Washington Post*, June 16, 2015.

Lindee, Susan. "Genetic Disease in the 1960s: A Structural Revolution." *American Journal of Medical Genetics (Seminars in Medical Genetics)* 115 (2002): 75–82.

———. *Moments of Truth in Genetic Medicine*. Baltimore: Johns Hopkins University Press, 2005.

Linder, Beth. "On the Borderland of Medical and Disability History: A Survey of the Fields." *Bulletin of the History of Medicine* 87, no. 4 (2013): 499–535.

Lippman, Abby. "Prenatal Genetic Testing and Screening: Constructing Needs and Reinforcing Inequities." *American Journal of Law and Medicine* 17 (1991): 15–50.

Littlewood, Jennifer. "From the Invisibility of Miscarriage to an Attribution of Life." *Anthropology and Medicine* 6 (1999): 217–30.

Livingston, Julie. *Improvising Medicine: An African Oncology Ward in an Emerging Cancer Epidemic*. Durham, NC: Duke University Press, 2012.

Livret d'inormation à l'usage des parents: Interruption médicale de la grossesse. Information leaflet. Paris: Assistance publique des hopitaux de Paris, n.d.

Lloyd, Amy. "Duchenne Muscular Dystrophy and Reproductive Decision Making: Implications for Newborn Screening." PhD diss., Cardiff University, 2009.

Lohr, Patricia. "Surgical Abortion in the Second Trimester." *Reproductive Health Matters* 16, suppl. (2008): S151–S161.

Lomas, David, and Rosemary Howell. "Medical Imagery in the Art of Frida Kahlo." *British Medical Journal* 299 (1989): 23–30.

Lopez-Beltran, Carlos. "The Medical Origins of Heredity." In *Heredity Produced: At the Crossroad of Biology, Politics and Culture, 1500–1870*, edited by Muller-Wille Staffan and Hans Joerg Rheinberger, 105–32. Cambridge MA: MIT Press, 2007.

Lotte, Lynda, and Isabelle Ville. "Histoire de la prénatalité et de la prévention des handicaps de la naissance en France." In final report, contract ANR 09-SSOC-026, Paris: Agence nationale de la recherche, diagnostic prénatal et prévention des handicaps, 2013.

Loudon, Irvine. *Death in Childbirth: An International Study of Maternal Care and Maternal Mortality, 1800–1950*. Oxford: Clarendon Press, 1992.

Löwy, Ilana. "Epidemiology, Immunology and Yellow Fever: The Rockefeller Foundation in Brazil, 1923–1939." *Journal of the History of Biology* 30 (1997): 397–417.

———. "How Genetics Came to the Unborn: 1960–2000." *Studies in History and Philosophy of Biological and Biomedical Sciences* 47 (2014): 154–62.

———. *L'emprise du genre Masculinité, Féminité, Inégalité*. Paris: La Dispute, 2005.

———. "Microscope Slides in the Life Sciences: Material, Epistemic and Symbolic Objects." *History and Philophy of Life Sciences* 35 (2013): 309–18.

———. "Prenatal Diagnosis: The Irresistible Rise of the 'Visible Fœtus.'" *Studies in History and Philosophy of Biological and Biomedical Sciences* 47 (2014): 154–62.

———. *Preventive Strikes: Women, Precancer and Preventive Surgery*. Baltimore: Johns Hopkins University Press, 2009.

———. "Review. Down's: The History of a Disability." *New Genetics and Society* 33 (2014): 229–31.

Löwy, Ilana, and Jean Paul Gaudillière. "Localizing the Global: Testing for Hereditary Risks of Breast Cancer." *Science, Technobiology and Human Values* 33 (2008): 299–325.

Luker, Kristin. *Abortion and the Politics of Motherhood*. Berkeley: University of California Press, 1984.

Lupton, Deborah. "'Precious Cargo': Foetal Subjects, Risk and Reproductive Citizenship." *Critical Public Health* 22 (2012): 329–40.

———. *Risk*. New York: Routledge, 1999.

———. *The Social Worlds of the Unborn*. Basingstoke, UK: Palgrave Macmillan, 2013.

Lutz, Amy. *Each Day I Like It Better: Autism, ECT and the Treatment of Our Most Impaired Children*. Nashville: Varnderbilt University Press, 2014.

Lyus, Richard; Stephen Robson; John Parsons; Jane Fisher; and Martin Cameron. "Second Trimester Abortion for Fetal Abnormality." *British Medical Journal* 347 (2013): f4165.

Määttä, Tuomo; Tuula Tervo-Määttä; Anja Taanila; Markus Kaski; and Matti Iivanainen. "Mental Health, Behaviour and Intellectual Abilities of People with Down Syndrome." *Down Syndrome Research and Practice* 11 (2006): 37–43.

Malacrida, Claudia. "Review of *Dangerous Pregnancies: Mothers, Disability and Abortion in Modern America,* by Leslie Reagan." *Disability and Society* 29 (2014): 336–38.

Manderbrodt, L., and G. Girard. "Aspects techniques des interruptions médicales de grossesse." *Mises à Jour en Gynécologie et Obstétrique* 32 (December 2008): 12–13.

Mangionea, R.; N Fries; P. Godard; M. Fontanges; G. Haddad; and V. Milresse. "Devenir des foetus présentant une ou plusieurs malformations découvertes avant 14 SA: La découverte d'une malformation à l'échographie du premier trimestre est-elle responsable d'interruption volontaire de grossesse?" *Journal de Gynécologie Obstétrique et Biologie de la Reproduction* 37 (2008): 154–62.

Marion, Robert. "Autism Spectrum Disorders and the Clinical Geneticist: An Approach to the Family." *Exceptional Parent Magazine,* April 2013, 33–36.

———. "The Christmas Present." *American Journal of Medical Genetics,* part A, 66 (1996): 247–49.

———. "The Insolvable Puzzles." *American Journal of Medical Genetics* 62 (1996): 327–29.

———. "The Skeleton in Mr. Anderson's Closet." *American Journal of Medical Genetics* 59 (1995): 1–3.

Marris, Claire; Catherine Jefferson; and Filippa Lentzos. "Negotiating the Dynamics of Uncomfortable Knowledge: The Case of Dual Use and Synthetic Biology." *Biosocieties* 9 (2014): 393–420.

Maurer, Kathryn; Janet Jacobson; and David Turok. "Same-Day Cervical Preparation with Misoprostol prior to Second Trimester D&E: A Case Series." *Contraception* 88 (2013): 116–21.

M'charek, Amanda. "Genetics and Its Others: On Three Versions of the Savage." *Etnofoor* 22 (2010): 127–38.

McCoyd, Judith. "'I'm Not a Saint': Burden Assessment as an Unrecognized Factor in Prenatal Decision Making." *Qualitative Health Research* 18 (2008): 1489–500.

———. "Pregnancy Interrupted: Loss of a Desired Pregnancy after a Diagnosis of Fetal Anomaly." *Journal of Psychosomatic Obstetrics and Gynecology* 28 (2007): 37–48.

———. "What Do Women Want? Experiences and Reflections of Women after Prenatal Diagnosis and Termination for Anomaly." *Health Care for Women International* 30 (2009): 507–35.

———. "Women in No Man's Land: The Abortion Debate in the USA and Women Terminating Desired Pregnancies Due to Fetal Anomaly." *British Journal of Social Work* 40 (2010): 133–53.

McIntosh, J.; N. Meriki; A. Joshi; V. Biggs; A. W. Welsh; D. Challis; and K. Lui. "Long Term Developmental Outcomes of Pre-school Age Children following Laser Surgery for Twin-to-Twin Transfusion Syndrome." *Early Human Development* 90 (2014): 837–42.

McIntosh, Neil, chair. *Report: Fetal, Perinatal and Paediatric Pathology; A Critical Future.* London: Royal College of Paediatric and Child Health, 2002.

McKeown, Thomas. *The Modern Rise of Population.* New York: Academic Press, 1976.

McKinney, Clare. "Selective Abortion as Moral Failure? Revaluation of the Feminist Case for Reproductive Rights in a Disability Context." *Disability Studies Quarterly* 36 (2016). http://dsq-sds.org/article/view/3885/4213. Accessed July 24, 2017.

McLaren, Angus. *A History of Contraception: From Antiquity to the Present*. Oxford: Blackwell, 1990.

Mefford, Heather; Mark Batshaw; and Eric Hoffman. "Genomics, Intellectual Disability and Autism." *New England Journal of Medicine* 366 (2012): 733–43.

Memmi, Dominique. "Archaïsme et modernité de la biopolitique contemporaine: L'interruption médicale de la grossesse." *Raisons Politiques* 1 (2003): 125–39.

———. *La deuxième vie des bébés morts*. Paris: Editions EHESS, 2011.

Menzel, Paul; Paul Dolan; Jeff Richardson; and Jan Abel Olsen. "The Role of Adaptation to Disability and Disease in Health State Valuation: A Preliminary Normative Analysis." *Social Science and Medicine* 55 (2002): 2149–58.

Merge, Dominique, and Patrick Schmoll. "Éthique de l'interruption médicale de grossesse en France: Questions posées par le dispositif des Centres pluridisciplinaires de diagnostic prénatal." *Les Dossiers de l'Obstétrique* 343 (2005): 21–29.

Meunier, Elisabeth. "Entretien préparatoire à l'interruption de grossesse." In *Interruption de grossesse pour pathologie fœtale*, edited by Veronique Mirlesse, 43–48. Paris: Flammarion Médecine-Sciences, 2002.

Miller, Alice Fiona. "A Blueprint for Defining Health: Making Medical Genetics in Canada." PhD diss., York University, 2000.

Miller, Callie Clark. "Discovering the Upside of Down: Children and Adults with Down Syndrome." *Exceptional Parent Magazine*, May 2012, 21–23.

Mirlesse, Veronique. "Diagnostic prénatal et médecine fœtale: Du cadre des pratiques à l'anticipation du handicap; Comparaison France-Brésil." PhD diss., University of Paris XI, 2014.

Mirlesse, Veronique; Frederique Perrotte; Francois Kieffer; and Isabelle Ville. "Women's Experience of Termination of Pregnancy for Fetal Anomaly: Effects of Socio-political Evolutions in France." *Prenatal Diagnosis* 31 (2011): 1021–28.

Mirlesse, Veronique, and Isabelle Ville. "The Uses of Ultrasonography in Relation to Foetal Malformations in Rio de Janeiro, Brazil." *Social Sciences and Medicine* 87 (2013): 168–75.

Mitchell, Lisa. "Time with Babe: Seeing Fetal Remains after Pregnancy Termination for Impairment." *Medical Anthropology Quarterly* 30 (2016): 168–85.

Mol, Annemarie. *The Body Multiple*. Durham, NC: Duke University Press, 2002.

———. *The Logic of Care: Health and the Problem of Patient Choice*. London: Routledge, 2008.

Molina, George; Thomas Weiser; Stuart Lipsitz; Micaela Esquivel; Tarsicio Uribe-Leitz; Tej Azad; Neel Shah; et al. "Relationship between Cesarean Delivery Rate and Maternal and Neonatal Mortality." *Journal of the American Medical Association* 314 (2015): 2263–70.

Monckton, Rosa. " 'When Will I Be Normal?': Heartbreaking Question of Diana's Downs Syndrome Goddaughter and Her Mother's Fears of How She Will Cope Alone." *Guardian* (Manchester), March 10, 2012.

Morgan, Lynn. *Icons of Life: A Cultural Study of Human Embryos*. Berkeley: University of California Press, 2009.

———. "Imagining the Unborn in the Ecuadoran Andes." *Feminist Studies* 23 (1997): 322–50.

———. " 'Properly Disposed Of': A History of Embryo Disposal and the Changing Claims on Fetal Remains." *Medical Anthropology* 21 (2002): 247–74.

———. "The Social Biography of Carnegie Embryo no. 836." *Anatomical Record*, part B, 276B (2004): 3–7.

————. "Strange Anatomy: Gertrude Stein and the Avant-Garde Embryo." *Hypatia* 21 (2006): 15–34.

Morgan, Lynn, and Carole Browner. "Why Worry about Embryos" (letter). *Social Science and Medicine* 40 (1995): 1015.

Morris, Joan; Mary Grinsted; and Anna Springetti. "Accuracy of Reporting Abortions with Down Syndrome in England and Wales: A Data Linkage Study." *Journal of Public Health* 38 (2016): 170–74.

Mukherjee, Siddhartha. *The Emperor of All Maladies: A Bigraphy of Cancer*. New York: Simon and Schuster, 2010.

Murphy, Michelle. "Unsettling Care: Troubling Transnational Itineraries of Care in Feminist Health Practices." *Social Studies of Science* 45 (2015): 717–37.

Myers, Amanda; Particia Lohr; and Naomi Pfeffer. "Disposal of Fetal Tissue following Elective Abortion: What Women Think." *Family Planning Reproductive Health Care* 41 (2015): 84–89.

Myers, B. A., and S. M. Pueschel. "Psychiatric Disorders in Persons with Down Syndrome." *Journal of Nervous and Mental Disease* 179 (1991): 609–13.

Nadler, Henry. "Antenatal Detection of Hereditary Disorders." *Pediatrics* 42 (1968): 912–18.

Navon, Daniel. "Genomic Designation: How Genetics Can Delineate New, Phenotypically Diffuse Medical Categories." *Social Studies of Science* 41 (2011): 203–26.

Nay, Olivier; Sophie Béjean; Daniel Benamouzig; Henri Bergeron; Partick Castel; and Bruno Ventelou. "Achieving Universal Health Coverage in France: Policy Reforms and the Challenge of Inequalities." *Lancet* 387 [10034] (2016): 2236–49.

Nelson, Jennifer. *More than Medicine: A History of the Feminist Women's Health Movement*. New York: New York University Press, 2015.

Nelson, Katherine; Hexem Kari; and Chris Feudtner. "Inpatient Hospital Care of Children with Trisomy 13 and Trisomy 18 in the United States." *Pediatrics* 129 (2012): 869–87.

Neuman, Karen. *Fetal Positions: Individualism, Science, Visuality*. Stanford, CA: Stanford University Press, 1996.

Nicolaides, Kypros; G. Azar; D. Byrne; C. Mansur; and K. Marks. "'Fetal Nuchal Translucency: Ultrasound Screening for Chromosomal Defects in the First Semester of Pregnancy." *British Medical Journal* 304 (1992): 867–69.

Nicolson, Malcolm, and John Fleming. *Imaging and Imagining the Fetus: The Development of Obstetric Ultrasound*. Baltimore: John Hopkins University Press, 2013.

Noble, Joan. "Natural History of Down's Syndrome: A Brief Review for Those Involved in Antenatal Screening." *Journal of Medical Screening* 5 (1998): 172–77.

Nomura, Rosalie Mieko Yamamoto; Maria de Lourdes Brizot; Adolfo Wenjaw Liao; and Wagner Rodriguez Hernandes. "Conjoined Twins and Legal Authorization for Abortion." *Revista de Associacion Medica Brasiliera* 57 (2011): 205–10.

North, Campbell. "New Down Syndrome Law Ensures Moms-to-Be Greater Access to Available Help." *Pittsburgh Post-Gazette*, July 18, 2014.

Novaes Machado, Heloisa. "A necropsia prenatal no campo dos defeitos congênitos e aconselhamento genético." PhD diss., Instituto Fernandes Figueira-Fiocruz, 2012.

Nowaczyk, M. J. M., and J. S. Waye. "The Smith-Lemli-Opitz Syndrome: A Novel Metabolic Way of Understanding Developmental Biology, Embryogenesis, and Dysmorphology." *Clinical Genetics* 59 (2001): 375–86.

Nuffield Council of Bioethics. *Human Bodies: Donation for Research and Medicine*. London: Nuffield Council of Bioethics Publications, 2011.

Nunez, Maria das Dores; Alberto Madeiro; and Debora Diniz. "Histórias de aborto provocado entre adolescentes em Teresina, Piauí, Brasil." *Ciência e Saúde Colectiva* 18 (2013): 2311–18.

Oakley, Ann. *The Captured Womb: A History of the Medical Care of Pregnant Women.* Oxford: Blackwell, 1986.

Oaks, Laury. "Smoke Filled Wombs and Fragile Fetuses: The Social Politics of Fetal Representation." *Signs* 26 (2000): 63–108.

Olsson, M. B., and C. P. Hwang. "Depression in Mothers and Fathers of Children with Intellectual Disability. *Journal of Intellectual Disability Research* 45 (2001): 535–43.

Olszynko-Gryn, Jesse. "Pregnancy Testing in Britain, 1900–1967: Laboratories, Animals and Demand from Doctors, Patients and Consumers." PhD diss., Cambridge University, 2014.

Osteen, Marc. Introduction to *Autism and Representation,* edited by Marc Osteen, 1–7. New York: Routledge, 2008.

———. "Urinatown: A Chronicle of the Potty Wars." In *Autism and Representation,* edited by Marc Osteen, 212–25. New York: Routledge, 2008.

O'Toole, Emer. "What Hope Has Pope Francis Offered to Women Exposed to Zika? None." *Guardian* (London), February 22, 2016.

Paillet, Anne. "The Ethnography of 'Particularly Sensitive' Activities: How 'Social Expectations of Ethnography' May Reduce Sociological and Anthropological Scope." *Etnography* 14 (2012): 126–42.

———. *Sauver la vie, donner la mort: Une sociologie de l'éthique en réanimation néonatale.* Paris: La Dispute, 2007.

Paim, Jairnilson; Claudia Travassos; Celia Almeida; Ligia Bahia; and James Macinko. "The Brazilian Health System: History, Advances, and Challenges." *Lancet* 377 (2011): 1778–97.

Palmer, Elisabeth Emma, and David Mowat. "Agenesis of the Corpus Callosum: A Clinical Approach to Diagnosis." *American Journal of Medical Genetics,* part C (Seminars in Medical Genetics), 166C (2014): 184–97.

Palomaki, Glen; Cosmin Deciu; Edward Kloza; Geralyn Lambert-Messerilan; James Haddow; Louis Neveux; Mathais Ehrich; et al. "DNA Sequencing of Maternal Plasma Reliably Identifies Trisomy 18 and Trisomy 13, as Well as Down Syndrome: An International Collaborative Study." *Genetics in Medicine* 14 (2012): 296–305.

Panzzanese, Christina. "Disarray at the VA: Interview with Linda Bilmes." *Harvard Gazette,* June 13, 2014.

Parens, Eric, and Adrienne Asch. "The Disability Rights Critique of Prenatal Genetic Testing: Reflections and Recommendations." In Parens and Asch, *Prenatal Testing and Disability Rights,* 3–43.

———, eds. *Prenatal Testing and Disability Rights.* Washington, DC: Georgetown University Press, 2000.

Parker, Rozsika. *Torn in Two: The Experience of Maternal Ambivalence.* London: Virago Press, 1995.

Parra, Flavia; Roberto Amado; Jose Lambertucci; Jorge Rocha; Carlos Antunes; and Sergio Pena. "Color and Genomic Ancestry in Brazilians." *Proceedings of the National Academy of Sciences of the United States of America* 100, no. 1 (January 7, 2003): 177–82.

Parry, Brownyn. "The Afterlife of a Slide: Exploring Emotional Attachment to Artifactualized Body Parts." *History and Philosophy of Life Sciences* 3 (2013): 431–48.

Paul, Diane. *Controlling Human Heredity, 1865 to the Present.* New Jersey: Humanities Press, 1995.

———. "From Eugenics to Medical Genetics." *Journal of Policy History* 9 (1997): 96–116.

———. *The Politics of Heredity: Essays on Eugenics, Biomedicine and the Nature-Nurture Debate.* Albany: State University of New York Press, 1998.

———. "Reflections on the Historiography of American Eugenics: Trends, Fractures, Tensions." *Journal of the History of Biology* 49 (2016): 641–58.

Paul, Diane, and Jeffrey Brosco. *The PKU Paradox: A Short History of a Genetic Disease.* Baltimore: Johns Hopkins University Press, 2013.

Paul, Lynn; Warren Brown; Ralph Adolph; J. Michael Tyszka; Linda Richards; Pratik Mukherjee; and Elliott Sherr. "Review: A Genesis of the Corpus Callosum; Genetic, Developmental and Functional Aspects of Connectivity." *Nature Neurosciences* 8 (2007): 287–97.

Penrose, Lionel. "The Relative Effects of Paternal and Maternal Age in Mongolism." *Journal of Genetics* 27 (1933): 219–21.

Pereira dos Santos, Fausto; Deborah Carvalho Malta; and Emerson Elias Mehry. "A regulação na saúde suplementar: Uma análise dos principais resultados alcançados." *Ciência e Saúde Coletiva* 13 (2008): 1463–75.

Perry, David. "Down Syndrome Isn't Just Cute: How the Down Community Sugarcoats Difficult Realities about the Condition." Editorial, Al Jazeera America, October 15, 2014.

Petchesky, Rosalind Pollack. "Fetal Images: The Power of Visual Culture in the Politics of Reproduction." *Feminist Studies* 13 (1987): 263–92.

Pfeffer, Naomi. "What British Women Say Matters to Them about Donating an Aborted Fetus to Stem Cell Research: A Focus Group Study." *Social Science and Medicine* 66 (2008): 2544–54.

Pfeffer, Naomi, and Julie Kent. "Framing Women, Framing Fetuses: How Britain Regulates Arrangements for the Collection and Use of Aborted Fetuses in Stem Cell Research and Therapies." *BioSocieties* 2 (2007): 429–47.

Pickstone, John. *Ways of Knowing: A New History of Science, Technology, and Medicine.* Chicago: University of Chicago Press, 2001.

Piepmeier, Alison. "Choosing to Have a Child with Down Syndrome." *New York Times,* March 2, 2012.

———. "The Inadequacy of 'Choice': Disability and What's Wrong with Feminist Framings of Reproduction." *Feminist Studies* 39 (2013): 159–86.

———. "Outlawing Abortion Won't Help Children with Down Syndrome." *New York Times,* April 1, 2013.

———. "Would It Be Better for Her Not to Be Born? Down Syndrome, Prenatal Testing and Reproductive Decision-Making." *Feminist Formations* 27 (2015): 1–24.

Pinkel, D.; J. Langedent; C. Colins; J. Fuscoe; R. Sergraves; J. Lucas; and J. Gray. "Fluorescence In Situ Hybridization with Human Chromosome–Specific Libraries: Detection of Trisomy 21 and Translocations of Chromosome 4." *Proceedings of the National Academy of Sciences of the United States of America* 85, no. 23 (1988): 9138–42.

Place, Fiona. "Motherhood and Genetic Screening: A Personal Perspective." *Down Syndrome Research and Practice* 12 (2008): 118–26.

Pollitt, Katha. *Pro: Reclaiming Abortion Rights.* New York: Picador, 2014.

Potter, Edith. *Pathology of the Fetus and the Newborn.* Chicago: Year Book Publishers, 1952.

Powell, Anne Martin. "On the Precarious Cusp of Genetic Medicine." *Journal of Genetic Counseling* 21 (2012): 793–802.

Powledge, Tabitha, and Sharmon Sollitto. "Prenatal Diagnosis: The Past and the Future." *Hastings Center Report* 4 (1974): 11–13.

Press, Nancy. "Assessing the Expressive Character of Prenatal Testing: The Choices Made or the Choices Made Available." In Parens and Asch, *Prenatal Testing and Disability Rights*, 214–33.

Quindlen, Anna. "Public and Private: A Public Matter." *New York Times*, June 17, 1992.

Rabinow, Paul. "Artificiality and the Enlightenment: From Sociobiology to Biosociality." In *Essays on the Anthropology of Reason*, 91–111. Princeton, NJ: Princeton University Press, 1996.

Rabinow, Paul, and Nikolas Rose. "Biopower Today." *BioSocieties* 1 (2006): 195–217.

Radkowska-Walkowicz, Magdalena. "The Creation of 'Monsters': The Discourse of Opposition to In Vitro Fertilization in Poland." *Reproductive Health Matters* 20 (2012): 30–37.

———. *Doświadczenie in vitro: Niepłodność i nowe technologie reprodukcyjne w perspektywie antropologicznej* [The in vitro experience: Infertility and new reproductive technologies in an anthropological perspective]. Warsaw: Wydawnictwo Uniwersytetu Warszawskiego, 2013.

———. "Frozen Children and Despairing Embryos in the 'New' Post-Communist State: Debate on IVF in the Context of Poland's Transition." *European Journal of Women's Studies* 21 (2014): 399–414.

Rajao, Raoni, and Ricardo Duque. "Between Purity and Hybridity: Technoscientific and Ethnic Myths of Brazil." *Science, Technology and Human Values* 39 (2014): 844–74.

Ramalho, Carla; Alexandra Matias; Otilia Brandao; and Nuno Montenegro. "Critical Evaluation of Elective Termination of Pregnancy in a Tertiary Fetal Medicine Center during 43 Months: Correlation of Prenatal Diagnosis Findings and Postmortem Examination." *Prenatal Diagnosis* 26 (2006): 1084–88.

Rapp, Emily. "Dear Dr. Frankenstein: Creation Up Close." In Boesky, *The Story Within*, 226–37.

———. *Poster Child: A Memoir*. New York: Bloomsbury, 2007.

———. *The Still Point of the Turning World*. New York: Penguin Press, 2013.

Rapp, Rayna. "Constructing Amniocentesis: Maternal and Medical Discourses." In *Uncertain Terms: Negotiating Gender in American Culture*, edited by Faye Ginsburg and Anna Lowenhaupt Tsing, 28–42. Boston: Beacon Press, 1990.

———. "Moral Pioneers: Women, Men and Fetuses on the Frontier of Reproductive Technology." *Women and Health* 13 (1987): 101–16.

———. "The Power of 'Positive' Diagnosis: Medical and Maternal Discourse on Amniocentesis." In *Representation of Motherhood*, edited by Donna Bassin, Margaret Honey, and Merele Maher Kaplan, 204–19. New Haven, CT: Yale University Press, 1994.

———. *Testing Women, Testing the Fetus: The Social Impact of Amniocentesis in America*. New York: Routledge, 1999.

———. "XYLO: An Amniocentesis Story." In *Test-Tube Women*, edited by Rita Arditti, Renate Duelli-Klein, and Shelley Minden, 313–28. New York: Routledge and Kegan Paul, 1984.

Rapp, Rayna, and Faye Ginsburg. "Enabling Disability, Rewriting Kinship, Reimagining Citizenship." *Public Culture* 13 (2001): 533–56.

Rasella, D.; R. Aquino; and M. L. Barreto. "Reducing Childhood Mortality from Diarrhea and Lower Respiratory Tract Infections in Brazil." *Pediatrics* 126 (2010): 534–40.

Raspberry, Kelly, and Debra Skinner. "Enacting Genetic Responsibility: Experiences of Mothers Who Carry the Fragile X Gene." *Sociology of Health and Illness* 33 (2010): 1–14.

———. "Experiencing the Genetic Body: Parents' Encounters with Pediatric Clinical Genetics." *Medical Anthropology* 26 (2007): 355–91.

Ratner, Mark. "Roche Swallows Ariosa, Grabs Slice of Prenatal Market." *Nature Biotechnology* 33 (2015): 113–14.

Ravitsky, Vardit. "Non Invasive Prenatal Testing (NIPT): Identifying Key Clinical, Ethical, Social, Legal and Policy Issues." Background paper, Nuffield Council on Bioethics, London, November 2015.

Raz, Aviad. " 'Important to Test, Important to Support': Attitudes toward Disability Rights and Prenatal Diagnosis among Leaders of Support Groups for Genetic Disorders in Israel." *Social Science and Medicine* 59 (2004): 1857–66.

Raz, Michal. "Médecins israeliens face au diagnostique prénatal des fetus intersexués." *Sciences Sociales et Santé* 33 (2015): 5–34.

Razavi, Férechté, and Dominique Carles, coordinators. *Pathologie foetale et placentaire pratique.* Montpellier: Sauramps Médical, 2008.

Reagan, Leslie. *Dangerous Pregnancies: Mothers, Disabilities and Abortion in Modern America.* Berkeley: University of California Press, 2010.

———. "From Hazard to Blessing to Tragedy: Representations of Miscarriage in Twentieth Century America." *Feminist Studies* 29 (2003): 357–78.

Reardon, William. "Professor Robin M. Winter: Clinical Geneticist Par Excellence." *Independent* (London), January 17, 2004.

Reardon, William, and Dian Donnai. "Dysmorphology Demystified." *Archives of Disease in Childhood, Fetal and Neonatal Edition* 92 (2007): F225–F229.

Rheinberger, Hans Joerg. "Preparations." In *An Epistemology of the Concrete: Twentieth-Century Histories of Life*, 233–43. Durham, NC: Duke University Press, 2010.

Riis, Powl, and Fritz Fuchs. "Antenatal Determination of Fœtal Sex in Prevention of Hereditary Diseases." *Lancet* 275 (1960): 180–82.

Roberts, Sam. "Helen Harrison, Authority on the Trials of Premature Births, Dies at 68." *New York Times*, July 8, 2015. https://www.nytimes.com/2015/07/09/us/helen-harrison-authority-on-the-trials-of-premature-births-dies-at-68.html?_r=0. Accessed July 20, 2017.

Robin, N. H., and R. J. Shprintzen. "Defining the Clinical Spectrum of Deletion 22q11.2." *Journal of Pediatrics* 147 (2005): 90–96.

Robinson, Gail Erlick. "Pregnancy Loss." *Best Practice and Research Clinical Obstetrics and Gynaecology* 28 (2014): 169–78.

Robyr, R.; J. P. Bernard; J. Roume; and Y. Ville. "Familial Diseases Revealed by a Fetal Anomaly." *Prenatal Diagnosis* 26 (2006): 1124–34.

Rodeck, Charles. "Sampling Pure Foetal Blood by Foetoscopy in Second Semester of Pregnancy." *British Medical Journal* 2, no. 6139 (1978): 728–30.

Romero, Simon. "Brazil Enacts Affirmative Action Law for Universities." *New York Times*, August 30, 2012.

Rose, Jacqueline. "Mothers." *London Review of Books* 36 (2014): 17–22.

Rose, Nikolas. *Politics of Life Itself: Biomedicine, Power and Subjectivity in the Twenty-First Century.* Princeton, NJ: Princeton University Press, 2006.

Rosenberg, Charles. "Managed Fear." *Lancet* 373 (2009): 802–3.

———. "The Tyranny of Diagnosis: Specific Entities and Individual Experience." *Milbank Quarterly* 80 (2002): 237–60.

Rosenthal, Mark. "Diego and Frida: High Drama in Detroit." In *Diego Rivera and Frida Kahlo in Detroit*, edited by Mark Rosenthal, 16–122. New Haven, CT: Yale University Press, 2015.

Rosman, Sophia. "Down Syndrome Screening Information in Midwifery Practices in the Netherlands: Strategies to Integrate Biomedical Information." *Health* (London) 20 (2016): 94–109.

Rothschild, Joan. *The Dream of the Perfect Child*. Bloomington: Indiana University Press, 2005.

Rothstein, William. *Public Health and the Risk Factor: A History of an Uneven Medical Revolution*. Rochester, NY: University of Rochester Press, 2003.

Rough, Bonnie. "Three Phone Calls: A Carrier's Journey into Motherhood." *American Journal of Medical Genetics*, part A, 161A (2013): 2119–21.

Rousseau, Jean-Jacques. *Émile, or Treatise on Education*. Translated by Barbara Foxley. www .online-literature.com/rousseau/emile, accessed July 25, 2017. First published in 1762.

Rowbotham, Sheila. *Woman's Consciousness, Man's World*. Harmondsworth: Penguin Books, 1973.

Rowland, Robyn. *Living Laboratories: Women and Reproductive Technologies*. Bloomington: Indiana University Press, 1992.

Rubin, Eric; Michael Greene; and Lindsey Baden. "Zika Virus and Microcephaly." *New England Journal of Medicine* 374 (2016): 984–85.

Ruhl, Lealle. "Liberal Governance and Prenatal Care: Risk and Regulation in Pregnancy." *Economy and Society* 28 (2009): 95–117.

Sacker, A.; Y. Kelly; M. Iacovou; N. Cable; and M. Bartley. "Breast Feeding and Intergenerational Social Mobility: What Are the Mechanisms?" *Archives of Diseases of Childhood* 98 (2013): 666–71.

Saint Louis, Catherine. "When Caregivers Need Healing." *New York Times*, July 28, 2014.

Saldanha, Fátima Aparecida Targino; Maria de Lourdes Brizot; Lilian M. Lopes; Adolfo Wenjaw Liao; and Marcelo Zugaib. "Anomalias e prognóstico fetal associados à translucência nucal aumentada e cariótipo anormal." *Revista da Associacao Médica Brasileira* 55 (2009): 54–59.

Samerski, Silja. "Genetic Counseling and the Fiction of Choice: Taught Self-Determination as a New Technique of Social Engineering." *Signs: Journal of Women in Culture and Society* 34 (2009): 735–61.

Sampaio Rodrigues, Claudia. "Sentidos, limites e potentialidades da medecina fetal: A visao dos specialistas." Master's thesis, Instituto Fernandes Figueira-Fiocruz, 2010.

Sanabria, Emilia. "From Sub- to Super-Citizenship: Sex Hormones and the Body Politic in Brazil." *Ethnos* 75 (2010): 377–401.

———. *Plastic Bodies: Sex Hormones and Menstrual Suppression in Brazil*. Durham, NC: Duke University Press, 2016.

Sandelowski, Margaret, and Julie Barroso. "The Travesty of Choosing after Positive Prenatal Diagnosis." *Journal of Obstetric, Gynecologic, and Neonatal Nursing* 34 (2005): 307–18.

Sanger, Carol. *About Abortion: Terminating Pregnancy in Twenty First Century America*. Cambridge, MA: Harvard University Press, 2017.

Sänger, Eva. "Obstetrical Care as a Matter of Time: Ultrasound Screening, Temporality and Prevention." *History and Philosophy of Life Sciences* 37 (2015): 105–20.

Santesmases, Maria Jesus. "Size and the Centromere: Translocations and Visual Cultures in Early Human Genetics." In *Making Mutations: Objects, Practices, Contexts*, edited by Luis Campos and Alexander von Schwerin, 189–208. Preprint 393. Berlin: Max Plank Institute for the History of Science, 2010.

Santo, S.; F. D'Antonio; T. Homfray; P. Rich; G. Pilu; A. Bhide; B. Thilaganathan; and A. T. Papagheorgiou. "Counseling in Fetal Medicine: A Genesis of the Corpus Callosum." *Ultrasound in Obstetrics and Gynecology* 40 (2012): 513–21.

Santo, Susana; Sahar Mansour; Basky Thilaganathan; Tessa Homfray; Aris Papageorghiou; Sandra Calvert; and Amar Bhide. "Prenatal Diagnosis of Non-immune Hydrops Fetalis: What Do We Tell the Parents?" *Prenatal Diagnosis* 31 (2011): 186–95.

Santos, Ricardo Ventura, and Maio Marcos Chor. "Race, Genomics, Identities and Politics in Contemporary Brazil." *Critique of Anthroplogy* 24 (2004): 347–78.

Sass, Nelson, and Susane Mei Hwang. "Dados epidemiológicos, evidências e reflexões sobre a indicação de cesariana no Brasil." *Diagnosis e Tratamento* 14 (2009): 133–37.

Saxton, Marsha. "Disability Rights and Selective Abortion." In *Abortion Wars: A Half Century of Struggle, 1950 to 2000*, edited by Rickie Solinger, 374–93. Berkeley: University of California Press, 1998.

———. "Prenatal Screening and Discriminatory Attitudes towards Disability." In *Embryos, Ethics and Women's Rights: Exploring the New Reproductive Technologies*, edited by Elaine Hoffman Baruch, Amadeo F. D'Adamo, and Joni Saeager, 217–24. New York: Haworth Press, 1987.

Scambler, P. J. "The 22q11 Deletion Syndromes." *Human Molecular Genetics* 9 (2000): 2421–26.

Scherter, Simon. "The Importance of Social Intervention in Britain's Mortality Decline, 1850–1914: A Re-interpretation of the Role of Public Health." *Social History of Medicine* 1 (1988): 1–37.

Schrad, Mark Lawrence. "Does Down Syndrome Justify Abortion?" *New York Times*, September 4, 2015.

Schwartz, Lisa; Steven Wolochin; Floyd Fowler; and Gilbert Welch. "Enthusiasm for Cancer Screening in the United States." *Journal of the American Medical Association* 291 (2004): 71–78.

Schwennesen, Nete, and Lene Koch. "Representing and Intervening: 'Doing' Good Care in First Trimester Prenatal Knowledge Production and Decision-Making." *Sociology of Health and Illness* 34 (2012): 283–98.

Schwennesen, Nete; Mette Nordahl Svendsen; and Lene Koch. "Beyond Informed Choice: Prenatal Risk Assessment, Decision-Making and Trust." *Clinical Ethics* 5 (2010): 207–16.

Scully, Jackie Leach. "What Is a Disease? Disease, Disability and Their Definitions." *EMBO Reports* 5 (2004): 650–53.

Senat, M. V., and R. Frydman. "Hyperclarté nucale a caryotype normal." *Gynécologie, Obstétrique et Fertilité* 35 (2007): 507–15.

Serres, Michel, and Bruno Latour. *Conversations on Science, Culture, and Time*. Ann Arbor: University of Michigan Press, 1995.

Shaffer, L. G.; C. D. Kashork; R. Saleki; E. Rorem; K. Sundin; B. C. Baiff; and B. A. Bejjani. "Target Genomic Microarray Analysis for Identification of Chromosome Abnormalities in 1500 Consecutive Clinical Cases." *Journal of Pediatrics* 149 (2006): 98–102.

Shaffer, Lisa G., on behalf of the American College of Medical Genetics (ACMG) Professional Practice and Guidelines Committee. "American College of Medical Genetics Guideline on the Cytogenetic Evaluation of the Individual with Developmental Delay or Mental Retardation." *Genetics in Medicine* 7 (2005): 650–54.

Shaffer, Lisa; Jill Rosenfeld; Mindy Dabell; Justine Coppinger; Anne M. Bandholz; Jay Ellison; J. Britt Ravnan; et al. "Detection Rates of Clinically Significant Genomic Alterations by Microarray Analysis for Specific Anomalies Detected by Ultrasound." *Prenatal Diagnosis* 32 (2012): 986–95.

Shakespeare, Tom. "The Content of Individual Choice." In *Quality of Life and Human Difference: Genetic Testing and Disability*, edited by David Wasserman, Jerome Bickenbach, and Robert Wachbroit, 217–36. Cambridge: Cambidge Univerity Press, 2005.

———. *Disabilities Rights and Wrongs*. London: Routledge, 2006.

———. "Losing the Plot? Medical and Activist Discourses of Contemporary Genetics and Disability." *Sociology of Health and Illness* 21 (1999): 669–88.

Shaw, Alison. "Rituals of Baby Death: Defining Life and Islamic Personhood." *Bioethics* 28 (2014): 84–95.

Shaw, Alison; Joanna Latimer; Paul Atkinson; and Katie Featherstone. "Surveying 'Slides': Clinical Perception and Clinical Judgment in the Construction of a Genetic Diagnosis." *New Genetics and Society* 22 (2003): 3–19.

Shaw-Smith, C.; R. Redon; L. Rickman; M. Rio; L. Wilatt; H. Fiegler; H. Firth; et al. "Microarray Based Comparative Genomic Hybridisation (Array-CGH) Detects Submicroscopic Chromosomal Deletions and Duplications in Patients with Learning Disability/Mental Retardation and Dysmorphic Features." *Journal of Medical Genetics* 41 (2004): 241–48.

Sheldon, Trevor, and John Simpson. "Appraisal of a New Scheme for Prenatal Screening for Down's Syndrome." *British Medical Journal* 302 (1991): 1133–36.

Shepard, T. "Obituary: David W. Smith, 1926–1981." *Teratology* 24 (1981): 111–12.

Shur, Natasha. "The Real Tiger Mothers: From a Clinical Genetics Perspective." *American Journal of Clinical Genetics*, part A, 155 (2011): 2088–90.

Siffredi, Vanessa; Vicki Anderson; Richard J. Leventer; and Megan M. Spencer-Smith. "Neuropsychological Profile of Agenesis of the Corpus Callosum: A Systematic Review." *Developmental Neuropsychology* 38 (2013): 36–57.

Silberman, Steve. *Neurotribes*. New York: Penguin Random House, 2015.

Silverman, Chloe. *Understanding Autism: Parents, Doctors and the History of a Disorder*. Princeton, NJ: Princeton University Press, 2012.

Silverman, William. "Incubator-Baby Side Shows." *Pediatrics* 64 (1979): 127–41.

Simon, Rafaelle. "Interview avec Eugène Green." *Trois Couleurs*, March 4, 2015, 25–27.

Simplican, Stacy Clifford. "Care, Disability and Violence: Theoretizing Complex Dependency." *Hypatia* 30 (2015): 217–33.

Simpson, Joe Leigh; Mitchell Golbus; Alice Martin; and Gloria Sarto. *Genetics in Obstetrics and Gynecology*. New York: Grune and Stratton, 1982.

Skotko, Brian; Susan Levine; and Richard Goldstein. "Having a Brother or Sister with Down Syndrome: Perspectives from Siblings." *American Journal of Medical Genetics*, part A, 155 (2011): 2348–59.

———. "Having a Son or Daughter with Down Syndrome: Perspectives from Mothers and Fathers." *American Journal of Medical Genetics*, part A, 155 (2011): 2335–47.

Sloan, Eileen; Sharon Kirsh; and Mary Mowbray. "Viewing the Fetus following Termination of Pregnancy for Fetal Anomaly." *Journal of Obstetrics, Gynecology and Neonatal Nursing* 37 (2008): 395–404.

Smith, David. "Dysmorphology (Teratology)." *Journal of Pediatrics* 69 (1966): 1150–69.

———. *Recognizable Patterns of Human Malformation: Genetic, Embryonic and Clinical Aspects*. Philadelphia: Saunders, 1970.

Smith, Dylan; George Loewenstein; Alexandra Jankovich; and Peter Ubel. "Happily Hopeless: Adaptation to a Permanent, but Not to a Temporary, Disability." *Health Psychology* 28 (2009): 787–91.

Smith, J. P. "Risky Choices: The Dangers of Teens Using Self-Induced Abortion Attempts." *Journal of Pediatric Health Care* 12 (1998): 147–51.

Snochowska-Gonzales, Claudia, ed. *A jak hipokryzja: Antologia tekstów o aborcji, władzy, pieniądzach i sprawiedliwości* [A like hypocrisy: An anthology of texts on abortion, power, money and justice]. Warsaw: Wydawnictwo O Matko!, 2011.

Sobrinho, D. Fonseca. *Estado e população: Uma história do planejamento familiar no Brasil.* Rio de Janeiro: Rosa dos Tempos; FNUAP, 1993.

Solomon, Andrew. *Far from the Tree: Parents, Children and the Search for Identity.* New York: Scribner, 2012.

———. *The Noonday Demon: An Anatomy of Depression.* London: Vintage Books, 2002.

Souka, Athena; Constantin von Kaisenberg; Jonathan Hyett; Jiri Sonek; and Kypros Nicolaides. "Increased Nuchal Translucency with Normal Karyotype." *American Journal of Obstetrics and Gynecology* 192 (2005): 1005–21.

Star, Leigh S. "Craft vs. Commodity, Mess vs. Transcendence: How the Right Tool Became the Wrong One in the Case of Taxidermy and Natural History." In *The Right Tools for the Job: At Work in Twentieth-Century Life Sciences,* edited by Adele Clarke and Joan Fujimura, 257–86. Princeton, NJ: Princeton University Press, 1992.

Statham, Helen. "Prenatal Diagnosis of Fetal Abnormality: The Decision to Terminate the Pregnancy and Its Consequences." *Fetal and Maternal Medicine Review* 13 (2002): 213–47.

Stein, Zena; Mervyn Susser; and Andrea Guterman. "Screening Programme for Down Syndrome." *Lancet* 301 (1973): 305–10.

Steinbock, Bonnie. "Disability, Prenatal Testing and Selective Abortion." In Parens and Asch, *Prenatal Testing and Disability Rights,* 108–23.

Stepan, Nancy. *The Hour of Eugenics: Race, Gender, and Nation in Latin America.* Ithaca, NY: Cornell University Press, 1991.

Stern, Alexandra Mina. *Eugenic Nation: Faults and Frontiers of Better Breeding in Modern America.* Berkeley: University of California Press, 2005.

———. *Telling Genes: The Story of Genetic Counseling in America.* Baltimore: Johns Hopkins University Press, 2012.

"Stillbirth: Your Stories." *New York Times,* June 26, 2016. https://www.nytimes.com/interactive /2015/health/stillbirth-reader-stories.html.

Stillwell, Devon. "'Pretty Pioneering-Spirited People': Genetic Counsellors, Gender Culture, and the Professional Evolution of a Feminised Health Field, 1947–1980." *Social History of Medicine* 28, no. 1 (2015): 172–93.

Sunday Times Insight Team. *Suffer the Children: The Story of Thalidomide.* London: André Deutsch, 1979.

Szostak, Violetta. "Justyna Bargielska: I Miscarried." *Gazeta Wyborcza* (Warsaw), May 23, 2011.

Tansey, E. M., and D. A. Christie, eds. *Looking at the Unborn: Historical Aspects of Obstetrics Ultrasound; A Witness Seminar Held at the the Wellcome Institute for the History of Medicine, London, on 10 March 1998.* Wellcome Witnesses to Twentieth Century Medicine, vol. 5. London: Wellcome Trust, 2000.

Taussig, Karen Sue. *Ordinary Genomes: Science, Citizenship and Genetic Identies.* Durham, NC: Duke University Press, 2009.

Taylor, Janelle. *The Public Life of the Fetal Sonogram: Technology, Consumption and the Politics of Reproduction.* New Brunswick, NJ: Rutgers University Press, 2008.

Teixiera, Maria; Maria da Conceição Costa; Wanderson de Oliveira; Marilia Lavocat Nunes; and Laura Rodrigues. "The Epidemic of Zika Virus–Related Microcephaly in Brazil: Detection, Control, Etiology, and Future Scenarios." *American Journal of Public Health* 106 (2016): 601–5.

Teixiera, Maria, and Laura Rodriguez. "Response." *American Journal of Public Health* 106, no. 8 (2016): e9.

Thayyil, Sudhin; Neil Sebire; Lyn Chitty; Angie Wade: Oystein Olsen, Roxana Gunny; and Amaka Offioah for the MARIAS Collaborative Group. "Post-mortem MRI versus

Conventional Autopsy in Fetuses and Children: A Prospective Validation Study." *Lancet* 382 (2013): 223–33.

Tibol, Raquel. *Frida Kahlo: Crônica, testimonios y aproximaciones.* Mexico City: Oasis, 1977. Translated by Elinor Randall as *Frida Kahlo: An Open Life* (Albuquerque: University of New Mexico Press, 1993).

Tredgold, Alfred Frank. *Mental Deficiency (Amentia).* London: Ballière, Tindall and Cox, 1908.

Trulsson, Otti, and Ingela Radestad. "The Silent Child: Mothers' Experiences Before, During, and After Stillbirth." *Birth* 31 (2004): 189–95.

Tsai-Goodmann, Beverly. "Towards a Safety Net for Management of 22q11.2 Deletion Syndrome: Guidelines for Our Times." *European Journal of Pediatrics* 173 (2014): 757–65.

Turton, P.; P. Hughes; C. D. Evans; and D. Fainman. "Incidence, Correlates and Predictors of Post Traumatic Stress Disorder in the Pregnancy after Stillbirth." *British Journal of Psychiatry* 178 (2001): 556–60.

United Kingdom. Department of Health. *Statistical Bulletin, Abortion Statistics for England and Wales 2009.* London: Department of Health, 2010.

United Kingdom. Parliament, Commons debate statutory instrument on mitochondrial donation, February 3, 2015. http://www.parliament.uk/business/news/2015/february/commons-debate-statutory-instrument-on-mitochondrial-donation/. Accessed July 28, 2017.

Valenti, Jessica. "My 28-Week Pregnancy and a 20-Week Abortion Ban: Why Choice Still Matters." *Guardian* (London), May 1, 2014.

Van der Sijpta, Erica. "The Unfortunate Sufferer: Discursive Dynamics around Pregnancy Loss in Cameroon." *Medical Anthropology: Cross-Cultural Studies in Health and Illness* 33 (2014): 395–410.

Van Dijck, José. *The Transparent Body: A Cultural Analysis of Medical Imaging.* Seattle: University of Washington Press, 2005.

Vassy, Carine. "From a Genetic Innovation to Mass Health Programmes: The Diffusion of Down's Syndrome Prenatal Screening and Diagnostic Techniques in France." *Social Science and Medicine* 63 (2006): 2041–51.

———. "How Prenatal Diagnosis Became Acceptable in France." *Trends in Biotechnology* 23 (2005): 246–49.

Vassy, Carine; Sophia Rosman; and Bénédicte Rousseau. "From Policy Making to Service Use: Down's Syndrome Antenatal Screening in England, France and the Netherlands." *Social Science and Medicine* 106 (2014): 67–74.

Viaux-Savelon, Sylvie; Marc Dommergues; Ouriel Rosenblum; Nicolas Bodeau; Elizabeth Aidane; Odile Philippon; Philippe Mazet; et al. "Prenatal Ultrasound Screening: False Positive Soft Markers May Alter Maternal Representations and Mother-Infant Interaction." *PloS One* 7 (2012): e30935.

Ville, Isabelle. "Politiques du handicap et périnatalité: La difficile conciliation de deux champs d'intervention sur le handicap." *ALTER—European Journal of Disability Research/Revue Européenne de Recherche sur le Handicap* 5 (2011): 16–25.

Ville, Isabelle, and Lynda Lotte. "Évolution des politiques publiques: handicap, périnatalité, avortement." In Final Report of the ANR project-09-SSOC-026, *Les enjeux du diagnostic prénatal dans la prévention des handicaps: L'usage des techniques entre progrès scientifiques et action publique.* Unpublished. Paris, 2013.

Ville, Isabelle, and Veronique Mirlesse. "Prenatal Diagnosis: From Policy to Practice; Two Distinct Ways of Managing Prognostic Uncertainty and Anticipating Disability in Brazil and in France." *Social Science and Medicine* 141 (2015): 19–26.

Vimercati, Antonella; Silvana Grasso; Marinella Abruzzese; Annarosa Chincoli; Alessandra de Gennaro; Angela Miccolis; Gabriella Serio; et al. "Correlation between Ultrasound

Diagnosis and Autopsy Findings of Fetal Malformations." *Journal of Prenatal Medicine* 6 (2012): 13–17.

Votino, C.; B. Bessieres; V. Segers; H. Kadhim; F. Razavi; M. Condorelli; R. Votino; et al. "Minimally Invasive Fetal Autopsy Using Three-Dimensional Ultrasound: A Feasibility Study." *Ultrasound Obstetrics and Gynecology* 44, S1 (2014): 43–44.

Wahlberg, Ayo. "Serious Disease as a Kind of Living." In *Contested Categories: Life Sciences in Society*, edited by Susan Bauer and Ayo Wahlberg, 89–112. Farnham, Surrey: Ashgate, 2009.

Wald, N. J.; L. George; J. W. Densem; K. Petterson; and International Prenatal Screening Research Group. "Serum Screening for Down's Syndrome between 8 and 14 Weeks of Pregnancy." *British Journal of Obstetrics and Gynecology* 103 (1996): 407–12.

Wald, N. J., and A. K. Hackshaw. "Combining Ultrasound and Biochemistry in First-Trimester Screening for Down's Syndrome." *Prenatal Diagnosis* 17 (1997): 821–29.

Waldman, Ayelet. "Rocketship." In *Bad Mother: A Chronicle of Maternal Crimes, Minor Calamities, and Occasional Moments of Grace*, 122–36. New York: Anchor Books, 2010.

Wapner, Ronald; Christa Lese Martin; Brynn Levy; Blake Ballif; Christine Eng; Julia Zachary; Melissa Savage; et al. "Chromosomal Microarray versus Karyotyping for Prenatal Diagnosis." *New England Journal of Medicine* 367 (2012): 2176–84.

Ward, Peter. *Birth, Weight and Economic Growth: Women's Living Standards in the Industrializing West*. Chicago: University of Chicago Press, 1973.

Warkany, Joseph. "The Medical Profession and Congenital Malformations, 1900–1979." *Teratology* 20 (1979): 201–4.

Watts, Georg, coordinator. *Novel Techniques for Mitochondrial DNA Disorders: An Ethical Review*. London: Nuffield Council on Bioethics, 2015,

Weber, Jean Christophe; Catherine Allamel-Ruffin; Thierry Rustenholtz; Isabelle Pons; and Isabelle Gobatto. "Les soignants et la décision d'interruption de grossesse pour motif médical: Entre indications cliniques et embarras éthique." *Sciences Sociales et Santé* 26 (2008): 93–119.

Weiner, Noga. "The Intensive Medical Care of Sick, Impaired and Preterm Newborns in Israel and the Production of Vulnerable Neonatal Subjectivities." *Medical Anthropology Quarterly* 23 (2009): 320–41.

Weintraub, Karen. "Three Biological Parents and a Baby." *New York Times*, December 16, 2013.

Weir, Lorna. *Pregnancy, Risk and Biopolitics: On the Threshold of Living Subject*. New York: Routledge, 2006.

Weisz, George. *Chronic Disease in the Twentieth Century: A History*. Baltimore: Johns Hopkins University Press, 2014.

Welch, H. Gilbert. *Should I Be Tested for Cancer? Maybe Not and Here's Why*. Berkeley: University of California Press, 2006.

Welch, H. Gilbert; Lisa Schwartz; and Steven Woloshin. *Overdiagnosed: Making People Sick in the Pursuit of Health*. Boston: Beacon Press, 2011.

Wendell, Susan. "Unhealthy Disabled: Treating Chronic Illnesses as Disabilities." *Hypatia* 16 (2001): 17–33.

Wertz, Dorothy, and John Fletcher. "A Critique of Some Feminist Challenges to Prenatal Diagnosis." *Journal of Women's Health* 2 (1993): 173–88.

———. "Ethical and Social Issues in Prenatal Sex Selection: A Survey of Geneticists in 37 Nations." *Social Sciences and Medicine* 46 (1998): 255–57.

———. "Ethics and Genetics: An International Survey." *Hastings Center Report* 19 (1989): 20–24.

———. "Geneticists Approach Ethics: An International Survey." *Clinical Genetics* 43 (1993): 104–10.

Whitley, Kari; Kevin Trinchere; Wendy Prutsman; Joanne Quiñones; and Meredith Rochon. "Midtrimester Dilation and Evacuation versus Prostaglandin Induction: A Comparison of Composite Outcomes." *American Journal of Obstetrics and Gynecology* 205 (2011): 386.e1–e7.

Whitmarsh, Ian; Arlene Davis; Debra Skinner; and Donald Bailey. "A Place for Genetic Uncertainty: Parents Valuing an Unknown in the Meaning of Disease." *Social Sciences and Medicine* 65 (2007): 1082–93.

"Why Isn't My Child as Clever as Me?" *Guardian* (London), April 22, 2013.

Whyte, Susan, and Benedicte Ingstad. "Disability and Culture: An Overview." In *Disability and Culture*, edited by Susan Whyte and Benedicte Ingstad, 3–37. Berkeley: University of California Press, 1995.

Wilkinson, Stephen. "Prenatal Screening, Reproductive Choice and Public Health." *Bioethics* 29 (2015): 26–35.

Willems, D. L.; A. A. Verhagen; and E. van Wijlick. "Infants' Best Interests in End-of-Life Care for Newborns." *Pediatrics* 134 (2014): e1163–e1168.

Williams, Clare. "Framing the Fetus in Clinical Work: Rituals and Practices." *Social Science and Medicine* 60 (2005): 2085–95.

Williams, Clare; Priscilla Alderson; and Bobbie Farsides. "Too Many Choices? Hospital and Community Staff Reflect on the Future of Prenatal Screening." *Social Science and Medicine* 55 (2002): 743–53.

———. "What Constitutes Balanced Information in the Practitioners' Portrayals of Down's Syndrome?" *Midwifery* 18 (2002): 230–37.

Williams, Daniel. "The Partisan Trajectory of the American Pro-life Movement: How a Liberal Catholic Campaign Became a Conservative Evangelical Cause." *Religions* 6 (2015): 451–75.

Williams, Zoe. "Mary Beard: 'The Role of the Academic Is to Make Everything Less Simple.'" *Guardian* (London), April 23, 2016.

Wilson, Emilie. "Ex Utero: Live Human Fetal Research and the Films of Davenport Hooker." *Bulletin of the History of Medicine* 88 (2014): 132–60.

Wolf, Joan. "Is Breast Really Best? Risk and Total Motherhood in the National Breastfeeding Awareness Campaign." *Journal of Health Politics, Policy and Law* 32 (2007): 596–635.

Wolfberg, Adam. *Fragile Beginnings: Discoveries and Triumphs in the Newborn ICU*. Harvard Health Publications. Boston: Beacon Press, 2012.

Woods, Robert. *Death before Birth: Fetal Health and Mortality in Historical Perspective*. Oxford: Oxford University Press, 2007.

World Health Organization. 2015. WHO's data on C-sections worldwide, 2015. http://www.who.int/mediacentre/news/releases/2015/caesarean-sections/en/. Accessed July 25, 2017.

Wright, David. *Downs: The History of a Disability*. Oxford: Oxford University Press, 2011.

Wthycombe, Shannon. "From Women's Expectations to Scientific Specimens: The Fate of Miscarriage Materials in Nineteenth Century America." *Social History of Medicine* 28 (2015): 245–62.

Wynn, Margaret, and Arthur Wynn. *Prevention of Handicap and the Health of Women*. London: Routledge and Kegan Paul, 1979.

Yeo, Lami; Edwin Guzman; Susan Shen-Schwarz; Christine Walters; and Anthony Vintzileos. "Value of a Complete Sonographic Survey in Detecting Fetal Abnormalities: Correlation with Perinatal Autopsy." *Journal of Ultrasound Medicine* 21 (2002): 501–10.

Young, Courtney. "Overselling Breast-Feeding." *New York Times*, October 16, 2015.

Yurkiewicz, Ilana R.; Bruce R. Korf; and Lisa Soleymani Lehmann. "Prenatal Whole-Genome Sequencing—is the Quest to Know a Fetus's Future Ethical?" *New England Journal of Medicine* 379 (2014): 195–97.

Zallen, Doris; Daphne Christie; and Elisabeth Tansey, eds. *The Rhesus Factor and Disease Prevention: The Transcript of a Witness Seminar Held by the Wellcome Trust Centre for the History of Medicine at UCL, London, on 3 June 2003*. Wellcome Witnesses to Twentieth Century Medicine, vol. 22. London: Wellcome Trust Centre for the History of Medicine, 2004.

Zamora, Martha. *Frida: El pincel de la angustia*. México DF: Bosque del Castillo, 1987. Translated by Marilyn Sode Smith as *Frida Kahlo: The Brush of Anguish* (San Francisco: Chronicle Books, 1990).

Zetterman, Eva. "Frida Kahlo's Abortions: With Reflections from a Gender Perspective on Sexual Education in Mexico." *Konsthistorisk Tidskrift/Journal of Art History* 75 (2006): 230–43.

Zihni, Lilian Serife. "The History of the Relationship between the Concept and the Treatment of People with Down Syndrome in Britain and America from 1866 to 1967." PhD diss., University College, London, 1989.

Zlot, Renata. "Anomalias congênitas em natimortos e neomortos: O papel do aconselhamento genético." Master's thesis, Instituto Fernandes Figueira-Fiocruz, 2008.

Zlotogora, Joel; Rivka Carmi; Boaz Lev; and Stavit Shalev. "A Targeted Population Carrier Screening Program for Severe and Frequent Genetic Diseases in Israel." *European Journal of Human Genetics* 17 (2009): 591–97.

Zlotogora, Joel, and Alex Lewenthal. "Screening for Genetic Disorders among Jews: How Should the Tay Sachs Screening Program Be Continued?" *Israeli Journal of Medical Science* 2 (2000): 665–67.

INDEX